Antenna Handbook

Antenna Handbook

VOLUME IV RELATED TOPICS

Edited by
Y. T. Lo

Electromagnetics Laboratory
Department of Electrical and Computer Engineering
University of Illinois–Urbana

S. W. Lee

Electromagnetics Laboratory
Department of Electrical and Computer Engineering
University of Illinois–Urbana

CHAPMAN & HALL
I(T)P An International Thomson Publishing Company

New York • Albany • Bonn • Boston • Cincinnati • Detroit • London • Madrid • Melbourne •
Mexico City • Pacific Grove • Paris • San Francisco • Singapore • Tokyo • Toronto • Washington

Copyright © 1993 by Van Nostrand Reinhold

This edition published by Chapman & Hall, New York, NY

Printed in the United States of America

For more information contact:

Chapman & Hall
115 Fifth Avenue
New York, NY 10003

Chapman & Hall
2-6 Boundary Row
London SE1 8HN
England

Thomas Nelson Australia
102 Dodds Street
South Melbourne, 3205
Victoria, Australia

Chapman & Hall GmbH
Postfach 100 263
D-69442 Weinheim
Germany

Nelson Canada
1120 Birchmount Road
Scarborough, Ontario
Canada M1K 5G4

International Thomson Publishing Asia
221 Henderson Road #05-10
Henderson Building
Singapore 0315

International Thomson Editores
Campos Eliseos 385, Piso 7
Col. Polanco
11560 Mexico D.F.
Mexico

International Thomson Publishing - Japan
Hirakawacho-cho Kyowa Building, 3F
1-2-1 Hirakawacho-cho
Chiyoda-ku, 102 Tokyo
Japan

All rights reserved. No part of this book covered by the copyright hereon may be reproduced or used in any form or by any means--graphic, electronic, or mechanical, including photocopying, recording, taping, or information storage and retrieval systems--without the written permission of the publisher.

2 3 4 5 6 7 8 9 XXX 01 00 99 98 97 96

Library of Congress Cataloging-in-Publication Data

The antenna handbook/edited by Y. T. Lo and S. W. Lee
     p.   Cm.
    Includes bibliographical references and indexes.
    Contents: v. 1. Fundamentals and mathematical techniques--v. 2. Antenna theory--v. 3. Applications--v. 4. Related topics.
    ISBN 0-442-01592-5 (v. 1).--ISBN 0-442-01593-3 (v.2).--ISBN 0-442-01594-1 (v. 3).--ISBN 0-442-01596-8 (v. 4)
    1. Antennas (Electronics) I. Lo, Y.T. II. Lee, S. W.
TK7871.6.A496   1993   93-6502
621.382'4--dc20   CIP

Visit Chapman & Hall on the Internet http://www.chaphall.com/chaphall.html

To order this or any other Chapman & Hall book, please contact **International Thomson Publishing, 7625 Empire Drive, Florence, KY 41042.** Phone (606) 525-6600 or 1-800-842-3636. Fax: (606) 525-7778. E-mail: order@chaphall.com.

For a complete listing of Chapman & Hall titles, send your request to **Chapman & Hall, Dept. BC, 115 Fifth Avenue, New York, NY 10003**.

Contents

Volume IV RELATED TOPICS

28. **Transmission Lines and Waveguides** — 28-1
    Y. C. Shih and T. Itoh

29. **Propagation** — 29-1
    C. H. Liu and D. J. Fang

30. **Antenna Response to Electromagnetic Pulses** — 30-1
    K. S. H. Lee

31. **Random Electromagnetic Design** — 31-1
    G. P. Tricoles

32. **Measurement of Antenna Radiation Characteristics on Far-Field Ranges** — 32-1
    E. S. Gillespie

33. **Near-Field Far-Field Antenna Measurements** — 33-1
    Jørgen Appel-Hansen

Appendices

A. Physical Constants, International Units, Conversion of Units, and Metric Prefixes — A-3
B. The Frequency Spectrum — B-1
C. Electromagnetic Properties of Materials — C-1
D. Vector Analysis — D-1
E. VSWR Versus Reflection Coefficient and Mismatch Loss — E-1
F. Decibels Versus Voltage and Power Ratios — F-1

Index — I-1

Preface

During the past decades, new demands for sophisticated space-age communication and remote sensing systems prompted a surge of R & D activities in the antenna field. There has been an awareness, in the professional community, of the need for a systematic and critical review of the progress made in those activities. This is evidenced by the sudden appearance of many excellent books on the subject after a long dormant period in the sixties and seventies. The goal of this book is to compile a reference to *complement* those books. We believe that this has been achieved to a great degree.

A book of this magnitude cannot be completed without difficulties. We are indebted to many for their dedication and patience and, in particular, to the forty-two contributing authors. Our first thanks go to Mr. Charlie Dresser and Dr. Edward C. Jordan, who initiated the project and persuaded us to make it a reality. After smooth sailing in the first period, the original sponsoring publisher had some unexpected financial problems which delayed its publication three years. In 1988, Van Nostrand Reinhold took over the publication tasks. There were many unsung heroes who devoted their talents to the perfection of the volume. In particular, Mr. Jack Davis spent many arduous hours editing the entire manuscript. Mr. Thomas R. Emrick redrew practically all of the figures with extraordinary precision and professionalism. Ms. Linda Venator, the last publication editor, tied up all of the loose ends at the final stage, including the preparation of the Index. Without their dedication and professionalism, the publication of this book would not have been possible.

Finally, we would like to express our appreciation to our teachers, students, and colleagues for their interest and comments. We are particularly indebted to Professor Edward C. Jordan and Professor George A. Deschamps for their encouragement and teaching, which have had a profound influence on our careers and on our ways of thinking about the matured field of electromagnetics and antennas.

This Preface was originally prepared for the first printing in 1988. Unfortunately, it was omitted at that time due to a change in the publication schedule. Since many readers questioned the lack of a Preface, we are pleased to include it here, and in all future printings.

Preface to the Second Printing

Since the publication of the first printing, we have received many constructive comments from the readers. The foremost was the bulkiness of a single volume for this massive book. The issue of dividing the book into multivolumes had been debated many times. Many users are interested in specific topics and not necessarily the entire book. To meet both needs, the publisher decided to reprint the book in multivolumes. We received this news with great joy, because we now have the opportunity to correct the typos and to insert the original Preface, which includes a heartfelt acknowledgment to all who contributed to this work.

We regret to announce the death of Professor Edward C. Jordan on October 18, 1991.

PART D
Related Topics

Chapter 28

Transmission Lines and Waveguides

Y. C. Shih
U.S. Naval Postgraduate School

T. Itoh
The University of Texas

CONTENTS

1. Introduction   28-5
2. Transmission Line Equations   28-5
3. TEM Transmission Lines   28-10
   Two-Wire Line   28-11
   Circular Coaxial Line   28-20
   Triplate Stripline   28-22
4. Planar Quasi-TEM Transmission Lines   28-30
   Microstrip Lines   28-30
   Coplanar Waveguides   28-33
   Coplanar Strips   28-35
5. TE/TM Waveguides   28-35
   Rectangular Waveguides   28-38
   Circular Waveguides   28-42
   Ridged Waveguides   28-45
6. Hybrid-Mode Waveguides   28-47
   Circular Dielectric Waveguides and Image Guides   28-50
   Rectangular Dielectric Waveguides and Image Guides   28-51
   Slot Lines   28-53
   Fin Lines   28-57
7. References   28-58

Yi-Chi Shih received the B Engr degree from National Taiwan University, Taiwan, R.O.C., in 1976, the MSc degree from the University of Ottawa, Ontario, Canada, in 1980, and the PhD degree from the University of Texas at Austin in 1982, all in electrical engineering.

From September 1982 to April 1984 he served as an adjunct professor at the Naval Postgraduate School, Monterey, California, where he offered courses in the areas of electromagnetic field theory, antennas, and propagation. From April 1984 to May 1986 he was with Hughes Aircraft Company as a senior member of technical staff. In May 1986 he joined MM-Wave Technology, Inc., as Technical Director. He is responsible for the development of CAD tools for hybrid and monolithic integrated circuits, including amplifiers, phase shifters, mixers, and filters.

Dr. Shih is a member of the Institute of Electrical and Electronics Engineers. He has published more than 20 papers on the analysis and design of transmission lines, waveguide discontinuities, and millimeter-wave filters.

Tatsuo Itoh received the PhD degree in electrical engineering from the University of Illinois, Urbana, in 1969.

From September 1966 to April 1976 he was with the Electrical Engineering Department, University of Illinois. From April 1976 to August 1977 he was a senior research engineer in the Radio Physics Laboratory, SRI International, Menlo Park, California. From August 1977 to June 1978 he was an associate professor at the University of Kentucky, Lexington. In July 1978 he joined the faculty of the University of Texas at Austin, where he is now a professor of electrical engineering. During the summer of 1979, he was a guest researcher at AEG-Telefunken, Ulm, West Germany. Since September 1983 he has held the Hayden Head Centennial Professorship of Engineering at the University of Texas.

Dr. Itoh is a fellow of the IEEE and a member of the Institute of Electronics and Communication Engineers of Japan, Sigma Xi, and Commission B of USNC/URSI. He serves on the administrative committee of the IEEE Microwave Theory and Techniques Society and is the editor of *IEEE Transactions on Microwave Theory and Techniques*. He is a professional engineer registered in the State of Texas.

1. Introduction

In antenna applications it is necessary to use some form of transmission line to connect the antenna to a transmitter or receiver. The purpose of this chapter is to provide the essential propagation characteristics of the more common forms of transmission lines.

The useful transmission line equations are summarized in Section 2, where most of the quantities are defined and tabulated in tables for convenience. In Sections 3 to 6, four classes of transmission lines are discussed. The classifications, based on the type of the fundamental-mode propagation, are transverse-electromagnetic (TEM) lines, planar quasi-TEM lines, transverse-electric/transverse-magnetic (TE/TM) waveguides, and hybrid-mode waveguides.

A TEM transmission line can support a TEM wave in which both the electric and magnetic field vectors are always perpendicular to the direction of propagation. In the planar quasi-TEM class the transmission lines usually have mixed dielecric boundaries and, therefore, cannot support a pure TEM wave. However, the longitudinal component of the fields is small, so it is neglected under the "quasi-TEM" approximations.

In the transverse-electric (TE) waves, sometimes called H-waves, the electric field vector is always perpendicular to the direction of propagation. In the transverse-magnetic (TM) waves, sometimes called E-waves, the magnetic field vector is always perpendicular to the direction of propagation.

In the case of hybrid-mode waveguides the modal field possesses relatively strong longitudinal components of electric and magnetic fields. Therefore, the quasi-TEM approximation technique cannot be applied.

In the following discussion, quantities are defined in SI units. The formulas are given under the assumption that all the materials are nonmagnetic and loss-free, unless specified.

2. Transmission Line Equations

In conventional analysis a transmission line is often represented by a series of equivalent networks. Fig. 1 shows the equivalent network for a short section Δx. The series impedance and the shunt admittance per unit length are

$$Z = R + j\omega L \tag{1a}$$

$$Y = G + j\omega C \tag{1b}$$

where ω is the angular frequency and R, L, G, and C are distributed resistance, inductance, conductance, and capacitance per unit length, respectively. The volt-

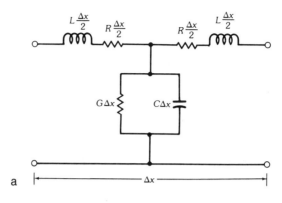

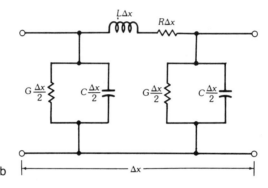

Fig. 1. Equivalent T and π networks of a short section Δx of a transmission line. (*After Jordan [1]*) (a) T network. (b) π network.

age (V) and current (I) waves along the transmission line are described by the following equations:

$$\frac{dV}{dz} = -ZI = -\gamma Z_0 I \tag{2a}$$

$$\frac{dI}{dz} = -YV = -\gamma Y_0 V \tag{2b}$$

From these equations expressions for important transmission line parameters can be derived and they are summarized in Tables 1 and 2. Fig. 2 defines the reference planes for impedance relations and reflection coefficients. The expressions for voltage, impedance, etc., are generally for the quantities at the input terminal of the line in terms of those at the output terminal. If it is desired to find the quantities at the output in terms of those at the input, it is simply necessary to interchange the subscripts 1 and 2 in the equations and to place a minus sign before ℓ wherever it appears.

The equations summarized here are primarily for transmission lines operating

Table 1. Summary of Transmission Line Equations (*After Ragan [2], © 1948 McGraw-Hill Book Company*)

Item No.	Quantity	General Line	Expression* Ideal Line	Approximation, Low-Loss Line
1	Propagation constant $\gamma = \alpha + j\beta$	$\sqrt{(R + j\omega L)(G + j\omega C)}$	$j\omega\sqrt{LC}$	See items 2 and 3
2	Phase constant β	$\mathrm{Im}\{\gamma\}$	$\omega\sqrt{LC} = \dfrac{\omega}{v_p} = \dfrac{2\pi}{\lambda_g}$	$\beta'\left[1 + \dfrac{1}{2}\left(\dfrac{\alpha_c}{\beta'} - \dfrac{\alpha_d}{\beta'}\right)^2\right]$
3	Attenuation constant α	$\mathrm{Re}\{\gamma\} = -\dfrac{1}{2P}\dfrac{dP}{d\ell}$	0	$\alpha_c + \alpha_d = \dfrac{R}{2Z_0'} + \dfrac{G}{2Y_0'}$
4	Characteristic impedance $Z_0 = 1/Y_0$	$\sqrt{\dfrac{R + j\omega L}{G + j\omega C}}$	$\sqrt{\dfrac{L}{C}}$	$Z_0'\left[1 + \dfrac{1}{2}\left(\dfrac{\alpha_c}{\beta'} - \dfrac{\alpha_d}{\beta'}\right)\left(\dfrac{\alpha_c}{\beta'} + \dfrac{3\alpha_d}{\beta'}\right)\right] - jZ_0'\left(\dfrac{\alpha_c}{\beta'} - \dfrac{\alpha_d}{\beta'}\right)$
5	Input impedance Z_1	$Z_0\dfrac{Z_2 + Z_0\tanh\gamma\ell}{Z_0 + Z_2\tanh\gamma\ell}$	$Z_0\dfrac{Z_2 + jZ_0\tan\beta\ell}{Z_0 + jZ_2\tan\beta\ell}$	
6	Impedance of short-circuited line ($Z_2 = 0$)	$Z_0\tanh\gamma\ell$	$jZ_0\tan\beta\ell$	$Z_0\dfrac{\alpha\ell + j\tan\beta\ell}{1 + j\alpha\ell\tan\beta\ell}$
7	Impedance of open-circuited line ($Z_2 = \infty$)	$Z_0\coth\gamma\ell$	$-jZ_0\cot\beta\ell$	$Z_0\dfrac{1 + j\alpha\ell\tan\beta\ell}{\alpha\ell + j\tan\beta\ell}$
8	Impedance of a line an odd number of quarter-wavelengths long ($\beta\ell = n\pi + \pi/2$)	$Z_0\dfrac{Z_2 + Z_0\coth\alpha\ell}{Z_0 + Z_2\coth\alpha\ell}$	$\dfrac{Z_0^2}{Z_2}$	$Z_0\dfrac{Z_0 + Z_2\alpha\ell}{Z_2 + Z_0\alpha\ell}$
9	Impedance of a line an integral number of half-wavelengths long ($\beta\ell = n\pi$)	$Z_0\dfrac{Z_2 + Z_0\tanh\alpha\ell}{Z_0 + Z_2\tanh\alpha\ell}$	Z_2	$Z_0\dfrac{Z_2 + Z_0\alpha\ell}{Z_0 + Z_2\alpha\ell}$
10	Voltage V_1 along line	$V_i(1 + \Gamma_0 e^{-2\gamma\ell})$	$V_i(1 + \Gamma_0 e^{-2j\beta\ell})$	
11	Current I_1 along line	$I_i(1 - \Gamma_0 e^{-2\gamma\ell})$	$I_i(1 - \Gamma_0 e^{-2j\beta\ell})$	
12	Voltage reflection coefficient at $z = 0$, Γ_0	$\dfrac{Z_2 - Z_0}{Z_2 + Z_0}$	$\dfrac{Z_2 - Z_0}{Z_2 + Z_0}$	

*The terms not defined above are as follows: $\beta' = \sqrt{LC}$ = phase constant, neglecting losses, $Z_0' = \sqrt{L/C}$ = characteristic impedance, neglecting loss, $Y_0' = 1/Z_0'$, V_i = incident voltage, I_i = incident current, λ_g = wavelength measured along line (guide wavelength), v_p = phase velocity of line, equals velocity of light in dielectric of line for an ideal line, α_c = conductor loss, and α_d = dielectric loss. Other notations refer to Fig. 2.

Table 2. Some Miscellaneous Relations in Low-Loss Transmission Lines
(After Ragan [2], © 1948 McGraw-Hill Book Company)

Item	Equation	Explanation
1a	$r = \dfrac{1 + |\Gamma|}{1 - |\Gamma|}$	r = voltage standing-wave ratio
1b	$|\Gamma| = \dfrac{r - 1}{r + 1}$	$|\Gamma|$ = magnitude of reflection coefficient
2a	$\Gamma = \dfrac{R - Z_0}{R + Z_0}$	Γ = reflection coefficient (real) at a point in a line where the impedance is real (R). Point may be at an actual resistive load or at a voltage max. or min. in standing-wave pattern
2b	$r = \dfrac{R}{Z_0};\ R = rZ_0$	Conditions for $R > Z_0$, i.e., at voltage maximum
2c	$r = \dfrac{Z_0}{R};\ R = \dfrac{1}{r}Z_0$	Conditions for $R < Z_0$, i.e., at voltage minimum
3a	$\dfrac{P_r}{P_i} = |\Gamma|^2 = \left(\dfrac{r-1}{r+1}\right)^2$	P_r = power in wave reflected by discontinuity or mismatched load
3b	$\dfrac{P_t}{P_i} = 1 - |\Gamma|^2 = \dfrac{4r}{(r+1)^2}$	P_i = power in incident wave P_t = power in transmitted (or absorbed) wave
4	$\dfrac{P_b}{P_m} = \dfrac{1}{r}$	P_b = *net power* transmitted to load at onset of breakdown in a line in which a vswr = r exists P_m = same when line is matched, $r = 1$
5	$\dfrac{\alpha_r}{\alpha_m} = \dfrac{1 + \Gamma^2}{1 - \Gamma^2} = \dfrac{r^2 + 1}{2r}$	α_m = ordinary attenuation constant; matched line, $r = 1$ α_r = attenuation constant allowing for increased *ohmic* loss caused by standing waves (vswr = r)
6a	$r_{\max} = r_1 r_2$	Resultant vswr when two separate mismatches combine in worst phase
6b	$r_{\min} = \dfrac{r_2}{r_1};\ r_1 < r_2$	Resultant when they combine in best phase
6c	$r_{\max} = r_1 r_2 r_3 \cdots r_n$	Resultant for n mismatches, worst phase
6d	$r_{\min} = \dfrac{r_n}{r_1 r_2 \cdots r_{n-1}}$ $r_1 < r_2 < r_2 < \cdots < r_n$	Resultant for n mismatches, best phase. If this gives $r_{\min} < 1$, then $r_{\min} = 1$
7a	$|\Gamma| = \dfrac{|X|}{\sqrt{X^2 + 4}}$	Relations for the case of a normalized reactance X in series with resistance Z_0
7b	$r = \dfrac{\sqrt{X^2 + 4} + |X|}{\sqrt{X^2 + 4} - |X|}$	
7c	$|X| = \dfrac{r - 1}{\sqrt{r}}$	

Transmission Lines and Waveguides

Table 2, *continued.*

Item	Equation	Explanation
8a	$\|\Gamma\| = \dfrac{\|B\|}{\sqrt{B^2 + 4}}$	Relations for the case of a normalized susceptance B shunting conductance Y_0
8b	$r = \dfrac{\sqrt{B^2 + 4} + \|B\|}{\sqrt{B^2 + 4} - \|B\|}$	
8c	$\|B\| = \dfrac{r - 1}{\sqrt{r}}$	

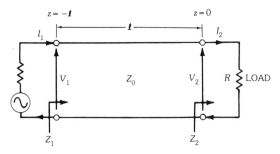

Fig. 2. Transmission line with generator and load.

in the TEM and quasi-TEM modes. The equations are accurate, according to conventional transmission line theory, and are applicable from the lowest power and communication frequencies, from zero frequency up to the frequency where the higher-order mode begins to appear on the line. The information is also valid for non-TEM waveguides under single-mode operation in general, except for those expressions in which the quantities R, L, G, and C are involved.

In addition to the exact equations, low-loss approximation yields simple expressions that are specially adapted for use in radio-frequency problems. In this case the quadratic $(\alpha^2 z^2)$ and higher powers in the expansion of $\exp(\alpha z)$, etc., are neglected. Thus, when $\alpha z = 0.1$ Np, the error in the approximate expressions is on the order of 1 percent.

In practice, the line loss is small and may be neglected in calculation of the characteristic impedance and phase velocity. The attenuation constant, including conductor loss and dielectric loss, may then be obtained by a perturbation method. In all cases it is assumed that the skin depth is small compared with the dimensions of the conductors and that the constructing materials are nonmagnetic. The conductor loss is normally a complicated function of frequency, geometry, and material constants, whereas the dielectric loss in nepers per meter takes the simple form

$$\alpha_d = \frac{\pi}{\lambda_g}(\tan \delta)_{\text{eff}} \qquad (3)$$

where the guide wavelength $\lambda_g = 2\pi/\beta'$ is obtained by neglecting the loss in the media and $(\tan \delta)_{\text{eff}}$ is the effective loss tangent of the structure. For TEM lines surrounded by a homogeneous dielectric, the effective loss tangent is simply the loss tangent of the dielectric. The conductor loss and the effective loss tangent for non-TEM lines will be given later for each case.

The peak power-handling capacity of a transmission line is determined by the maximum allowable electric field intensity E_b in the dielectric. The maximum E_b, or dielectric strengths, for some commonly used dielectrics are given in Table 3. From this table 3.0 kV/mm is the theoretical maximum for air dielectric at normal temperature and atmosphere pressure, but for proper derating a value of 2.0 kV/mm is more practical. A solid dielectric may be used to obtain higher power level. However, any air pockets within the dielectric may not be able to support the increased **E**-field intensity. A common method of increasing the power limit is to use pressurized air, for which the maximum power is proportional to the square of the absolute pressure. It should be pointed out that in an unmatched transmission line, the maximum power is inversely proportional to the standing-wave ratio (refer to item 4 of Table 2).

3. TEM Transmission Lines

The characteristics of lossless uniform TEM lines are completely described by two parameters, the phase velocity v_p and the characteristic impedance Z_0. The latter is a function of the line cross-sectional dimensions and dielectric constant. The phase velocity in meters per second is given simply by

$$v_p = c/\sqrt{\epsilon_r} \qquad (4)$$

where c is the free-space speed of light and ϵ_r is the relative dielectric constant of the filling material. Equation 4 is valid for all TEM structures. Consequently, the characteristic impedance becomes the only parameter necessary to characterize the transmission line.

Table 3. Properties of Commonly Used Dielectric Materials

Material	Dielectric Constant ϵ_r	Loss Tangent at 10 GHz ($\times 10^{-4}$)	Thermal Conductivity K (W/m/°C)	Dielectric Strength (kV/mm)
Air	1.0	≈ 0	0.024	3.0
Alumina	9.7	2.0	30.0	400.0
Berylium oxide	6.6	1.0	250.0	
Cuflon	2.1	5.0		78.7
Gallium arsenide	12.9	16.0	30.0	35.0
Polystyrene	2.53	4.7	0.15	28.0
Quartz (fused)	3.78	1.0	1.0	1000.0
RT/Duroid 5880	2.2	10.0	0.26	11.8
Sapphire	11.7	1.0	40.0	400.0
Silicon	11.7	50.0	90.0	30.0

Transmission Lines and Waveguides

Table 4 is a summary of the impedance formulas for common TEM lines.* They are roughly grouped into three families, namely, open-wire lines, shielded-wire lines, and strip lines. Open-wire lines may consist of either a pair of conductors carrying the going and returning currents (nos. 1 and 2), or of a multiplicity of conductors interconnected in different manners (nos. 3, 4, and 5). The arrangement is sometimes used in conjunction with a ground plate to which the wires are parallel (nos. 6, 7, and 8). Generally, the wires have small dimensions compared with the spacing between the wires and between the wires and the ground plate. Furthermore, the spacing is assumed to be very much less than a wavelength at the operating frequency. This type of transmission line has the advantage of simplicity and economy. However, the open-wire lines become unusable at high frequency because of excessive radiation loss.

Surrounding the wires of a transmission line with a shielding conductor (e.g., nos. 9 through 19) effectively overcomes the radiation problem. The popular coaxial line (no. 16) is essentially a self-shielded transmission, and is widely used for propagation of microwave power.

Strip transmission lines (nos. 20 through 28) are those in which the conductors have the form of flat plates or strips that are amenable to photolithographic techniques for mass production of circuit components. Both open and shielded types exist, but open structures suffer radiation problems at higher frequencies.

For a pair of wires near ground or in a shield, two kinds of modes may be excited:

(a) both "positive" with respect to ground; this is called "even-mode" excitation; or
(b) one "positive" and one "negative" with respect to ground; this is called "odd-mode" excitation.

These modes of excitation are illustrated schematically in Fig. 3, for the case of simple two-wire line near ground. The characteristic impedances, Z_{0e} and Z_{0o}, associated with these modes of excitation are defined as the input impedance of an infinite length of one line, in the presence of the second line, when both are excited in the appropriate manner [(a) or (b) above].

Two-Wire Line

Fig. 4 shows the cross section and field configuration of a two-wire line with wires of diameter d, spaced at a center-to-center distance D. The characteristic impedance in ohms of this line is

$$Z_0 = \frac{119.9}{\sqrt{\epsilon_r}} \cosh^{-1}\left(\frac{D}{d}\right) = \frac{119.9}{\sqrt{\epsilon_r}} \ln\left(\frac{D + \sqrt{D^2 - d^2}}{d}\right) \tag{5}$$

and is plotted in Fig. 5.

The conductor loss in nepers per meter is

---

*Most of the materials in this table are from [1] and [3]. No. 19 is a quasi-TEM structure; it is included here for convenience of comparison.

Table 4. Impedance Formulas for Common TEM Transmission Lines

No.	Structure	Cross Section	Impedance Formulas*	Range of Validity
1	Open two-wire line		$Z_0 = 119.9 \ln(x + \sqrt{x^2 - 1})$ $x = D/d$	Exact; no limit
2	Balanced two-wire unequal diameters		$Z_0 = 59.95 \ln(x + \sqrt{x^2 - 1})$ $x = \dfrac{1}{2}\left[\dfrac{4D^2}{d_1 d_2} - \dfrac{d_1}{d_2} - \dfrac{d_2}{d_1}\right]$	Exact; no limit
3	Three-wire line		$Z_0 = 476.6 \ln(1.59\, D/d)$	$d \ll D$
4	Balanced four-wire		$Z_0 = 59.95 \ln\left\{\dfrac{2D_2}{d}\left[1 + \left(\dfrac{D_2}{D_1}\right)^2\right]^{-1/2}\right\}$	$d \ll D_1, D_2$

5	Five-wire line		$Z_0 = 74.95 \ln(D/0.993 d)$	$d \ll D$
6	Single wire near ground		$Z_0 = 59.95 \ln(x + \sqrt{x^2 - 1})$ $x = 2h/d$	Exact; no limit
7	Open two-wire near ground		$Z_{0e} = 29.98 \ln\{(4h/d)[1 + (2h/D)^2]^{1/2}\}$ $Z_{0o} = 119.9 \ln\{(2D/d)[1 + (D/2h)^2]^{-1/2}\}$	$d \ll D, h$
8	Balanced two-wire near ground		$Z_0 = 119.9 \ln\left\{\dfrac{2D}{d}\left[1 + \dfrac{D^2}{4h_1 h_2}\right]^{-1/2}\right\}$	$d \ll D, h_1, h_2$
9	Single wire between two grounded planes		$Z_0 = 15\ln[1 + 1.314x + \sqrt{(1.314x)^2 + 2x}]$ $x = (1 + 2h/d)^4 - 1$	Less than 0.1%; no limit

Table 4, *continued.*

No.	Structure	Cross Section	Impedance Formulas*	Range of Validity
10	Two-wire line between grounded planes		$Z_{0e} = 59.95 \ln\left\{\coth\left(\dfrac{\pi D}{2h}\right) \coth\left(\dfrac{\pi d}{4h}\right)\right\}$ $Z_{0o} = 59.95 \ln\left\{\tanh\left(\dfrac{\pi D}{2h}\right) \coth\left(\dfrac{\pi d}{4h}\right)\right\}$	$d/h < 0.25$ $D/h > 3d/h$
11	Balanced line between grounded planes		$Z_0 = 119.9 \ln(2h/\pi d)$	$d \ll h$
12	Single wire in trough		$Z_0 = 59.95 \ln\left\{\tanh\left(\dfrac{\pi h}{W}\right) \coth\left(\dfrac{\pi d}{4W}\right)\right\}$	$d/W < 0.25$ $h/W > 3d/2W$
13	Balanced line in trough		$Z_{0o} = 119.9 \ln[2W/\pi d(A^{1/2})]$ $A = \operatorname{cosec}^2(\pi D/W) + \operatorname{cosech}^2(2\pi h/W)$	$d \ll D, W, h$

#	Name	Figure	Formula	Accuracy
14	Single wire in square enclosure		$Z_0 = 59.95\ln(1.0787 D/d)$	Less than 1.5%; $d/D < 0.8$
15	Two-wire line in rectangular enclosure	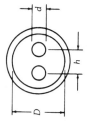	$Z_{0e} = 119.9\left[\ln\dfrac{4h\coth\left(\dfrac{\pi D}{2h}\right)}{\pi d} + \displaystyle\sum_{m=1}^{\infty}(-1)^m \ln\dfrac{1+\dfrac{\cosh^2(\pi D/2h)}{\sinh^2(m\pi W/2h)}}{1-\dfrac{\cosh^2(\pi D/2h)}{\cosh^2(m\pi W/2h)}}\right]$ $Z_{0o} = 119.9\left[\ln\dfrac{4h\tanh\left(\dfrac{\pi D}{2h}\right)}{\pi d} + \displaystyle\sum_{m=1}^{\infty}\ln\dfrac{1+\dfrac{\sinh^2(\pi D/2h)}{\cosh^2(m\pi W/2h)}}{1-\dfrac{\sinh^2(\pi D/2h)}{\sinh^2(m\pi W/2h)}}\right]$	$d \ll D, W, h$
16	Coaxial line		$Z_0 = 59.95\ln(D/d)$	Exact; no limit
17	Shielded two-wire line		$Z_{0e} = 29.98\ln[(v/2\sigma^2)(1-\sigma^4)]$ $Z_{0o} = 119.9\ln\{2v[(1-\sigma^2)/(1+\sigma^2)]\}$ $v = h/d;\ \sigma = h/D$	$d \ll D, h$
18	Eccentric line		$Z_0 = 119.9\ln(x + \sqrt{x^2-1})$ $x = \dfrac{1}{2}[(D/d)+(d/D)-(4c^2/dD)]$	Exact; no limit

Table 4, *continued.*

No.	Structure	Cross Section	Impedance Formulas*	Range of Validity
19	Air coaxial with dielectric supporting wedge		$Z_0 \cong \dfrac{59.95 \ln(D/d)}{[1 + (\epsilon_r - 1)(\theta/2\pi)]^{1/2}}$ θ: in rad	—
20	Broadside-coupled stripline		$Z_0 = 2(Z_0 \text{ of No. 22})$	Approx. 0.1%; $W/h < 1000$
21	Edge-coupled stripline		$Z_0 = 376.7\, K(x)^\dagger$ $x = S/(S + 2W)$	Exact for $t = 0$
22	Single-dielectric microstrip line		$Z_0 = 59.95 \ln\left\{\dfrac{6}{u} + \dfrac{2\pi - 6}{u} \exp\left[-\left(\dfrac{30.666}{u}\right)^{0.7528}\right] + \sqrt{1 + \left(\dfrac{2}{u}\right)^2}\right\}$ $u = \dfrac{W}{h} + \dfrac{t}{\pi h} \ln\left[1 + \dfrac{h}{t}\dfrac{4e}{\coth^2\sqrt{6.517\,W/h}}\right]$	Approx. 0.1%; $W/h < 1000$

#	Type	Diagram	Formulas	Accuracy
23	Triplate stripline		$Z_0 = 30 \ln\left\{1 + \dfrac{8h}{\pi W'}\left[\dfrac{16h}{\pi W'} + \sqrt{\left(\dfrac{16h}{\pi W'}\right)^2 + 6.27}\right]\right\}$ $W' = W + \dfrac{t}{\pi}\ln\left\{e\left[\left(\dfrac{1}{4h/t + 1}\right)^2 + \left(\dfrac{1/4\pi}{W/t + 1.1}\right)^m\right]^{-1/2}\right\}$ $m = 2/(1 + t/3h)$	0.1% or less $W'/h < 20$
24	Rounded-edge triplate stripline		$Z_0 = \dfrac{188.34 \ln x}{\pi + 2[(W/D - t/D)/(1 - t/D)]\ln x}$ $x = 4/(\pi t/b)$	Less than 1%; $t/D \leqq 0.5$
25	Edge-coupled triplate stripline		$Z_{0e} = 29.98\pi K(k_e)$ $Z_{0o} = 29.98\pi K(k_0)$ $k_e = \tanh\left(\dfrac{\pi W}{2D}\right)\tanh\left(\dfrac{\pi}{2}\dfrac{W+S}{D}\right)$ $k_0 = \tanh\left(\dfrac{\pi W}{2D}\right)\coth\left(\dfrac{\pi}{2}\dfrac{W+S}{D}\right)$	Exact; for $t = 0$
26	Broadside-coupled horizontal stripline		$Z_{0e} = 59.95\pi/[(W/D)/(1 - h/D) + C_e]$ $Z_{0o} = 59.95\pi/[(W/D)/(1 - h/D) + (W/h) - C_0]$ $C_e = 0.4413 - H/\pi;\ C_0 = HD/h\pi$ $H = \ln(1 - h/D) + \ln(h/D)/(D/h - 1)$	$W/h > 0.35$ $W/D \geqq 0.35(1 - h/D)$

Table 4, *continued.*

No.	Structure	Cross Section	Impedance Formulas*	Range of Validity
27	Square coaxial line		$Z_0 = \dfrac{47.086(1 - W/D)}{0.279 + 0.721 W/D}$; $\quad W/D > 0.25$ $Z_0 = 59.37 \ln(0.9259 \, D/W)$; $W/D \leqq 0.5$	Less than 1%; no limit
28	Rectangular coaxial line		$Z_0 = 59.95 \ln\left(\dfrac{1 + W'/D}{W/D + t/D}\right)$	10% for $t/D < 0.3$ $W/W' < 0.8$

*Air dielectric is assumed for the formulas; for transmission lines with material of relative dielectric constant ϵ_r, the impedance is $Z_0' = Z_0/\sqrt{\epsilon_r}$.

† $K(x) = \begin{cases} \pi^{-1} \ln[2(1 + \sqrt{x})/(1 - \sqrt{x})]; & 1/\sqrt{2} \leqq x \leqq 1 \\ \pi/\ln\{2[1 + (1 - x^2)^{1/4}]/[1 - (1 - x^2)^{1/4}]\}; & 0 \leqq x \leqq 1/\sqrt{2} \end{cases}$

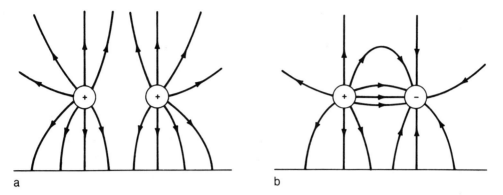

Fig. 3. Schematic representation of even-mode and odd-mode excitation for a typical pair of coupled transmission lines. (*a*) Even-mode excitation. (*b*) Odd-mode excitation.

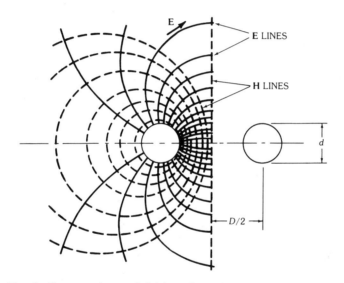

Fig. 4. Cross section and field configuration of a two-wire line.

$$\alpha_c = 5.274 \times 10^{-6} \sqrt{\frac{\epsilon_r f}{\sigma}} \frac{1}{d[\cosh^{-1}(D/d)]} \frac{D}{\sqrt{D^2 - d^2}} \qquad (6)$$

where σ is the conductivity of the conductors.

The power-handling capacity in watts is

$$P = \frac{E_b^2 d^2 \sqrt{\epsilon_r}}{239.81} \frac{D - d}{D + d} \cosh^{-1}\left(\frac{D}{d}\right) \qquad (7)$$

where E_b, the maximum allowable field intensity, occurs at the innermost surface of each conductor.

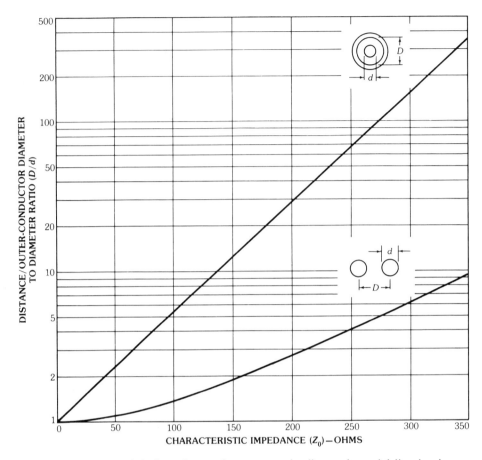

Fig. 5. Characteristic impedance of open two-wire line and coaxial line in air.

Circular Coaxial Line

The circular coaxial line consists of two cylindrical conductors which are located coaxially, one within the other, as shown in Fig. 6. For a line with inner-conductor diameter d and outer-conductor diameter D, the characteristic impedance in ohms is

$$Z_0 = \frac{59.95}{\sqrt{\epsilon_r}} \ln\left(\frac{D}{d}\right) \tag{8}$$

This relation is also plotted in Fig. 5.

The cutoff wavelength in meters of the first higher-order mode is given approximately by

$$\lambda_c = \frac{\pi}{2}\sqrt{\epsilon_r}(D + d) \tag{9}$$

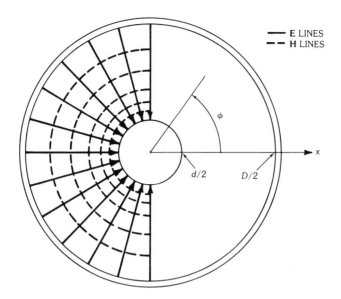

Fig. 6. Cross section and field configuration of a coaxial line.

The conductor loss in nepers per meter for the fundamental mode is

$$\alpha_c = 5.274 \times 10^{-6} \sqrt{\frac{\epsilon_r f}{\sigma}} \left(\frac{1}{D} + \frac{1}{d}\right) \frac{1}{\ln(D/d)} \tag{10}$$

The minimum conductor loss occurs when $D/d = 3.5911$ and is given in nepers per meter by

$$\alpha_{c,\min} = 1.894 \times 10^{-5} \frac{1}{D} \sqrt{\frac{\epsilon_r f}{\sigma}} \tag{11}$$

The power-handling capacity in watts is

$$P = \frac{E_b^2 d^2 \sqrt{\epsilon_r}}{479.6} \ln\left(\frac{D}{d}\right) \tag{12}$$

where E_b is the electric field intensity at the surface of the center conductor. For a fixed frequency the maximum power-handling capacity occurs when $D/d = \sqrt{e} = 1.6487$ and is

$$P_{\max} = 1.043 \times 10^{-3} E_b^2 d^2 \sqrt{\epsilon_r} \tag{13}$$

For air-dielectric coaxial lines, insulating beads are often used to support the center conductor, as shown in Fig. 7. The use of the beads, with relative dielectric constant ϵ_1, changes the line constants. For thin beads at frequent intervals, i.e., $W \ll S \ll \lambda/4$, the new characteristic impedance Z_0' in ohms is

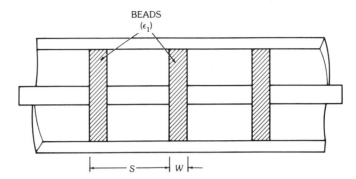

Fig. 7. Bead-supported coaxial line.

$$Z_0' = Z_0/[1 + (\epsilon_1 - 1)(W/S)]^{1/2} \qquad (14)$$

"Flexible" coaxial line is probably the most common means of connecting together many separated components of a radio-frequency (rf) system. It consists of a solid or stranded inner conductor, a plastic-dielectric support, and a braided outer conductor. Although the transmission loss is relatively high because of the dielectric loss, the convenience often outweighs this factor in applications where some loss is tolerable. Table 5 summarizes the characteristics of the more common lines.

Figs. 8 and 9 illustrate the approximate attenuation and power-handling capacity of general-purpose rf coaxial cables up to their practical upper frequency limit. For the RG-type cables, only the number is given (for instance, the curve for RG-218/U is labeled 218). The curves for rigid copper coaxial lines are labeled with the diameter of the line only, as ⅞ in C. Typical curves are also shown for three sizes of 50-Ω semirigid cables labeled by size in inches, as ⅞ in S. In Fig. 9 the curves are for unity voltage standing-wave ratio. For polyethylene cables an inner-conductor maximum temperature of 80°C is specified. For high-temperature cables (types 211, 228, 225, and 227) the inner-conductor temperature is 250°C. An ambient temperature of 40°C is assumed.

Triplate Stripline

The cross section and field configuration of a triplate stripline is shown in Fig. 10a, where ϵ_r is the relative dielectric constant of the medium filling the entire space between the two ground plates. Practical stripline circuits are frequently constructed from copper-clad printed-circuit boards.

When $t = 0$, the characteristic impedance and a design formula are given by [4]

$$Z_0 = \frac{29.98}{\sqrt{\epsilon_r}} \ln\{1 + \tfrac{1}{2}\chi[\chi + \sqrt{\chi^2 + 6.27}\,]\} \qquad (15a)$$

$$\frac{W}{h} = \frac{16}{\pi} \frac{\sqrt{\exp(4\pi r) + 0.568}}{\exp(4\pi r) - 1} \qquad (15b)$$

Table 5. List of Radio-Frequency Cables (Adapted from Jordan [1], © 1985 Howard W. Sams & Company; reprinted with permission)

Class of Cable		JAN Type RG-	Inner Conductor*	Dielectric Material†	Nominal Diameter of Dielectric (in)	Shielding Braid*	Protective Covering‡	Nominal Overall Diameter (in)	Weight (lb/ft)	Nominal Capacitance (pF/ft)	Maximum Operating Voltage (rms)	Remarks
50 Ω	Single-braid	58C/U	19/0.0071 tc	A	0.116	tc	II	0.195	0.029	28.5	1900	
		213/U	7/0.0296 c	A	0.285	c	II	0.405	0.120	29.5	5000	
		215/U	7/0.0296 c	A	0.285	c	IIa	0.475max	0.160	29.5	5000	RG-213/U with armor
		218/U	0.195 c	A	0.680	c	II	0.870	0.491	29.5	11000	Low-attenuation high-power
		219/U	0.195 c	A	0.680	c	IIa	0.945max	0.603	29.5	11000	RG-218/U with armor
		220/U	0.260 c	A	0.910	c	II	1.120	0.745	29.5	14000	Low-attenuation high-power
		221/U	0.260 c	A	0.910	c	IIa	1.195max	0.925	29.5	14000	RG-220/U with armor
	Double-braid	55B/U	0.032 sc	A	0.116	sc	III	0.206	0.032	28.5	1900	
		212/U	0.0556 sc	A	0.185	sc	II	0.332	0.093	28.5	3000	
		214/U	7/0.0296 sc	A	0.285	sc	II	0.425	0.158	30.0	5000	
		217/U	0.106 c	A	0.370	c	II	0.545	0.236	29.5	7000	
		223/U	0/035 sc	A	0.116	sc	II	0.216	0.036	28.5	1900	
		224/U	0.106 c	A	0.370	c	IIa	0.615max	0.282	29.5	7000	RG-217/U with armor
75 Ω	Single-braid	11A/U	7/26 AWG tc	A	0.285	c	II	0.412	—	20.5	5000	
		12A/U	7/26 AWG tc	A	0.285	c	IIa	0.475	—	20.5	5000	RG-11A/U with armor
		34B/U	7/0.0249 c	A	0.460	c	II	0.630	0.231	21.5	6500	High-power low-attenuation
		35B/U	0.1045 c	A	0.680	c	IIa	0.945max	0.480	21.5	10000	High-power low-attenuation
		59B/U	0.0230 ccs	A	0.146	c	II	0.242	—	21.0	2300	
		84A/U	9.1045 c	A	0.680	c	II*	1.000	1.325	21.5	10000	*Same as RG-35B/U with lead sheath
		85A/U	0.1045 c	A	0.680	c	II*	1.565max	2.910	21.5	10000	*Same as RG-84A/U, with special armor for underground installations
		164/U	0.1045 c	A	0.680	c	II	0.870	0.490	21.5	10000	RG-35B/U without armor
		307A/U	17/0.0058 sc	A*	0.029	sc	III	0.270	—	20.0	400	*Foamed
	Double-braid	6A/U	21 AWG ccs	A	0.185	In: sc	II	0.332	—	20.0	2700	
		216/U	7/0.0159 tc	A	0.285	Out: c	II	0.425	0.121	20.5	5000	

Table 5, *continued.*

Class of Cable		JAN Type RG-	Inner Conductor*	Dielectric Material†	Nominal Diameter of Dielectric (in)	Shielding Braid*	Protective Covering‡	Nominal Overall Diameter (in)	Weight (lb/ft)	Nominal Capacitance (pF/ft)	Maximum Operating Voltage (rms)	Remarks
High temperature	Single-braid	144/U	7/0.179 sccs	F1	0.285	sc	X	0.410	0.120	20.5	5000	$Z = 75\ \Omega$
		178B/U	7/0.004 sccs	F1	0.034	sc	IX	0.075max	—	29.0	1000	$Z = 50\ \Omega$
		179B/U	Same as above	F1	0.063	sc	IX	0.105	—	20.0	1200	$Z = 95\ \Omega$
		180B/U	Same as above	F1	0.102	sc	IX	0.145	—	15.5	1500	$Z = 95\ \Omega$
		187A/U	7/0.004 asccs	F1	0.060	sc	VII	0.110	—	—	1200	$Z = 75\ \Omega$
		195A/U	Same as above	F1	0.102	sc	VII	0.155	—	—	1500	$Z = 95\ \Omega$
		196A/U	Same as above	F1	0.034	sc	VII	0.080	—	—	1000	$Z = 50\ \Omega$
		211A/U	0.190 c	F1	0.620	c	X	0.730	0.450	29.0	7000	Semiflexible; operating at −55 to +200°C. $Z = 50\ \Omega$
		228A/U	0.190 c	F1	0.620	c	Xa	0.795	0.600	29.0	7000	RG-211A/U with armor. $Z = 50\ \Omega$
		302/U	0.025 sc	F1	0.146	sc	IX	0.206	—	21.0	2300	$Z = 75\ \Omega$
		303/U	0.039 sccs	F1	0.116	sc	IX	0.170	—	28.5	1900	$Z = 50\ \Omega$
		304/U	0.059 sccs	F1	0.185	sc	IX	0.280	—	28.5	3000	$Z = 50\ \Omega$
		316/U	7/0.0067 asccs	F1	0.060	sc	IX	0.102	—	—	1200	$Z = 50\ \Omega$
	Double-braid	115/U	7/0.028 sc	F2	0.250	sc	X	0.375	—	29.5	5000	Slightly expandable. $Z = 50\ \Omega$
		142B/U	0.039 sccs	F1	0.116	sc	IX	0.195	—	28.5	1900	$Z = 50\ \Omega$
		225/U	7/0.0312 sc	F1	0.285	sc	X	0.430	0.176	29.5	5000	Semiflexible; operating at −55 to +200°C. $Z = 50\ \Omega$
		226/U	19/0.0254 sc	F2	0.370	c	X	0.500	0.247	29.0	7000	Slightly expandable. $Z = 50\ \Omega$
		227/U	7/0.0312 sc	F1	0.285	sc	Xa	0.490	0.224	29.5	5000	RG-225/U with armor. $Z = 50\ \Omega$
Pulse	Single-braid	26A/U	19/0.0117 tc	E	0.288	tc	IVa	0.505	0.189	50.0	10000	$Z = 48\ \Omega$
		27A/U	19/0.0185 tc	D	0.455	tc	IVa	0.670	0.304	50.0	15000 peak	$Z = 48\ \Omega$

Category	Type	Inner conductor	Dielectric	Diel. OD	Shield	Jacket	Overall OD	(in.)	(pF/ft)	V (peak)	Notes
Double-braid	25A/U	19/0.0117 tc	E	0.288	tc	IV	0.505	0.205	50.0	10000	Z = 48 Ω
	28B/U	19/0.0185 tc	D	0.455	In: tc Out: gs	IV	0.750	0.370	50.0	15000 peak	Z = 48 Ω
	64A/U	19/0.0117 tc	E	0.288	tc	IV	0.475max	0.205	50.0	10000	Z = 48 Ω
	156/U	7/21 AWG tc	AH*	0.285	In: tc Out: gs	II	0.540	0.211	30.0	10000	*1st layer A; 2nd layer H. Z = 50 Ω
	157/U	19/24 AWG tc	HAH*	0.455	↑	II	0.725	0.317	38.0	15000	*1st layer H; 2nd layer A; 3rd layer H. Z = 50 Ω
	158/U	37/21 AWG tc	HAH	0.455	↑	II	0.725	0.380	78.0	15000	Z = 25 Ω
	190/U	19/0.0117 tc	HJH*	0.380	↑	VIII	0.700	0.353	50.0	15000	Z = 50 Ω. *1st layer H; 2nd layer J; 3rd layer H.
	191/U	30 AWG c	HJH	1.065	↑	VIII	1.460	1.469	85.0	15000	Z = 25 Ω
Four braids	88/U	19/0.0117 tc	E	0.288	tc	III	0.515	—	50.0	10000	Z = 48 Ω
Low capacitance Single-braid	62A/U	0.0253 ccs	A	0.146	c	II	0.242	0.382	14.5	750	Z = 93 Ω
	63B/U	0.0253 ccs	A	0.285	c	II	0.405	0.082	10.0	1000	Low-capacitance. Z = 125 Ω
	79B/U	0.0253 ccs	A	0.285	c	IIa	0.475max	0.138	10.0	1000	RG-63B/U with armor. Z = 125 Ω
Double-braid	71B/U	0.0253 ccs	A	0.146	tc	III	0.250max	—	14.5	750	Low-capacitance. Z = 93 Ω
High attenuation Single-braid	301/U	7/0.0203 Karma wire	F1	0.185	Karma wire	IX	0.245	—	29.0	3000	High-attenuation. Z = 50 Ω
High delay Single-braid	65A/U	No. 32 Formex F. Helix diam. 0.128	A	0.285	c	II	0.405	0.096	44.0	1000	High-impedance high-delay line. Z = 950 Ω

*Diameter of strands given in inches (does not apply to AWG types). As, 7/0.0296 = 7 strands, each 0.0296-inch diameter. Conductors: c = copper, gs = galvanized steel, sc = silvered copper, tc = tinned copper, ccs = copper-covered steel, sccs = silvered copper-covered steel, ascs = annealed silvered copper-covered steel.

†Dielectric materials: A = polyethylene, D = layer of synthetic rubber between two layers of conducting rubber, E = layer of conducting rubber plus two layers of synthetic rubber, F1 = solid polytetrafluoroethylene (Teflon), F2 = semisolid or taped polytetrafluoroethylene (Teflon), H = conducting synthetic rubber, and J = insulating butyl rubber.

‡Jacket types: I = polyvinyl chloride (colored black), II = noncontaminating synthetic resin, III = noncontaminating synthetic resin (colored black), IV = chloroprene, V = Fiberglas silicone-impregnated varnish, VII = polytetrafluoroethylene, VIII = polychloroprene, IX = fluorinated ethylene propylene, and X = Teflon-tape moisture seal with double-braid type V jacket; the letter "a" means "with armor," e.g., Ia = I, with armor.

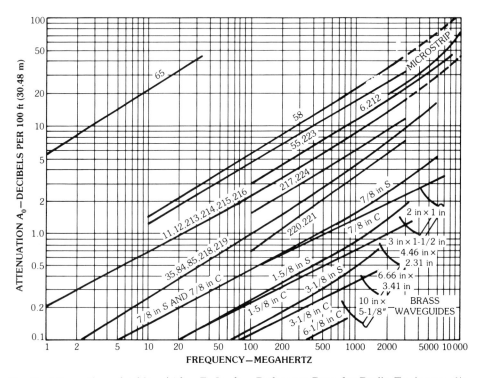

Fig. 8. Attenuation of cables. (*After E. Jordan*, Reference Data for Radio Engineers, *6/e*, © *1965 Howard W. Sams & Company, reproduced with permission*)

where

$$\chi = 16h/\pi W$$
$$r = Z_0\sqrt{\epsilon_r}/376.7$$

The relative error in Z_0 is less than 0.005 for $W/h < 20$.

When $t > 0$, the thickness effect may be taken into account by replacing the original strip with an equivalent infinitely thin strip of width $W' = W + \Delta W$, where

$$\Delta W = \frac{t}{\pi}\left\{1 - \frac{1}{2}\ln\left[\left(\frac{1}{4h/t + 1}\right)^2 + \left(\frac{1/4\pi}{W/t + 1.1}\right)^m\right]\right\} \quad (16a)$$

or

$$\Delta W = \frac{t}{\pi}\left\{1 - \frac{1}{2}\ln\left[\left(\frac{1}{4h/t + 1}\right)^2 + \left(\frac{1/4\pi}{W'/t - 0.26}\right)^m\right]\right\} \quad (16b)$$

in which

$$m = \frac{6}{3 + t/h}$$

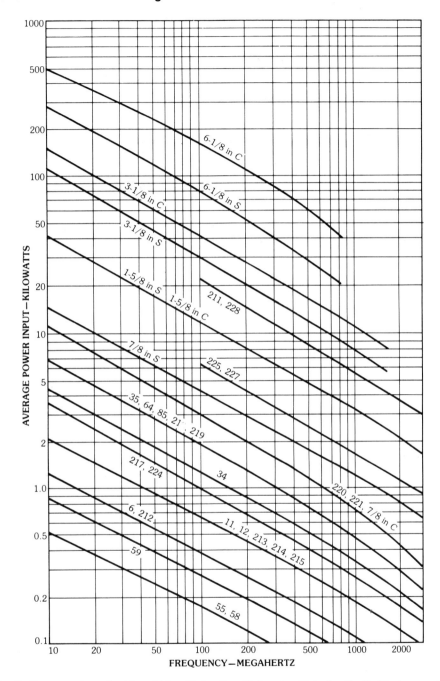

Fig. 9. Power rating of cables. (*After E. Jordan*, Reference Data for Radio Engineers, *6/e*, © *1965 Howard W. Sams & Company, reproduced with permission*)

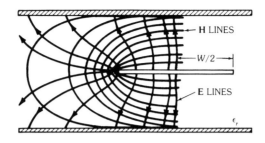

Fig. 10. Triplate stripline. (*a*) Cross section and field configuration. (*b*) Geometrical dimensions.

The relative error is less than 0.015. This adjustment enables a width conversion either way between strips with or without thickness. The small relative error makes a much smaller contribution to the relative error of the effective width and Z_0.

The conductor loss in nepers per meter is

$$\alpha_c = \frac{\beta'}{2Q} = \frac{\sqrt{\epsilon_r}}{Q}\frac{\pi}{\lambda_0} \tag{17}$$

with the dissipation factor ($1/Q$) given by

$$\frac{1}{Q} = P = 1 - Z_1/Z_\delta \tag{18}$$

where Z_1 is the characteristic impedance of the original structure, neglecting the loss, and Z_δ is the impedance of the modified structure as indicated by the dashed line in Fig. 10b, where δ is the skin depth defined by $\delta = \sqrt{2/\omega\mu_0\sigma}$.

The characteristic impedance and the dissipation factor of the triplate stripline are computed and plotted in Figs. 11 and 12.

The power-handling capacity of stripline is limited by field concentration at sharp corners of the conducting strip. Rounded-edge stripline (e.g., no. 24, Table 4) are necessary for high-power applications.

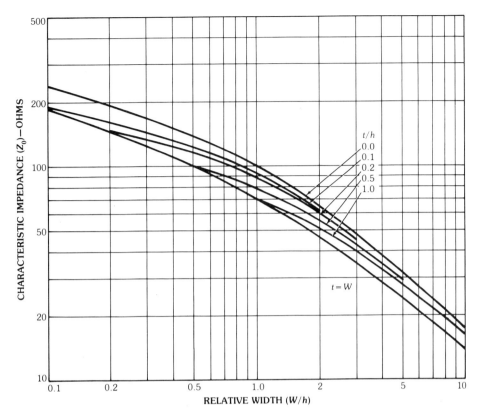

Fig. 11. Characteristic impedance of triplate stripline, showing the effect of thickness.

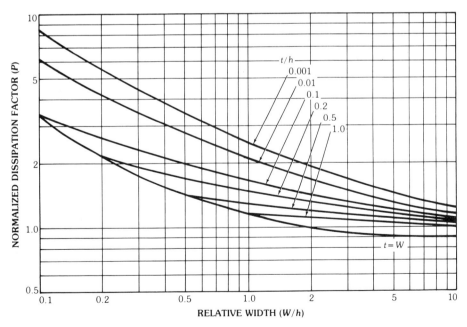

Fig. 12. The normalized dissipation factor of triplate stripline, showing the effect of thickness.

4. Planar Quasi-TEM Transmission Lines

Planar quasi-TEM transmission lines are transmission lines of metallic layers deposited on a dielectric substrate. The complete conductor pattern can be deposited and processed on a single dielectric substrate which may be supported by a metal ground plane. Beam-lead active and passive devices can be bonded directly to metal strips on the dielectric substrate. Such circuits can be fabricated at a substantially lower cost than waveguide or coaxial circuit configurations.

The electric parameters of these lines required for circuit design are the impedance, the attenuation, and the guide wavelength. These parameters can be approximated from the effective dielectric constant and the characteristic impedance of the corresponding air line. This section gives the line parameters for the most important cases of microstrip lines, coplanar waveguides, and coplanar strips.

Microstrip Lines

A microstrip line consists of a metal strip placed on a dielectric substrate supported by a metal ground plane as shown in Fig. 13. The upper dielectric is assumed to be a lossless air dielectric.

The quasi-TEM characteristic impedance Z_0 and the effective dielectric constant $\epsilon_{\text{eff}} = (\lambda_0/\lambda_g)^2$ are functions of structure and dielectric constant [5]:

$$Z_0\sqrt{\epsilon_{\text{eff}}} = Z_{01}(U_1) = 59.95 \ln\left\{\frac{6}{U_1} + \frac{2\pi - 6}{U_1}\exp\left[-\left(\frac{30.666}{U_1}\right)^{0.7528}\right] + \sqrt{1 + \left(\frac{2}{U_1}\right)^2}\right\} \quad (19\text{a})$$

$$\epsilon_{\text{eff}} = \frac{\epsilon_r + 1}{2} + \frac{\epsilon_r - 1}{2}\left(1 + \frac{10}{U}\right)^{-ab}[Z_{01}(U_1)/Z_{01}(U_2)]^2 \quad (19\text{b})$$

where

$$a = 1 + \frac{1}{49}\ln\left[\frac{U^4 + (U/52)^2}{U^4 + 0.432}\right] + \frac{1}{18.7}\ln\left[1 + \left(\frac{U}{18.1}\right)^3\right]$$

$$b = 0.564\left(\frac{\epsilon_r - 0.9}{\epsilon_r + 3}\right)^{0.053}$$

$$U = W/h$$

$$U_1 = \frac{W}{h} + \frac{t}{\pi h}\ln\left[1 + \frac{h}{t}\frac{4e}{\coth^2\sqrt{6.517\,W/h}}\right]$$

$$U_2 = \frac{W}{h} + \frac{t}{2\pi h}\ln\left[1 + \frac{h}{t}\frac{4e}{\coth^2\sqrt{6.517\,W/h}}\right]\left(1 + \frac{1}{\cosh\sqrt{\epsilon_r - 1}}\right)$$

For $t = 0$, the accuracy of these formulas is on the order of 0.1 percent. The computed data are plotted in Fig. 14.

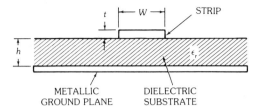

Fig. 13. Cross section of microstrip line. (*After Jordan [1], © 1985 Howard W. Sams & Company; reproduced with permission*)

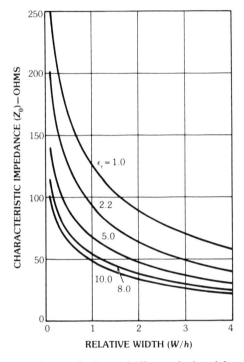

Fig. 14. Characteristic impedance of microstrip line, calculated from quasi-TEM formulas by Hammerstadt and Jansen. (*After Jordan [1], © 1985 Howard W. Sams & Company; reproduced with permission*)

The dielectric loss in a microstrip line in nepers per meter is [6]

$$\alpha_d = \frac{\pi}{\lambda_g}(\tan\delta)_{\text{eff}} = \frac{\pi}{\lambda_g}\left[\tan\delta \bigg/ \left(1 + \frac{1}{\epsilon_r}\frac{F-1}{F+1}\right)\right] \qquad (20)$$

where

$$F = \sqrt{1 + 10h/W}$$

and the guide wavelength λ_g is calculated from (19b).

The conductor loss in nepers per meter due to the dissipation in the strip and ground plane is

$$\alpha_c = \frac{\pi}{\lambda_g}(1 - Z_1/Z_\delta) \tag{21}$$

where Z_1 is the characteristic impedance of the corresponding air-filled lossless microstrip line obtained by (19), and Z_δ is calculated with the same formula by replacing W, h, and t with W', h', and t' such that

$$W' = W - \delta$$
$$h' = h + \delta \tag{22}$$
$$t' = t - \delta$$

where $\delta \ (= \sqrt{2/\omega\mu_0\sigma})$ is the skin depth. For accurate results the thickness of the conductors must be greater than about four times the skin depth.

At lower microwave frequencies the above quasi-TEM approximations are very accurate. As the frequency is increased, however, the characteristic impedance and the effective dielectric constant must be modified to be functions of frequency:

$$Z_0(f) = Z_0(0) \sqrt{\frac{\epsilon_{\text{eff}}(0)}{\epsilon_{\text{eff}}(f)}} \frac{\epsilon_{\text{eff}}(f) - 1}{\epsilon_{\text{eff}}(0) - 1} \tag{23a}$$

$$\epsilon_{\text{eff}}(f) = \epsilon_r - \frac{\epsilon_r - \epsilon_{\text{eff}}(0)}{1 + G \cdot (f/f_p)^2} \tag{23b}$$

where $f_p = Z_0(0)/(2\mu_0 h)$ and the factor G is

$$G = \frac{\pi^2}{12} \frac{\epsilon_r - 1}{\epsilon_{\text{eff}}(0)} \sqrt{\frac{Z_0(0)}{59.95}}$$

In the above, $Z_0(0)$ and $\epsilon_{\text{eff}}(0)$ are dc values obtained from (19).

The power-handling capability of the microstrip line is limited by heating due to ohmic and dielectric losses and by dielectric breakdown. The increase in temperature due to losses limits the average power of the line, while the breakdown between the strip conductor and ground plate limits the peak power. Some theoretical values of average power-handling capability for various substrates at 2 GHz, 10 GHz, and 20 GHz are given in Table 6.

The peak power is limited primarily by the sharp edges of the line due to electric field concentration. To increase the breakdown voltage, one may use thick strip conductors with rounded edges. For a microstrip line composed of a strip 7/32 in (5.55 mm) wide on a Teflon-impregnated Fiberglas base 1/16 in (1.58 mm) thick, corona effects appear at the edge of the strip conductor for pulse power of roughly 10 kW at 9 GHz.

Table 6. Average Power-Handling Capacity for 50-ohm Microstrip Lines

Substrate	Maximum Average Power (kW)*		
	2 GHz	10 GHz	20 GHz
Alumina	12.1	5.2	3.4
Beryllium oxide	174.5	75.7	51.5
Gallium arsenide	3.5	1.5	0.93
Polystyrene	0.32	0.12	0.075
Quartz	1.2	0.52	0.36
RT/Duroid 5880	0.29	0.15	0.049
Sapphire	11.6	5.1	3.5
Silicon	3.2	2.2	1.6

*Maximum temperature = 100°C; ambient temperature = 25°C.

Coplanar Waveguides

Coplanar waveguide consists of a center strip with two ground planes located on the top surface of a dielectric substrate, as shown in Fig. 15. Under the quasi-TEM approximation, the characteristic impedance and the effective dielectric constant are [6]

$$Z_0 = \frac{30\pi}{\sqrt{\epsilon_{\text{eff}}}} \frac{1}{K(k_e)} \tag{24a}$$

$$\epsilon_{\text{eff}} = \left(\frac{\lambda_0}{\lambda_g}\right)^2 = \epsilon_e - \frac{0.7(\epsilon_e - 1)t/W}{K(k) + 0.7\,t/W} \tag{24b}$$

where

$$K(x) = \begin{cases} \frac{1}{\pi} \ln\left(2\frac{1+\sqrt{x}}{1-\sqrt{x}}\right), & \sqrt{1/2} \leq x \leq 1 \\ \pi/\ln\left[2\frac{1+(1-x^2)^{1/4}}{1-(1-x^2)^{1/4}}\right], & 0 \leq x \leq \sqrt{1/2} \end{cases} \tag{24c}$$

$$\epsilon_e = \frac{\epsilon_r + 1}{2} \{\tanh[0.775 \ln(h/W) + 1.75] + (kW/h)[0.04 - 0.7k + 0.01(1 - 0.1\epsilon_r)(0.25 + k)]\}$$

$$k_e = \frac{S_e}{S_e + 2W_e} = \frac{S + \Delta}{S + 2W - \Delta}$$

$$k = \frac{S}{S + 2W}$$

$$\Delta = (1.25t/\pi)[1 + \ln(4\pi S/t)]$$

Note that as $t \to 0$, $\Delta \to 0$. For $t = 0$, the accuracy of these expressions is better than 1.5 percent for $h/W \geq 1$. The characteristic impedance computed from the above formulas is plotted in Fig. 16.

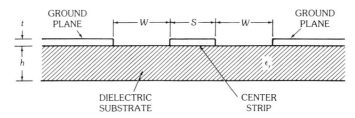

Fig. 15. Cross section of coplanar transmission line. (*After Jordan [1], © 1985 Howard W. Sams & Company; reproduced with permission*)

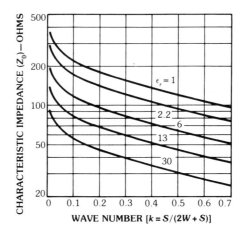

Fig. 16. Characteristic impedance of coplanar transmission line calculated from formula (24) for $t = 0$. (*After Jordan [1], © 1985 Howard W. Sams & Company; reproduced with permission*)

The dielectric loss of this line in nepers per meter is

$$\alpha_d = \frac{\pi \tan \delta}{\lambda_g} \frac{\epsilon_r}{\epsilon_{\text{eff}}} \frac{\epsilon_{\text{eff}} - 1}{\epsilon_r - 1} \tag{25}$$

The conductor loss in nepers per meter is

$$\alpha_c = \frac{\pi}{\lambda_g}(1 - Z_1/Z_\delta) \tag{26}$$

where Z_1 is the characteristic impedance of the line when losses are neglected, Z_δ is the impedance when the dimensions S, W, and t are replaced by S', W', and t' such that

$$\begin{aligned} S' &= S - \delta \\ W' &= W + \delta \\ t' &= t - \delta \end{aligned} \tag{27}$$

and δ is the skin depth of the conductor.

Coplanar Strips

A configuration of coplanar strips which is complementary to coplanar waveguide is shown in Fig. 17. It consists of two equal-width strips running parallel on the same surface of a dielectric slab.

The characteristic impedance in ohms and the effective dielectric constant of the line are

$$Z_0 = \frac{120\pi}{\sqrt{\epsilon_{\text{eff}}}} K(k_e) \quad (28a)$$

$$\epsilon_{\text{eff}} = \left(\frac{\lambda_0}{\lambda_g}\right)^2 = \epsilon_e - \frac{1.4(\epsilon_e - 1)t/S}{1/K(k) + 1.4t/S} \quad (28b)$$

where $K(x)$, ϵ_e, and k have the same expressions as defined in (24), and

$$k_e = \frac{S - \Delta}{S + 2W + \Delta}$$

$$\Delta = (1.25t/\pi)[1 + \ln(4\pi W/t)]$$

Notice that W is now the strip width and S is the spacing between the strips.

The expressions for the dielectric loss and conductor loss in coplanar strips are given in (25) and (26). But in this case S', W', and t' are

$$\begin{aligned} S' &= S + \delta \\ W' &= W - \delta \\ t' &= t - \delta \end{aligned} \quad (29)$$

5. TE/TM Waveguides

The most commonly used waveguides operating in TE/TM modes are the hollow-tube waveguides.

For propagation of energy through a hollow metal tube under fixed conditions, two types of waves, i.e., TE and TM waves, are available. (Note: the TEM waves cannot be propagated in a hollow-tube waveguide.) The possible configurations of the fields, commonly termed "modes," are characterized by the introduction of

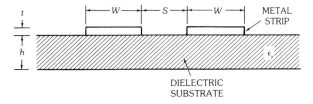

Fig. 17. Cross section of coplanar strips.

integer subscripts m and n, which can take on separate values from 0 or 1 to infinity.

A particular mode can be propagated in a hollow-tube guide only if the frequency is greater than an associated particular value called the cutoff frequency, f_c, for that mode; the corresponding free-space wavelength is called the cutoff wavelength, λ_c. The cutoff frequency is a function of the waveguide dimension, the mode indexes m and n, and the relative dielectric constant ϵ_r of the medium enclosed by the waveguide walls.

As mentioned in Section 2, the transmission line equations are valid for single-mode waveguides, except for those expressions involving the quantities R, L, G, and C. The transmission line description of a waveguide mode is based on the fact that the transverse electric field $\mathbf{E}_t$ and transverse magnetic field $\mathbf{H}_t$ of each mode can be expressed as

$$\mathbf{E}_t(x, y, z) = V(z)\,\mathbf{e}(x, y) \tag{30a}$$

$$\mathbf{H}_t(x, y, z) = I(z)\,\mathbf{h}(x, y) \tag{30b}$$

where $\mathbf{e}(x, y)$ and $\mathbf{h}(x, y)$ are vector functions indicative of the cross-sectional form of the mode field, and $V(z)$ and $I(z)$ are voltage and current functions that measure the rms amplitudes of the transverse electric and magnetic fields at any point z along the direction of propagation. The voltage and current are described by the transmission line equations given in (1), where, for a medium of uniform dielectric constant,

$$\gamma = \alpha + j\beta = \sqrt{k_c^2 - k^2} \tag{31}$$

$$Z_0 = \frac{1}{Y_0} = \begin{cases} \dfrac{j\omega\mu_0}{\gamma} & \text{for TE modes} \\ \dfrac{\gamma}{j\omega\epsilon} & \text{for TM modes} \end{cases} \tag{32}$$

The parameters k_c, k, and Z_0 are respectively termed the cutoff wave number, the free-space wave number, and the characteristic wave impedance of the mode in question.

When the structure is loss-free, $\alpha = 0$ and the following relationships hold:

$$\begin{aligned} k &= 2\pi/\lambda = 2\pi\sqrt{\epsilon_r}/\lambda_0, & k_c &= 2\pi/\lambda_c \\ \beta &= 2\pi/\lambda_g, & \lambda_g &= \lambda/\sqrt{1 - (\lambda/\lambda_c)^2} \end{aligned} \tag{33}$$

where λ, λ_c, and λ_g are respectively the free-space wavelength, the cutoff wavelength, and the guide wavelength. The guide wavelength is always greater than the wavelength in the unbounded medium, i.e., the free-space wavelength.

The phase velocity $v_p = f\lambda_g = c(\lambda_g/\lambda)$ is always greater than that in an unbounded medium. The group velocity, the velocity of energy propagation down the guide, is less than that in an unbounded medium.

Transmission Lines and Waveguides

The dielectric loss of the medium in a waveguide may be taken into account by introduction of a complex permittivity

$$\epsilon = \epsilon' - j\epsilon'' = (\epsilon'_r - j\epsilon''_r)\epsilon_0 \tag{34}$$

where ϵ' is the dielectric constant and ϵ'' is the loss factor. For a medium having unity relative permeability, the propagation constant of the waveguide is

$$\gamma = \sqrt{\left(\frac{2\pi}{\lambda_c}\right)^2 - \frac{\epsilon}{\epsilon_0}\left(\frac{2\pi}{\lambda_0}\right)^2} \tag{35}$$

In a waveguide having a cutoff wavelength $\lambda_c > \lambda_0$, the attenuation constant in nepers per meter is

$$\alpha_d = \frac{\pi \lambda_g \epsilon''_r}{\lambda_0^2} = \frac{2\pi}{\lambda_{g0}}\sqrt{\frac{-1 + (1 + x^2)^{1/2}}{2}} = \frac{2\pi}{\lambda_{g0}}\sinh\left(\frac{\sinh^{-1}x}{2}\right) \tag{36a}$$

$$\alpha_d \cong \frac{\pi x}{\lambda_{g0}}\left(1 - \frac{x^2}{8} + \cdots\right), \quad x \ll 1 \tag{36b}$$

and the phase constant in radians per meter is

$$\beta = \frac{2\pi}{\lambda_g} = \frac{2\pi}{\lambda_{g0}}\sqrt{\frac{1 + (1 + x^2)^{1/2}}{2}} = \frac{2\pi}{\lambda_{g0}}\cosh\left(\frac{\sinh^{-1}x}{2}\right) \tag{37a}$$

$$\beta \cong \frac{2\pi}{\lambda_{g0}}\left(1 + \frac{x^2}{8} + \cdots\right), \quad x \ll 1 \tag{37b}$$

where $x = \epsilon''(\lambda_{g0}/\lambda_0)^2$ and

$$\lambda_{g0} = \lambda_0/\sqrt{\epsilon''_r - (\lambda_0/\lambda_c)^2}$$

is the guide wavelength neglecting the loss. The approximations (36b) and (37b) are valid for $\epsilon''/\epsilon' \ll 1$, when λ_0 is not too close to the cutoff wavelength λ_c.

The characteristic wave impedance at a certain point has been defined in (32) as the ratio of the total transverse electric-field strength to the total transverse magnetic-field strength. The wave impedance is constant over the cross section of the guide.

In addition to the wave impedance, the "integrated" characteristic impedance, Z_g, of the line is a quantity of great usefulness in connection with ordinary two-conductor transmission lines. For such lines, Z_g can be defined in terms of the voltage-current ratio or in terms of the power transmitted for a given voltage or a given current. That is, for an infinitely long line (or matched line),

$$Z_{VI} = \frac{V}{I}, \quad Z_{PI} = \frac{2P}{|I|^2}, \quad Z_{PV} = \frac{|V|^2}{2P} \tag{38}$$

where V and I are the peak values of voltage and current. For TEM lines these definitions are equivalent, but for non-TEM waveguides they lead to three values that depend on the guide dimensions in the same way, but differ by a constant. Of the three, Z_{PV} is most widely used, but Z_{VI} is found to be more nearly correct in matching a coaxial line to a waveguide.

Rectangular Waveguides

A uniform waveguide of rectangular cross section is described by a Cartesian coordinate system shown in Fig. 18. The inner dimensions of the guide are a and b.

For TE$_{mn}$ modes in rectangular waveguides, m and n may take any integer value from 0 to infinity, except the case $m = n = 0$. For the TM$_{mn}$ modes, m and n may take any value from 1 to infinity. Field patterns for some of the lower-order modes are shown in Fig. 19. The dominant mode in the rectangular waveguide is the TE$_{10}$ mode.

The cutoff wavelength of the TE$_{mn}$ or TM$_{mn}$ mode is

$$\lambda_c = \frac{2\sqrt{ab}}{\sqrt{(b/a)m^2 + (a/b)n^2}} \tag{39}$$

The conductor loss for a propagating TE$_{mn}$ mode due to the dissipative waveguide walls is

$$\alpha_c = \frac{5.274 \times 10^{-6}}{b} \sqrt{\frac{\epsilon_r f}{\sigma}} \left[\frac{\epsilon_n m^2 (b/a) + \epsilon_m n^2}{m^2(b/a) + n^2(a/b)} \sqrt{1 - \left(\frac{\lambda}{\lambda_c}\right)^2} + \frac{[\epsilon_n + \epsilon_m(b/a)](\lambda/\lambda_c)^2}{\sqrt{1 - (\lambda/\lambda_c)^2}} \right] \tag{40}$$

where

$\epsilon_m = 1$ if $m = 0$, $\epsilon_m = 2$ if $m \neq 0$

$\epsilon_n = 1$ if $n = 0$, $\epsilon_n = 2$ if $n \neq 0$

σ = conductivity of conductor walls

ϵ_r = relative permittivity of filling dielectric

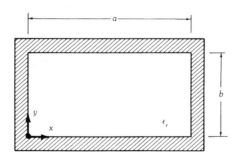

Fig. 18. Rectangular-waveguide cross section.

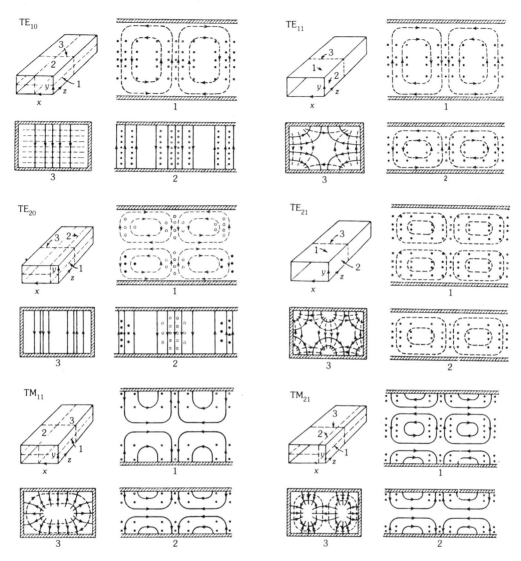

Fig. 19. Field patterns of rectangular-waveguide modes. (*After Blake [7], copyright © 1969 John Wiley & Sons, Inc., reprinted by permission of John Wiley & Sons, Inc.*)

The maximum power-handling capacity in watts for a TE$_{mn}$ mode in a matched nondissipative guide is, for $m \neq 0$, $n \neq 0$,

$$P = 3.318 \times 10^{-4} \sqrt{\epsilon_r}\, ab \left(\frac{mb}{na} + \frac{na}{mb}\right)^2 E_b^2 \sqrt{1 - \left(\frac{\lambda}{\lambda_c}\right)^2} \qquad (41)$$

where E_b is the maximum electric field intensity, which occurs at values of x and y such that

Table 7. Standard Waveguides (After Jordan [1], © 1985 Howard W. Sams & Company; reprinted with permission)

EIA Waveguide Designation (Standard RS-261-A)	JAN Waveguide Designation (MIL-HDBK-216, 4 January 1962)	Outer Dimensions and Wall Thickness (in)	Frequency for Dominant (TE$_{10}$) Mode (GHz)	Cutoff Wavelength λ_c for TE$_{10}$ Mode (cm)	Cutoff Frequency f_c for TE$_{10}$ Mode (GHz)	Theoretical Attenuation, Lowest to Highest Frequency (dB/100 ft)	Theoretical Power Rating for Lowest to Highest Frequency* (MW)
WR-2300	RG-290/U†	23.250 × 11.750 × 0.125	0.32–0.49	116.8	0.256	0.051–0.031	153.0–212.0
WR-2100	RG-291/U†	21.250 × 10.750 × 0.125	0.35–0.53	106.7	0.281	0.054–0.034	120.0–173.0
WR-1800	RG-201/U†	18.250 × 9.250 × 0.125	0.425–0.620	91.4	0.328	0.056–0.038	93.4–131.9
WR-1500	RG-202/U†	15.250 × 7.750 × 0.125	0.49–0.740	76.3	0.393	0.069–0.050	67.6–93.3
WR-1150	RG-203/U†	11.750 × 6.000 × 0.125	0.64–0.96	58.4	0.514	0.128–0.075	35.0–53.8
WR-975	RG-204/U†	10.000 × 5.125 × 0.125	0.75–1.12	49.6	0.605	0.137–0.095	27.0–38.5
WR-770	RG-205/U†	7.950 × 4.100 × 0.125	0.96–1.45	39.1	0.767	0.201–0.136	17.2–24.1
WR-650	RG-69/U	6.660 × 3.410 × 0.080	1.12–1.70	33.0	0.908	0.317–0.212	11.9–17.2
WR-510		5.260 × 2.710 × 0.080	1.45–2.20	25.9	1.16		
WR-430	RG-104/U	4.460 × 2.310 × 0.080	1.70–2.60	21.8	1.375	0.588–0.385	5.2–7.5
WR-340	RG-112/U	3.560 × 1.860 × 0.080	2.20–3.30	17.3	1.735	0.877–0.572	
WR-284	RG-48/U	3.000 × 1.500 × 0.080	2.60–3.95	14.2	2.08	1.102–0.752	2.2–3.2
WR-229		2.418 × 1.273 × 0.064	3.30–4.90	11.6	2.59		
WR-187	RG-49/U	2.000 × 1.000 × 0.064	3.95–5.85	9.50	3.16	2.08–1.44	1.4–2.0
WR-159		1.718 × 0.923 × 0.064	4.90–7.05	8.09	3.71		

WR	RG	Dimensions (in)	Freq (GHz)			Power/Atten
WR-137	RG-50/U	1.500 × 0.750 × 0.064	5.85–8.20	6.98	4.29	2.87–2.30	0.56–0.71
WR-112	RG-51/U	1.250 × 0.625 × 0.064	7.05–10.00	5.70	5.26	4.12–3.21	0.35–0.46
WR-90	RG-52/U	1.000 × 0.500 × 0.050	8.20–12.40	4.57	6.56	6.45–4.48	0.20–0.29
WR-75		0.850 × 0.475 × 0.050	10.00–15.00	3.81	7.88		
WR-62	RG-91/U	0.702 × 0.391 × 0.040	12.40–18.00	3.16	9.49	9.51–8.31	0.12–0.16
WR-51		0.590 × 0.335 × 0.040	15.00–22.00	2.59	11.6		
WR-42	RG-53/U	0.500 × 0.250 × 0.040	18.00–26.50	2.13	14.1	20.7–14.8	0.043–0.058
WR-34		0.420 × 0.250 × 0.040	22.00–33.00	1.73	17.3		
WR-28	RG-96/U‡	0.360 × 0.220 × 0.040	26.50–40.00	1.42	21.1	21.9–15.0	0.022–0.031
WR-22	RG-97/U‡	0.304 × 0.192 × 0.040	33.00–50.00	1.14	26.35	31.0–20.9	0.014–0.020
WR-19		0.268 × 0.174 × 0.040	40.00–60.00	0.955	31.4		
WR-15	RG-98/U‡	0.228 × 0.154 × 0.040	50.00–75.00	0.753	39.9	52.9–39.1	0.0063–0.0090
WR-12	RG-99/U‡	0.202 × 0.141 × 0.040	60.00–90.00	0.620	48.4	93.3–52.2	0.0042–0.0060
WR-10		0.180 × 0.130 × 0.040	75.00–110.00	0.509	59.0		
WR-8	RG-138/U§	0.140 × 0.100 × 0.030	90.00–140.00	0.406	73.84	152–99	0.0018–0.0026
WR-7	RG-136/U§	0.125 × 0.0925 × 0.030	110.00–170.00	0.330	90.84	163–137	0.0012–0.0017
WR-5	RG-135/U§	0.111 × 0.0855 × 0.030	140.00–220.00	0.259	115.75	308–193	0.00071–0.0010
WR-4	RG-137/U§	0.103 × 0.0815 × 0.030	170.00–260.00	0.218	137.52	384–254	0.00052–0.0007
WR-3	RG-139/U§	0.094 × 0.0770 × 0.030	220.00–325.00	0.173	173.28	512–348	0.00035–0.0004

*For these computations, the breakdown strength of air was taken as 15 000 V/cm. A safety factor of approximately 2 at sea level has been allowed.
†Aluminum, 2.83×10^{-6} Ω-cm resistivity.
‡Silver, 1.62×10^{-6} Ω-cm resistivity.
§JAN types are silver, with a circular outer diameter of 0.156 in and a rectangular bore matching EIA types. All other types are of a Cu–Zn alloy, 3.9×10^{-6} Ω-cm resistivity.

$$\tan\left(\frac{m\pi}{a}x\right) = \pm\frac{na}{mb}, \quad \tan\left(\frac{n\pi}{b}y\right) = \pm\frac{mb}{na} \tag{42}$$

For either $m = 0$ or $n = 0$, the maximum power in watts is

$$P = 6.636 \times 10^{-4}\sqrt{\epsilon_r}\, abE_b^2\sqrt{1-(\lambda/\lambda_c)^2} \tag{43}$$

where E_b occurs at integer multiples of $x = a/2m$ for $m \neq 0$, $n = 0$ and at integer multiples of $y = b/2n$ for $m = 0$, $n \neq 0$.

The conductor attenuation in nepers per meter for a propagating TM_{mn} mode due to dissipation in the guide walls is

$$\alpha_c = \frac{1.055 \times 10^{-6}}{a}\sqrt{\frac{\epsilon_r f}{\sigma}}\left(\frac{m^2 + n^2 a^3/b^3}{m^2 + n^2 a^2/b^2}\right)\frac{1}{\sqrt{1-(\lambda/\lambda_c)^2}} \tag{44}$$

For a matched nondissipative guide, the maximum power-handling capacity of a TM mode is given in watts by

$$P = 3.318 \times 10^{-4}\frac{\sqrt{\epsilon_r}}{\sqrt{1-(\lambda/\lambda_c)^2}}ab\left(\frac{mb}{na}+\frac{na}{mb}\right)^2 E_b^2 \tag{45}$$

where E_b occurs at values of x and y for which

$$\tan\left(\frac{m\pi}{a}x\right) = \pm\frac{mb}{na}, \quad \tan\left(\frac{n\pi}{b}y\right) = \pm\frac{na}{mb} \tag{46}$$

Some properties and dimensions of standard rectangular waveguides are listed in Table 7.

Circular Waveguides

A uniform waveguide of circular cross section is most conveniently described by a polar coordinate system as shown in Fig. 20. The cross section has an inner dimension of radius a.

For both TE_{mn} and TM_{mn} modes in circular waveguides, m may take any integer value from 0 to infinity, and n from 1 to infinity. Field patterns for some of the lower-order modes are shown in Fig. 21. The dominant mode in circular waveguide is the TE_{11} mode. Of the circularly symmetrical modes the TM_{01} mode has the lowest cutoff frequency.

The cutoff wavelength of the TE_{mn} mode is

$$\lambda_c = 2\pi a/x'_{mn} \tag{47}$$

where x'_{mn} is the nth-order positive root of the derivative of the mth-order Bessel function. Several of the lower-order roots are given in Table 8.

The TE_{mn} attenuation in nepers per meter due to dissipation in the guide walls is

Transmission Lines and Waveguides

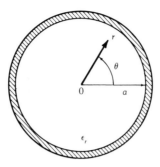

Fig. 20. Circular-waveguide cross section.

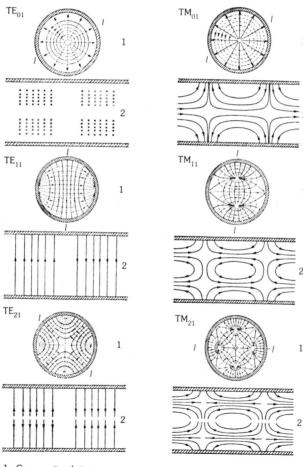

1. Cross-sectional view
2. Longitudinal view through plane *l-l*

Fig. 21. Cross-sectional and longitudinal views of field patterns of six lower-order modes in circular waveguides. (*After Marcuvitz [8]*, © *1951 McGraw-Hill Book Company*)

Table 8. Roots of $J'_m(x) = 0$ *(After Marcuvitz [8], © 1951 McGraw-Hill Book Company)*

n \ m	0	1	2	3	4	5	6	7
1	3.832	1.841	3.054	4.201	5.317	6.416	7.501	8.578
2	7.016	5.331	6.706	8.015	9.282	10.520	11.735	12.932
3	10.173	8.536	9.969	11.346	12.682	13.987		
4	13.324	11.706	13.170					

$$\alpha_c = \frac{5.274 \times 10^{-6}}{a\sqrt{1 - (\lambda/\lambda_c)^2}} \sqrt{\frac{\epsilon_r f}{\sigma}} \left[\frac{m^2}{(x'_{mn})^2 - m^2} + \left(\frac{\lambda}{\lambda_c}\right)^2 \right] \quad (48)$$

For the dominant TE_{11} mode in a matched nondissipative guide, the power-handling capacity in watts is

$$P = 1.985 \times 10^{-3} \sqrt{1 - (\lambda/\lambda_c)^2} \, a^2 E_b^2 \quad (49)$$

where E_b is the maximum electric field intensity, which occurs at the axis of the guide.

The cutoff wavelength of the TM_{mn} mode is

$$\lambda_c = 2\pi a / x_{mn} \quad (50)$$

where x_{mn}, the nth positive root of the mth-order Bessel function, is tabulated in Table 9 for several values of m and n.

The TM_{mn} mode attenuation in nepers per meter due to finite conductivity of the guide walls is

$$\alpha_c = \frac{5.274 \times 10^{-6}}{a\sqrt{1 - (\lambda/\lambda_c)^2}} \sqrt{\frac{\epsilon_r f}{\sigma}} \quad (51)$$

For the TM_{01} mode in a matched lossless guide, the power-handling capacity in watts is, for $a/\lambda < 0.761$,

$$P = 7.69 \times 10^{-3} (a/\lambda)^2 \sqrt{1 - (\lambda/\lambda_c)^2} \, a^2 E_b^2 \quad (52a)$$

Table 9. Roots of $J_m(x) = 0$ *(After Marcuvitz [8], © 1951 McGraw-Hill Book Company)*

n \ m	0	1	2	3	4	5	6	7
1	2.405	3.832	5.136	6.380	7.588	8.771	9.936	11.086
2	5.520	7.016	8.417	9.761	11.065	12.339	13.589	14.821
3	8.654	10.173	11.620	13.015	14.372			
4	11.792	13.323	14.796					

where E_b occurs at the axis of the guide. For $a/\lambda > 0.761$,

$$P = 3.33 \times 10^{-3} a^2 E_b^2 / \sqrt{1 - (\lambda/\lambda_c)^2} \qquad (52b)$$

In this case the maximum field intensity E_b occurs at $r = 0.765a$ and is independent of the angle.

Ridged Waveguides

A common ridged waveguide is a rectangular waveguide having a rectangular ridge projecting inward from one or both sides, as shown in Fig. 22. The ridge loading in the center of the guide lowers the cutoff frequency of the dominant TE_{10} mode so that a bandwidth ratio of over 4:1 exists between cutoff frequencies for the TE_{20} and TE_{10} modes.

In Fig. 23 the extension factors λ_c/a for the TE_{10} and TE_{20} modes are given as a function of the geometrical parameters. By comparing the factor λ_c/a for the TE_{10} and TE_{20} modes, the useful bandwidth can be easily obtained.

The attenuation in the ridged guides is greater than the equivalent rectangular waveguide of identical cutoff frequency. Furthermore, for ridged guides of identical cutoff frequency, more conductor loss is encountered in the guide with wider operating bandwidth.

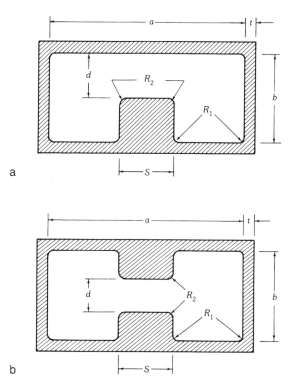

Fig. 22. Cross-sectional views of ridged waveguides. (*a*) Single-ridge waveguide. (*b*) Double-ridge waveguide.

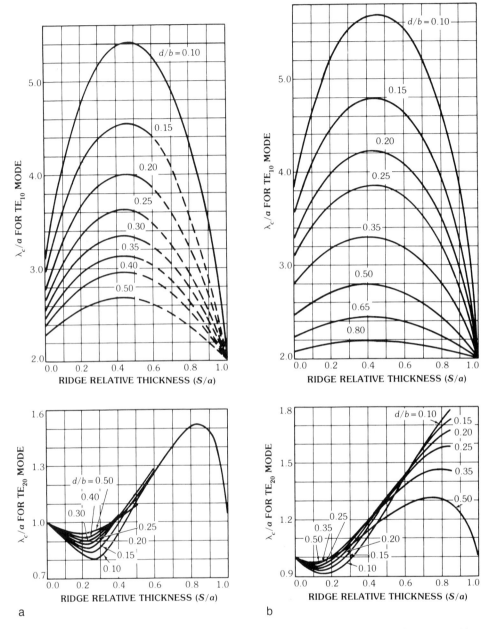

Fig. 23. TE_{10}- and TE_{20}-mode cutoff wavelength of ridged waveguides. (*a*) Double-ridge waveguide with $b/a = 0.5$. (*b*) Single-ridge waveguide with $b/a = 0.45$. (*After Hopfer [9]*, © *1955 IEEE*)

Based on a power-voltage definition, the ridge-guide characteristic admittance for the TE_{10} mode is plotted in Fig. 24.

Tables 10 and 11 give the essential characteristics of commonly used single- and double-ridge guides. The parameters are defined in Fig. 22, where R_1 and R_2 are the radii of curvature of the edges.

6. Hybrid-Mode Waveguides

The transmission lines to be discussed in this section include circular and rectangular dielectric waveguides and the corresponding image guides, slot lines, and fin lines. The fundamental mode in these waveguides propagates as a hybrid mode, having longitudinal components of both electric and magnetic fields.

The dielectric waveguides have their major applications in fiber optics and integrated optics, while the slot line is compatible with other planar circuits in microwave frequencies. The image guides and fin lines are proposed for millimeter-wave applications because of their low-loss advantage over the microstrip line and the ease of fabrication compared with the metal-tube waveguides.

The parameters required for circuit design are the guide wavelength, the attenuation, and the impedance. Since the fundamental modes are non-TEM, the definition of the impedance is not unique. The various definitions are given in (32) and (38).

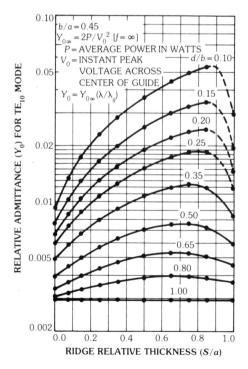

Fig. 24. Admittance of single-ridge waveguide. (*After Hopfer [9], © 1955 IEEE*)

Table 10. Characteristics of Single-Ridge Waveguides (*After Jordan [1], © 1985 Howard W. Sams & Company; reprinted with permission*)

Frequency Range (GHz)	f_{c10} (GHz)	λ_{c10} (in)	f_{c20} (GHz)	Dimensions in Inches							At $f = (3)^{1/2} f_{c10}$	
				a	b	d	s	t	R_1 (max)	R_2	Attenuation* (dB/ft)	Power Rating[†] (kW)
Bandwidth 2.4:1												
0.175–0.42	0.148	79.803	0.431	28.129	12.658	5.278	4.360	—	—	1.056	0.000 24	32 870.
0.267–0.64	0.226	52.260	0.658	18.421	8.289	3.457	2.855	—	—	0.691	0.000 45	14 100.
0.42–1.0	0.356	33.177	1.036	11.695	5.263	2.195	1.813	0.125	0.047	0.439	0.000 87	5 682.
0.64–1.53	0.542	21.792	1.577	7.682	3.457	1.442	1.191	0.125	0.047	0.288	0.001 64	2 451.
0.84–2.0	0.712	16.588	2.072	5.847	2.631	1.097	0.906	0.080	0.047	0.219	0.002 48	1 421.
1.5–3.6	1.271	9.293	3.699	3.276	1.474	0.615	0.508	0.080	0.047	0.123	0.005 91	445.8
2.0–4.8	1.695	6.968	4.933	2.456	1.105	0.461	0.381	0.080	0.047	0.092	0.009 08	250.6
3.5–8.2	2.966	3.982	8.632	1.404	0.632	0.264	0.218	0.064	0.031	0.053	0.0212	81.87
4.75–11.0	4.025	2.934	11.714	1.034	0.465	0.194	0.160	0.050	0.031	0.039	0.0333	44.43
7.5–18.0[‡]	6.356	1.858	18.498	0.655	0.295	0.123	0.1015	0.050	0.015	0.025	0.0661	17.82
11.0–26.5[‡]	9.322	1.267	27.130	0.4466	0.2010	0.0838	0.0692	0.040	0.015	0.017	0.117	8.285
18.0–40.0[‡]	15.254	0.7743	44.393	0.2729	0.1228	0.0512	0.0423	0.040	0.015	0.010	0.246	3.035
Bandwidth 3.6:1												
0.108–0.39	0.092	128.37	0.404	31.218	14.048	2.402	5.307	—	—	0.480	0.0016	14 550.
0.27–0.97	0.229	51.572	1.006	12.542	5.644	0.965	2.132	—	—	0.193	0.0065	2 348.
0.39–1.4	0.331	35.680	1.454	8.677	3.905	0.668	1.475	0.125	0.047	0.134	0.0112	1 124.
0.97–3.5	0.822	14.367	3.611	3.494	1.572	0.269	0.594	0.080	0.047	0.054	0.0438	182.2
1.4–5.0	1.186	9.958	5.210	2.422	1.090	0.186	0.412	0.080	0.047	0.037	0.0758	87.56
3.5–12.4	2.966	3.982	13.030	0.968	0.436	0.075	0.165	0.050	0.031	0.015	0.300	13.99
5.0–18.0[‡]	4.237	2.787	18.613	0.678	0.305	0.052	0.115	0.050	0.015	0.010	0.513	6.857
12.4–40.0[‡]	10.508	1.124	46.162	0.273	0.123	0.021	0.046	0.040	0.015	0.004	2.008	1.115

*Copper.
[†] Based on breakdown of air—15 000 V/cm (safety factor of approximately 2 at sea level). Corner radii considered.
[‡] Fig. 22a in these frequency ranges only.

Table 11. Characteristics of Double-Ridge Waveguides (*After Jordan [1], © 1985 Howard W. Sams & Company; reprinted with permission*)

Frequency Range (GHz)	f_{c10} (GHz)	λ_{c10} (in)	f_{c20} (GHz)	Dimensions in Inches								At $f = (3)^{1/2} f_{c10}$	
				a	b	d	s	t	R_1 (max)	R_2		Attenuation* (dB/ft)	Power Rating[†] (kW)
						Bandwidth 2.4:1							
0.175–0.42				29.667	13.795	5.863	7.417	—	—	1.173			
0.267–0.64				19.428	9.034	3.839	4.857	—	—	0.768			
0.42–1.0				12.333	5.737	2.437	3.083	0.125	0.050	0.487			
0.64–1.53				8.100	3.767	1.601	2.025	0.125	0.050	0.320			
0.84–2.0				6.167	2.868	1.219	1.542	0.125	0.050	0.244			
1.5–3.6				3.455	1.607	0.683	0.864	0.080	0.050	0.137			
2.0–4.8				2.590	1.205	0.512	0.648	0.080	0.050	0.102			
3.5–8.2				1.480	0.688	0.292	0.370	0.064	0.030	0.058			
4.75–11.0				1.090	0.506	0.215	0.272	0.050	0.030	0.043			
7.5–18.0				0.691	0.321	0.136	0.173	0.050	0.020	0.027			
11.0–26.5[‡]				0.471	0.219	0.093	0.118	0.040	0.015	0.019			
18.0–40.0[‡]				0.288	0.134	0.057	0.072	0.040	0.015	0.011			
						Bandwidth 3.6:1							
0.108–0.39	0.092	128.37	0.401	34.638	14.894	2.904	8.660	—	—	0.581		0.0014	28 830.
0.27–0.97	0.229	51.572	0.999	13.916	5.984	1.167	3.479	—	—	0.233		0.0055	4 653.
0.39–1.4	0.331	35.680	1.444	9.628	4.140	0.807	2.407	0.125	0.050	0.161		0.0097	2 227.
0.97–3.5	0.822	14.367	3.587	3.877	1.667	0.325	0.969	0.080	0.050	0.065		0.0378	361.2
1.4–5.0	1.186	9.958	5.176	2.687	1.155	0.225	0.672	0.080	0.050	0.045		0.0656	173.5
3.5–12.4	2.966	3.982	12.944	1.074	0.462	0.090	0.269	0.050	0.030	0.018		0.259	27.74
5.0–18.0	4.237	2.787	18.490	0.752	0.323	0.063	0.188	0.050	0.020	0.013		0.443	13.59
12.4–40.0[‡]	10.508	1.124	45.857	0.303	0.130	0.025	0.076	0.040	0.015	0.005		1.730	2.210

*Copper.
[†] Based on breakdown of air—15 000 V/cm (safety factor of approximately 2 at sea level). Corner radii considered.
[‡] Fig. 22b in these frequency ranges only.

Circular Dielectric Waveguides and Image Guides

This type of waveguide has applications in antenna structures, laser devices, fiber optics, and millimeter-wave techniques.

The field structures for the nonradiating modes fall into two classes, circularly symmetric and nonsymmetric modes. The symmetric modes are TE_{0m} and TM_{0m} modes whose cutoff wavelengths λ_c are

$$\lambda_c = \pi d \sqrt{\epsilon_r - 1}/x_{0m} \tag{53}$$

where d is the rod diameter, ϵ_r the relative dielectric constant, and x_{0m} the mth root of the zeroth-order Bessel function $J_0(x)$.

The nonsymmetric modes are hybrid modes that require the coexistence of an H(TE) wave with an E(TM) wave. These modes are described as HE if the H mode is predominant, and as EH if the E mode is predominant. Among the hybrid modes the HE_{11} mode is the dominant mode of the waveguide, with a zero cutoff frequency. The field pattern of the HE_{11} mode is shown in Fig. 25. Fig. 26 describes the relation between λ_g/λ_0 and d/λ_0 for dielectric guides of different ϵ_r. Fig. 27 is a plot showing the combined effect of ϵ_r on dielectric loss, field spread, and waveguide size, all normalized for frequency and loss tangent.

Since the HE_{11} mode exhibits a plane of symmetry containing the axis of the rod, an image plane may be used to replace half of the rod and surrounding space. This results in an image guide. The image plane reduces the required cross section by one-half, and also largely eliminates the support and shielding problems. In addition, it acts as a polarization anchor and reduces the mode conversion problem.

For the HE_{11} mode the dispersion (λ_0/λ_g) and dielectric attenuation (α_d) in the image guide are the same as those in the corresponding dielectric rod, given in Figs. 26 and 27. The conduction loss due to the dissipation in the image plane remains an order of magnitude less than α_d for reasonable values of $2a/\lambda_0$ and for wavelengths well into the millimeter region [10].

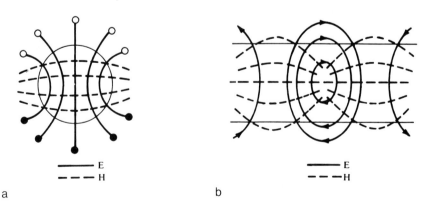

Fig. 25. Field distribution of the HE_{11} mode. (*a*) Cross section. (*b*) Longitudinal section.

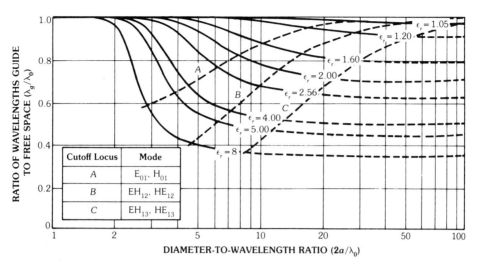

Fig. 26. Effect of dielectric constant on the dispersion characteristics of round rod. (*After Schlesinger and King [10], © 1958 IEEE*)

Rectangular Dielectric Waveguides and Image Guides

Rectangular dielectric waveguides (Fig. 28) find use in integrated optics and in millimeter-wave integrated circuits. In the latter case, image guides (Fig. 29) are more frequently used. Modes are classified into E^y_{pq} and E^x_{pq}. The former has a principal **E** field in the y direction, and the latter in the x direction. The subscripts p and q indicate the number of extrema of electric or magnetic field in the x and y directions.

For a well-guided mode, approximate expressions of phase constant for these modes in dielectric waveguides are [11]

$$\beta = \{\epsilon_1 k_0^2 - (p\pi/a)^2/(1 + 2A_3/\pi a)^2 - (q\pi/b)^2/[1 + (\epsilon_2 A_2 + \epsilon_3 A_3)/(\epsilon_1 \pi b)]^2\}^{1/2} \tag{54a}$$

for the E^y_{pq} mode, with $p, q = 1, 2, \ldots$, and

$$\beta = \{\epsilon_1 k_0^2 - (p\pi/a)^2/(1 + 2\epsilon_3 A_3/\epsilon_1 \pi a)^2 - (q\pi/b)^2/[1 + (A_2 + A_3)/\pi b]^2\}^{1/2} \tag{54b}$$

for the E^x_{pq} mode, with $p, q = 1, 2, \ldots$, where

$$A_2 = \lambda_0/2\sqrt{\epsilon_1 - \epsilon_2}$$
$$A_3 = \lambda_0/2\sqrt{\epsilon_1 - \epsilon_3}$$

in which λ_0 is the free-space wavelength, and ϵ_1, ϵ_2, ϵ_3 are relative dielectric constants of the materials involved.

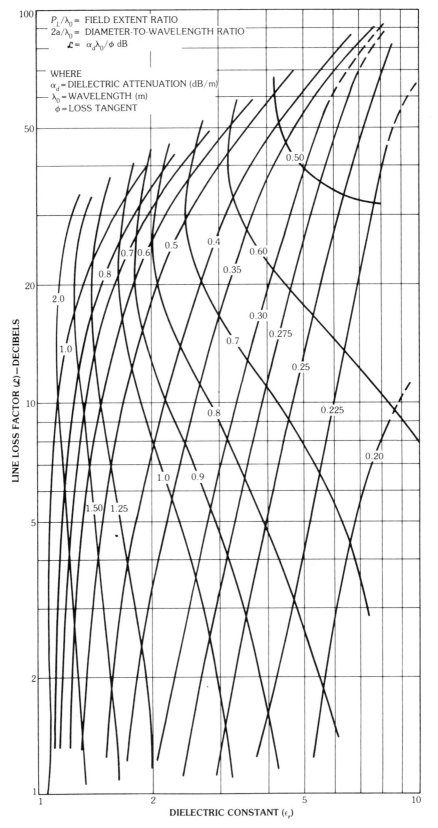

Fig. 27. Normalized loss factor $\mathscr{L}$ as a function of the dielectric constant for constant $2a/\lambda_0$ and ϱ_L/λ_0. (*After Schlesinger and King [10], © 1958 IEEE*)

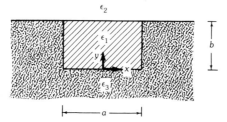

Fig. 28. Cross section of rectangular dielectric waveguide. (*After Jordan [1], © 1985 Howard W. Sams & Company; reprinted with permission*)

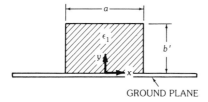

Fig. 29. Cross section of image guide. (*After Jordan [1], © 1985 Howard W. Sams & Company; reprinted with permission*)

The above formulas are not very accurate near the cutoff frequencies. More accurate data are obtainable by using a number of techniques, the simplest of which is the effective dielectric constant (EDC) method [12]. Fig. 30 shows the HE_{11} dispersion characteristics calculated by EDC and by the above formulas for a dielectric waveguide with $a/b = 1$ and $\epsilon_2 = \epsilon_3 = 1$. The dielectric constant ϵ_1 of the guide is varied as a parameter.

As the image guide is usually surrounded by air and is equivalent to a dielectric waveguide of height $2b'$ in free space for the dominant E^y_{11} mode, the approximation formula and the EDC can be used for dispersion characteristics. The results in Fig. 30 are valid for the image guide with the aspect ratio $a/b' = 2$.

Slot Lines

Slot line consists of a narrow gap (or slot) in a conductive coating on a dielectric substrate, as shown in Fig. 31. The geometry is planar and is well suited for its usage in microwave integrated circuits. Normally, the permittivity of the dielectric substrate in a slot line is sufficiently large (e.g., $\epsilon_r = 16$) so that the fields are closely confined to the slot and the radiation loss is minimized. The nature of the slot-mode configuration is such that the electric field extends across the slot while the magnetic field is in a plane perpendicular to the slot and forms closed loops at half-wave intervals. The slot mode is non-TEM and has no low-frequency cutoff.

Since the slot mode is non-TEM, the definition of characteristic impedance is not unique. The power-voltage definition, i.e., $Z_0 = |V|^2/2P$, is normally used.

Over the intervals given by $9.6 < \epsilon_r < 20$, $0.02 < W/H < 2$, and $0.015 < H/\lambda_0 < 0.08$ and $t = 0$, a closed-form approximation to λ_g/λ_0 is expressed as [13]

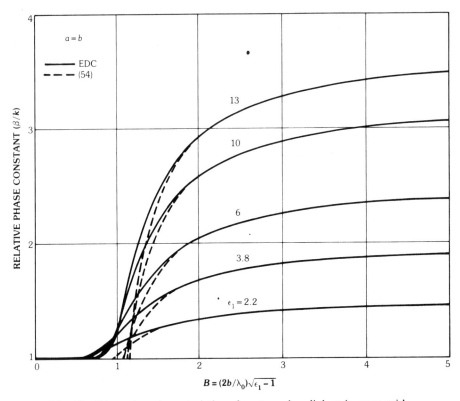

Fig. 30. Dispersion characteristics of rectangular dielectric waveguide.

$$\frac{\lambda_g}{\lambda_0} = f_1(\epsilon_r)\left[f_2\left(\frac{W}{H}\right) + f_3\left(\frac{W}{H}\right)\left(\frac{H}{\lambda_0}\right)^{f_4(W/H)} + f_5\left(\frac{W}{H}\right)\right] \quad (55)$$

where

$$f_1(\epsilon_r) = 3.549\epsilon_r^{-0.56}$$

$$f_2\left(\frac{W}{H}\right) = 0.5632\left(\frac{W}{H}\right)^{0.104(W/H)^{0.266}}$$

$$f_3\left(\frac{W}{H}\right) = -0.877\left(\frac{W}{H}\right)^{0.81} + 0.4233\left(\frac{W}{H}\right) - 0.2492$$

$$f_4\left(\frac{W}{H}\right) = -1.269 \times 10^{-2}\left[\ln\left(50\frac{W}{H}\right)\right]^{1.7} + 0.0674\ln\left(50\frac{W}{H}\right) + 0.20$$

$$f_5\left(\frac{W}{H}\right) = 1.906 \times 10^{-3}\left[\ln\left(50\frac{W}{H}\right)\right]^{2.9} - 7.203 \times 10^3 \ln\left(50\frac{W}{H}\right) + 0.1223$$

The accuracy of the expression is less than 3.7 percent.

A closed-form approximation for Z_0 in ohms is

$$Z_0 = \left(\frac{11}{\epsilon_r}\right)^{p(W/H, H/\lambda_0)} g\left(\frac{W}{H}, \frac{H}{\lambda_0}\right) \quad (56)$$

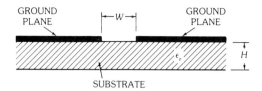

Fig. 31. Cross section of slot line. (*After Jordan [1]*, © *1985 Howard W. Sams & Company; reprinted with permission*)

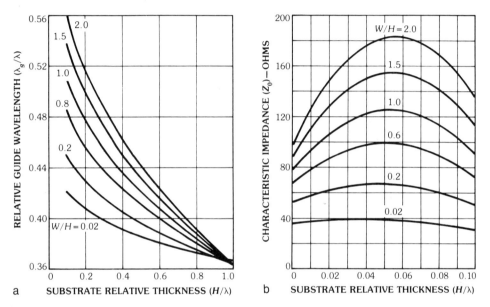

Fig. 32. Characteristics of slot line with $\epsilon_r = 13$. (*a*) Relative guide wavelength. (*b*) Characteristic impedance. (*After Jordan [1]*, © *1985 Howard W. Sams & Company; reprinted with permission*)

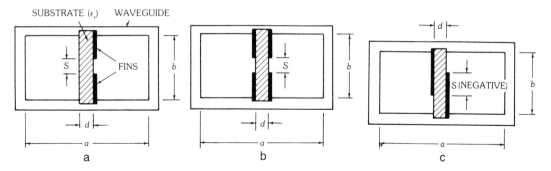

Fig. 33. Cross sections of fin lines. (*a*) Unilateral. (*b*) Bilateral. (*c*) Antipodal. (*After Jordan [1]*, © *1985 Howard W. Sams & Company; reprinted with permission*)

where

$$p\left(\frac{W}{H}, \frac{H}{\lambda_0}\right) = \left[-30.21 \ln\left(\frac{W}{H}\right) - 46.03\right]\left(\frac{H}{\lambda_0}\right)^2 + \left[0.5073 \ln\left(\frac{W}{H}\right) + 3.358\left(\frac{W}{H}\right)\right.$$
$$\left. + 6.492\right]\left(\frac{H}{\lambda_0}\right) - 2.013 \times 10^{-2} \ln\left(\frac{W}{H}\right) - 0.1374\left(\frac{W}{H}\right) + 0.2365$$

$$g\left(\frac{W}{H}, \frac{H}{\lambda_0}\right) = \left[-1.176 \times 10^4 \left(\frac{W}{H}\right)^{0.502} - 6.311 \times 10^3 \left(\frac{W}{H}\right) - 162.7\right]\left(\frac{H}{\lambda_0}\right)^2$$
$$+ \left[900.5\left(\frac{W}{H}\right)^{0.28} + 1262\left(\frac{W}{H}\right) - 123.8\right]\left(\frac{H}{\lambda_0}\right)$$
$$+ 1.637 \ln\left(50\frac{W}{H}\right) + 40.99\left(\frac{W}{H}\right)^{0.46} + 30.96$$

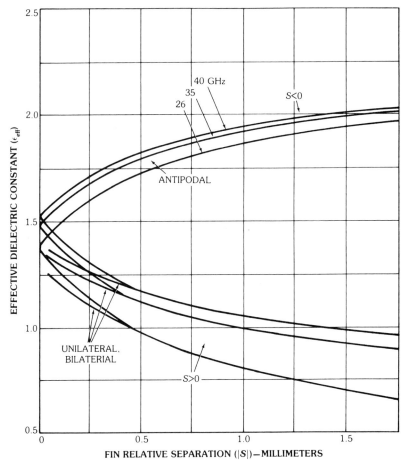

Fig. 34. Effect of fin separation on effective dielectric constant and characteristic impedance of fin lines. (*a*) Effective dielectric constant. (*b*) Characteristic impedance. (*After Hofmann [14], © 1977 IEEE*)

Transmission Lines and Waveguides

Over the same parameter ranges as above, the accuracy of Z_0 is ±14.5%. For $W/H > 0.2$, it is better than 4 percent. Data calculated by these formulas are plotted in Fig. 32.

Fin Lines

Fin lines consist of fins separated by a gap printed on one or both sides of a dielectric substrate, which is in turn placed at the center of a rectangular waveguide along its E-plane, as shown in Fig. 33. Therefore, fin lines are considered printed versions of ridged waveguides and have single-mode operating bandwidths wider than the one for the enclosing waveguide itself. Fin lines are widely used for millimeter-wave integrated circuits in the frequency range from 26.5 to 100 GHz and beyond.

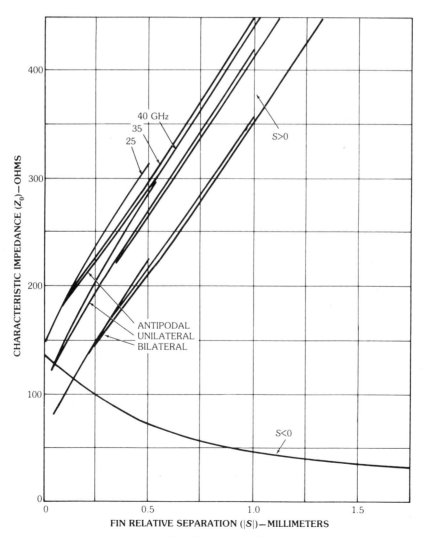

Fig. 34, *continued.*

The effective dielectric constant $\epsilon_{\text{eff}} = (\lambda_0/\lambda_g)^2$ and the characteristic impedance Z_0 for fin lines are plotted in Fig. 34 for the K_a-band (26.5 to 40 GHz) applications [14]. The characteristic impedance is defined as the ratio of the voltage across the slot to the current on the fins and is believed useful for small values of substrate thickness and gap width. Negative values of S indicate the overlap of fins in the antipodal fin lines. The waveguides supporting the fins are WR-28 ($a = 7.112$ mm, $b = 3.556$ mm). The substrate used for calculation is RT-Duroid ($\epsilon_r = 2.22$) of thickness $d = 0.254$ mm.

7. References

[1] E. C. Jordan, ed., *Reference Data for Engineers: Radio, Electronics, Computer, and Communications*, 7th ed., Indianapolis: Howard W. Sams & Co., 1985, pp. 29-3, 29-25, 29-27 to 29-37, 30-10 to 30-12.

[2] G. L. Ragan, *Microwave Transmission Circuits*, MIT Radiation Lab Series, vol. 8, New York: McGraw-Hill Book Co., 1948, pp. 32–35.

[3] M. A. Gunston, *Microwave Transmission-Line Impedance Data*, New York: Van Nostrand Reinhold Co., 1972.

[4] H. A. Wheeler, "Transmission-line properties of a strip between parallel planes," *IEEE Trans. Microwave Theory Tech.*, vol. MTT-28, pp. 866–876, November 1978.

[5] E. Hammerstadt and O. Jansen, "Accurate models for microstrip computer-aided design," *1980 IEEE Intl. Microwave Symp. Dig.*, pp. 407–409.

[6] K. C. Gupta et al., *Microstrip Lines and Slotlines*, Dedham: Artech House, 1979.

[7] L. V. Blake, *Transmission Lines and Waveguides*, New York: John Wiley & Sons, 1969, pp. 115–120.

[8] N. Marcuvitz, *Waveguide Handbook*, MIT Radiation Lab Series, vol. 10, New York: McGraw-Hill Book Co., 1951, pp. 68–71.

[9] S. Hopfer, "The design of ridged waveguides," *IRE Trans. Microwave Theory Tech.*, vol. MTT-3, pp. 20–29, October 1955.

[10] S. P. Schlesinger and D. D. King, "Dielectric image lines," *IRE Trans. Microwave Theory Tech.*, vol. MTT-6, pp. 291–299, 1958.

[11] E. A. J. Marcatili, "Dielectric rectangular waveguide and directional coupler for integrated optics," *Bell Syst. Tech. J.*, vol. 48, pp. 2071–2102, 1969.

[12] K. J. Button and J. C. Wiltse, eds., *Infrared and Millimeter Waves*, vol. 4, New York: Academic Press, 1981, pp. 195–273.

[13] C. M. Krowne, "Approximations to hybrid mode slot line behavior," *Electron Lett.*, vol. 14, pp. 258–259, April 13, 1978.

[14] H. Hofmann, "Calculation of quasi-planar lines for mm-wave application," *1977 IEEE MTT-S Microwave Symp. Dig.*, pp. 381–384, San Diego, June 1977.

Chapter 29

Propagation

C. H. Liu
University of Illinois

D. J. Fang
Telectronics International, Inc.

CONTENTS

1. Introduction — 29-5
2. Satellite-Earth Propagation — 29-7
    Free-Space Loss Along a Satellite-Earth Path 29-8
    Gaseous Attenuation 29-8
    Rain Attenuation 29-10
    Rain Depolarization 29-14
    Transionospheric Propagation 29-14
    Group Delay 29-15
    Scintillation 29-16
3. Propagation beyond the Horizon via the Ionosphere — 29-21
    Ionospheric Propagation at High Frequency 29-21
    Sky-Wave Propagation at Medium Frequency 29-25
    Ionospheric Waveguide-Mode Propagation at Very Low and Extremely Low Frequencies 29-27
    Scatter Propagation at Very High and Ultrahigh Frequencies 29-28
4. Tropospheric and Surface Propagation — 29-30
    Refractive Index of the Atmosphere 29-30
    Line-of-Sight Propagation 29-32
    Multipath Propagation over an Earth Surface 29-37
    Diffraction 29-41
    Surface-Wave Propagation 29-44
    Tropospheric Scatter Propagation 29-46

Chao-Han Liu was born in Kwangsi, China, on January 3, 1939. He received the BS degree in electrical engineering from National Taiwan University in 1960 and the PhD degree in electrical science from Brown University in 1965.

In 1965 he joined the faculty of the Department of Electrical Engineering, University of Illinois at Urbana-Champaign, where he is now a professor. In 1974 and 1977 he was a visiting scientist at the Max Planck Institut Für Aeronomie, Lindau, Germany. In 1981 he was National Science Council Chair Professor at National Taiwan University, Taipei, Taiwan. His research interests are wave propagation, ionospheric and atmospheric physics, and communication. He has published approximately 120 papers and is a coauthor of the book *Theory of Ionospheric Waves*. He served as an associate editor for *IEEE Transactions on Antennas and Propagation* from 1980 to 1987 and currently is the scientific secretary of the Scientific Committee on Solar Terrestrial Physics (SCOSTEP), ICSU.

He is a Fellow of IEEE and a member of the American Physical Society, the American Geophysical Union, US Commissions C, G, and H of the International Scientific Radio Union, Sigma X, and Tau Beta Pi.

Dah-Jeng "Dickson" Fang received his BSEE in 1962 from Taiwan University, Taipei, China, and his MSEE and PhD from Stanford University in 1964 and 1967, respectively. From 1967 to 1974, he was a radio physicist at Stanford Research Institute, studying electromagnetic-wave propagations, radio auroras, and ionospheric irregularities. While on leave from Stanford (1969–1970) he was a visiting professor at Taiwan University, where he published a graduate textbook on electromagnetic theory and five articles on the college education system.

Dr. Fang joined the Propagation Studies Department, COMSAT Laboratories, in 1974 and became manager in 1980. He was involved in both theoretical and experimental studies of ionospheric scintillations, microwave precipitation, multipath effects, and atmospheric optical transmission characterizations. He left COMSAT in 1985 and formed his own consultant company, Telectronics International, Inc. He is the author or coauthor of more than 50 technical papers and reports. He is a member of CCIR (Study Groups 5 and 6) and URSI (Commissions F and G).

5. Noise 29-47
   Atmospheric Noise   29-50
   Emission by Atmospheric Gases and Precipitation   29-50
   Extraterrestrial Noise   29-50
   Human-Made Noise   29-51
6. References 29-53

1. Introduction

In free space the propagation of radio waves from a transmitter to a receiver is governed by the fundamental equation

$$P_r = \frac{P_t G_t G_r \lambda^2}{(4\pi D)^2} \tag{1}$$

where the subscripts t and r refer to transmitting and receiving antennas, respectively, P is power in watts, G is gain over an isotropic antenna (numerical ratio), λ is the wavelength in meters, and D is distance between transmitter and receiver in meters. If the propagation path is not in free space, the medium's effect can be included by a correction factor F such that

$$P_r = \frac{P_t G_t G_r \lambda^2}{(4\lambda D)^2} |F|^2 \tag{2}$$

Since P is proportional to the square of the electric-field intensity, the factor F is obviously a numerical ratio of actual electric-field intensity E_m in the medium to the free-space electric-field intensity E_0, i.e.,

$$F = \frac{E_m}{E_0} \tag{3}$$

Both E_m and E_0 are in volts per meter.

For practical engineering applications one often writes equation (2) in the form

$$P_r = P_t + G_t + G_r - L_0 - L_p \tag{4}$$

where

P_r = received power in decibels referred to 1 W
P_t = transmitted power, in decibels referred to 1 W
G_t = transmitting antenna gain in decibels
G_r = receiving antenna gain in decibels
L_0 = free-space loss in decibels
L_p = loss due to the medium in decibels

The free-space loss is given by

$$L_0 = 10\log\left(\frac{4\pi D}{\lambda}\right)^2 \tag{5}$$

If we express D in kilometers and convert λ into the frequency f in megahertz, L_0 becomes

$$L_0 = 32.5 + 20\log f + 20\log D \tag{6}$$

The loss L_p due to the medium is related to the F factor by

$$L_p = -10\log|F|^2 \tag{7}$$

In radio-wave propagations the medium almost invariably behaves as an attenuator, so the factor F is less than unity in magnitude. It follows that L_p is generally a positive number.

The loss L_p due to the medium in general includes the effects due to absorption, scattering, refraction, diffraction, and reflection of the propagation medium as well as from the earth surface. The main text of this chapter is devoted to engineers who require readily useful information on either L_p or F for an assessment of system impact.

Before we move ahead, it is worthwhile to note the utility of the above approach. Generally speaking, in a real-world communications system design problem, the propagation issue is one of the many important issues in considering system constraints and trade-offs. However, the propagation medium is complicated to the extent that usually not a single effect, such as absorption per se, but multiple effects pose as propagation anomalies. System engineers can ill afford the luxury of achieving first an understanding of all the propagation effects before engaging in system design. Therefore, it is believed that the above L_p (or F) approach is a useful one since it essentially treats the propagation medium as a black box. Engineers can apply formulas, curves, or models of L_p or F for system design applications without much detailed knowledge of propagation.

There are, of course, cases where the black box approach is found to be unacceptable. In such cases some knowledge, or an assumption, of the propagation medium has to be provided. One can then proceed with an approach based on principles of physics and calculate the effects of the medium on propagation. For the simplest case, the basic propagation constant for a plane wave in a general isotropic medium is given by

$$k = \omega\sqrt{\mu_0\epsilon_0}\,n \tag{8}$$

where

k = propagation constant, or wave number in meters^{-1}

$\omega = 2\pi f$, where f is the frequency of the wave in hertz

ϵ_0 = free-space permittivity, 8.854×10^{-12} F/m

μ_0 = free-space permeability, $4\pi \times 10^{-7}$ H/m

Propagation

and n is the refractive index in the medium given by

$$n = \sqrt{\epsilon'_r} = \sqrt{\epsilon_r - j\sigma/\omega\epsilon_0} \qquad (9)$$

in which

ϵ_r = relative permittivity of the medium
σ = conductivity of the medium in siemens per meter

The speed of phase propagation is governed by the real part of k or n while the imaginary part of k or n determines the attenuation of the plane wave. Therefore, to study the propagation of radio waves in a material medium, the first step is to determine the refractive index of the medium for each characteristic mode. In the atmosphere, it turns out, n varies as a function of position as well as time, resulting in complications in propagation characteristics. Radio-wave propagation in such media has been studied by many authors [1–5].

As technology advances, the demands for precision and instantaneity increase. For modern communication and remote-sensing applications, these demands suggest that many of the conventional propagation models that are based on first-order approximations are no longer adequate. Instead of the deterministic models used to describe the media, the complexity and the randomness of the media require a stochastic approach. Phenomena such as scattering by hydrometeors, scintillation, and scattering by rough surfaces are examples in this category. New techniques developed in treating problems of wave propagation in random media are discussed in several books [6,7]. Specific applications to ionosphere and troposphere problems have been review [8,9].

In this chapter the authors will attempt to apply the results from the various sources, both theoretical and observational, to estimate the propagation effects for applications in system engineering. Many of the results in terms of empirical equations or graphs are derived by consensus, as documented by empirical modelings, FCC regulations, and/or CCIR documents.* For cases where consensus has not yet been reached, the authors will exercise their best judgment in recommending formulations based on theoretical consideration as well as limited observational data.

The chapter begins with a discussion of satellite-earth propagation with an emphasis on the application to modern satellite communications. Consideration of propagation via the ionosphere follows. Troposphere as well as ground waves are discussed next. A brief disscussion of noise and its effect on propagation concludes the chapter.

2. Satellite-Earth Propagation

Previously, satellite communications systems were largely below 10 GHz. Typical examples are the L-band (1.6 GHz) maritime systems and the C-band (4 to

---

*The CCIR documents, which are subjected to periodic reviews and revisions every two years, quoted in this chapter are the output of the XVth Plenary Assembly held in Geneva in 1982, since, at the time of preparing this chapter, formally approved later versions of the documents have not been released.

6 GHz) international communications systems. Due to technology advances and rapid increase of traffic, systems with frequencies higher than 10 GHz have begun to emerge for both fixed services and direct broadcasting services. The K_u-band (14/11 GHz) domestic and international systems are now widely available. The K_a-band (30/20 GHz) systems for the 1990s are topics of intense research.

Propagation anomalies for frequencies above 1 GHz cannot be scaled and estimated from those at lower frequencies. This section provides engineers with a concise summary of major propagation effects experienced at gigahertz frequencies for both satellite-earth and terrestrial paths. These include the gaseous-absorption loss due to atmospheric constituents; path loss due to atmospheric hydrometeors; and ionosphere-induced propagation anomalies.

Free-Space Loss Along a Satellite-Earth Path

A general geometry of wave propagation between two points over a spherical earth is shown in Fig. 1. The slant path distance R_d from an earth station pointing at the satellite with an elevation angle θ_d can be evaluated by the following great-circle equations:

$$\frac{\sin(90° + \theta_d)}{h_2 + a_e} = \frac{\sin \alpha}{h_1 + a_e} = \frac{\sin \Phi}{R_d} \qquad (10)$$

where $\Phi = 90° - \theta_d - \alpha$. For a synchronous satellite the altitude h_2 is a constant of 35 860 km and the earth radius a_e is 6376 km.

For example, if the elevation angle θ_d is 10°, pointing to a synchronous satellite, (10) implies

$$h_1 \cong 0, \qquad \alpha = 8.55°$$

$$\Phi = 71.45°, \qquad R_d = 40\,659.5 \text{ km}$$

At a transmission frequency of 4 GHz, the free-space loss L_0, according to (6), will be

$$L_0 = 32.5 + 20 \log 4000 + 20 \log 40\,659.5 = 196.72 \text{ dB} \qquad (11)$$

Gaseous Attenuation

Gaseous attenuation is caused by molecular absorption of gaseous constituents in the atmosphere. Absorptions generally have narrow spectrum peaks corresponding to resonance frequencies. Between the peaks, there are "windows," where transmission losses are relatively low. In the lower troposphere the spectral lines are broadened to wide absorption bands. An example of gaseous attenuation at sea level of standard atmosphere is shown in Fig. 2 [10]. To estimate gaseous attenuation along a slant path above the sea level, a CCIR procedure is available as documented in two Study Group 5 documents [11, 12].

For applications in modern satellite communications under 15 GHz, the CCIR procedure can be considerably simplified as below [13]:

Propagation

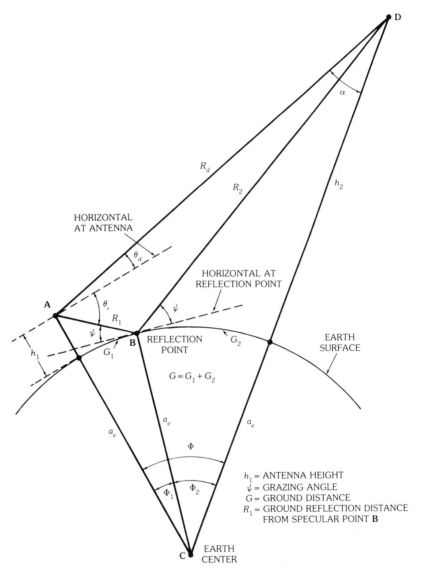

Fig. 1. General geometry of wave propagation between points A and D over a spherical earth. (*After Blake [58]*)

$$L_{\text{gas}} = \gamma_a \ell_a \tag{12}$$

$$\gamma_a = 0.00466\, e^{0.1362f} \tag{13}$$

$$\ell_a = \frac{2H_a}{(\sin^2\theta + 2H_a/R_e)^{1/2} + \sin\theta}, \quad \theta \leqq 10° \tag{14}$$

$$= \frac{H_a}{\sin\theta}, \quad \theta > 10°$$

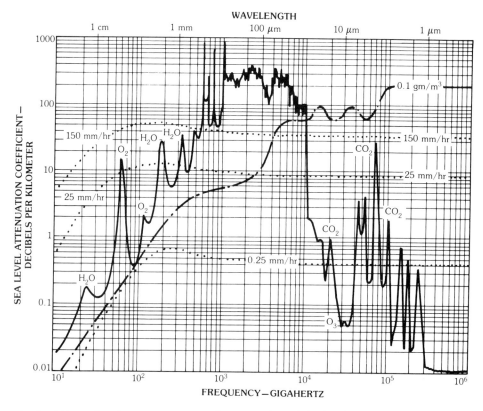

Fig. 2. Attenuation due to gaseous constituents and precipitation for propagation within the atmosphere. (*After Dougherty and Rush [10]*)

$$H_a = 6.01 e^{-0.0485 f} \tag{15}$$

where

- L_{gas} = gaseous attenuation in decibels
- γ_a = specific gaseous attenuation in decibels per kilometer
- ℓ_a = slant path length in kilometers
- f = frequency in gigahertz
- H_a = effective height of the absorptive atmosphere in kilometers
- R_e = effective earth radius, 8500 km

Rain Attenuation

The effects of rain on radio-wave propagation have been investigated by many authors both theoretically and experimentally [14–18]. These results have been used in constructing practical models for rain effects. In this section we discuss a procedure to compute rain attenuation based on these models.

Since precipitation is a statistical meteorological event, the attenuation due to

Propagation

rain is necessarily evaluated in reference to a percentage of time of an averaged year. The basic equation is still the same as (12), except it includes an additional reduction factor δ, i.e.,

$$L_R(P) = \gamma_R \ell_R \delta \tag{16}$$

where

L_R = attenuation in decibels at P percent of the time in a year
γ_R = specific rain attenuation in decibels per kilometer
ℓ_R = slant path length in kilometers
δ = reduction factor for P percent of the time

The specific rain attenuation γ_R is related to rain rate at P percent of the time, $R(P)$ in millimeters per hour, by the following power law equation:

$$\gamma_R = K_{H,V} R^{\alpha_{H,V}}(P) \tag{17}$$

where K and α are power law coefficients. This is based on theoretical scattering computations verified by experimental data. The subscripts H, V refer to horizontal or vertical polarizations of a radio-wave transmission, respectively. Numerical values of K and α as recommended by CCIR [19] are tabulated in Table 1.

To calculate $L_R(P)$, the input parameters are:

$R(P)$: the point rainfall rate in millimeters per hour for the location at a required P percent of the time. Global prediction models based on meteorological observations can be used to estimate this rate [19]
h_0: the height in kilometers above mean sea level of the earth station
θ: the elevation angle in degrees
ϕ: north or south latitude of the earth station, in degrees

According to the CCIR document [20], the calculation of $L_R(P)$ involves the following steps:

Step 1
Evaluate rain height h_R in kilometers:

$$h_R = \varrho_p(\phi)[5.1 - 2.15 \log\{1 + 10^{(\phi-27)/10}\}] \tag{18}$$

where

$$\varrho_p(\phi) = \begin{cases} 0.6 & \phi \leq 20° \\ 0.6 + 0.02(\phi - 20) & 20° < \phi \leq 40° \\ 1.0 & \phi < 40° \end{cases} \tag{19}$$

Step 2
Evaluate slant path length ℓ_R in kilometers:

Table 1. Values of Power Law Coefficients $K_{H,V}$ and $\alpha_{H,V}$ to Be Used for Evaluating Specific Rain Attenuation as Given by (17)

Frequency (GHz)	K_H	K_V	α_H	α_V
1	0.0000387	0.0000352	0.912	0.880
2	0.000154	0.000138	0.963	0.923
4	0.000650	0.000591	1.12	1.07
6	0.00175	0.00155	1.31	1.27
8	0.00454	0.00395	1.33	1.31
10	0.0101	0.00887	1.28	1.26
12	0.0188	0.0168	1.22	1.20
15	0.0367	0.0347	1.15	1.13
20	0.0751	0.0691	1.10	1.07
25	0.124	0.113	1.06	1.03
30	0.187	0.167	1.02	1.00
35	0.263	0.233	0.979	0.963
40	0.350	0.310	0.939	0.929
45	0.442	0.393	0.903	0.897
50	0.536	0.479	0.873	0.868
60	0.707	0.642	0.826	0.824
70	0.851	0.784	0.793	0.793
80	0.975	0.906	0.769	0.769
90	1.06	0.999	0.753	0.754
100	1.12	1.06	0.743	0.744
120	1.18	1.13	0.731	0.732
150	1.31	1.27	0.710	0.711
200	1.45	1.42	0.689	0.690
300	1.36	1.35	0.688	0.689
400	1.32	1.31	0.683	0.684

$$\ell_R = \frac{2(h_R - h_0)}{[\sin^2\theta + 2(h_R - h_0)/R_e]^{1/2} + \sin\theta}, \quad \theta < 10°$$

$$= \frac{h_R - h_0}{\sin\theta}, \quad \theta > 10° \quad (20)$$

where $R_e = 8500$ km.

Step 3

Evaluate reduction factor δ at P percent of the time.

The reduction factor is due to the fact that during heavier rain the rainfall rate is distributed nonuniformly over the path in the form of rain cells or rain front. The rainfall rate decreases from the center of the rain cell outwards. The rate of decrease is higher for heavier rains. The empirically determined formula for the reduction factor at P percent of the time for the satellite-earth propagation geometry is given by

$$\delta = \frac{90\,C}{90 + 4\ell_R \cos\theta}\left(\frac{P}{0.01}\right)^{-a}$$

where

$$a = 0.33, \quad C = 1; \quad 0.001 \leq P \leq 0.01 \tag{21}$$

$$a = 0.41, \quad C = 1; \quad 0.01 < P \leq 0.1 \tag{22}$$

$$a = 0.50, \quad C = 1.3; \quad 0.1 < P \leq 1.0 \tag{23}$$

Step 4
   Evaluate $L_R(P)$ by (16).

Among more than a dozen methods available, the above method is regarded to be the one that produces least overall errors against currently available global experimental data collected from more than 30 locations [20]. For an individual site, however, the method could make predictions which deviate considerably from experimental results. Typical comparisons between predicted and measured results [21, 22, 23] are provided in Table 2.

In many digital communications applications, rain attenuation statistics for a worst month in a year is required for system design. In the absence of reliable measurement data the statistics can be estimated from the annual statistics by the following conversion relationship:

$$P = 0.29 P_w^{1.15} \quad \text{or} \quad P_w = 2.93 P^{0.87} \tag{24}$$

where P_w is the percentage time in a worst month. It suggests that the worst month $L_R(P_w)$ is the same numeric decibel value as that of $L_R(P)$ in an averaged year.

Table 2. Comparison between Measured Rain Attenuation Data and Predictions

Location (Latitude, h_0)	Frequency (GHz)	Polarization	Elevation Angle (°)	Time (%)	Rain Rate R (mm/hr)	Measured Attenuation (dB)	Predicted Attenuation (dB)
Kjeller, Norway (60.0N, 0)	11.6	H	22	0.001	75.0	12.6	16.0
				0.01	40.0	6.9	7.5
				0.1	15.0	2.1	2.9
				1.0	3.0	0.8	1.0
Yonaguni, Japan (24.3N, 0.20)	12.1	V	58	0.001	117.1	16.7	23.3
				0.01	82.8	14.3	10.9
				0.1	37.8	8.1	4.2
				1.0	5.6	2.7	1.4
Clarksburg, Md. (39.2N, 0.18)	19	H	47.5	0.001	—	—	64.0
				0.001	68.0	22.0	29.9
				0.1	17.0	11.0	11.6
				1.0	2.0	3.0	3.9
Blacksburg, Va. (37.2N, 0.30)	28.6	V	45	0.001	100.0	—	48.5
				0.01	36.0	24.0	22.7
				0.1	8.0	11.8	8.8
				1.0	1.0	5.0	2.9

Rain Depolarization

The determination of depolarization is important for communications, interference considerations, and remote-sensing applications, particularly for dual-polarization antennas. At gigahertz frequencies, depolarization can be caused by raindrops, snow, and ice particles, with raindrops being the major contributor [24]. Engineering modelings for estimating depolarizations due to snow and ice particles have yet to be established. Even for rain, CCIR models are rather divergent. The basic reason is that depolarization results from the nonspherical geometry of raindrops as they tilt in the air while falling under the influence of turbulent wind shear. This is a highly random meteorological process [25]. For first-order applications, a simple equation is selected here for estimating cross polarization [26], XPD in decibels, from a signal at a known level of attenuation, $L_R(R)$, given in the previous section:

$$\text{XPD} = 30 \log f - 40 \log(\cos \theta) - 20 \log(\sin 2\tau) - 20 \log[L_R(P)] \qquad (25)$$

where

f = frequency in gigahertz

θ = elevation angle in degrees

τ = polarization tilt angle in degrees with respect to the horizon (for circular polarization $\tau = 45°$, the term $20 \log[\sin 2\tau]$ disappears)

Again, the XPD here refers to long-term averaged cross-polarization level at P percent of the time for an averaged year.

Transionospheric Propagation

Caused by solar radiation, the earth's ionosphere consists of several regions of ionizations located from about 50 km to 1000 km in altitude. For all practical communication purposes three regions of the ionosphere, D, E, and F (F1, F2), have been identified as shown in Fig. 3, where N_e is the electron density in electrons per cubic meter.

In each region the ionized medium is neither homogeneous in space nor stationary in time. Generally speaking, the background ionization has relatively regular diurnal, seasonal, and 11-year solar cycle variations, with many anomalies, and depends strongly on geographical location. An international reference ionosphere has been constructed [27] that can be used as a general reference. In addition, there are highly dynamic, small-scaled nonstationary structures known as *irregularities*. Both the background ionization and irregularities degrade radio waves. The charged particles in the ionosphere cause the refractive index to become frequency dependent, i.e., the medium becomes dispersive. For transionospheric propagation above the very high frequencies, when the effect of earth's magnetic field can be neglected, the refractive index n can be expressed as

$$n = \left(1 - \frac{80.6}{f^2} N_e\right)^{1/2} \qquad (26)$$

Propagation

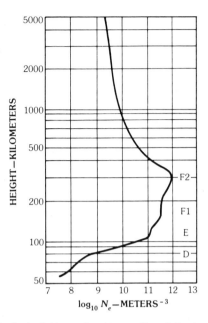

Fig. 3. Typical electron-density profile of the ionosphere.

where N_e is the electron density in electrons per cubic meter and f is the signal frequency in hertz. The density N_e has a regular part corresponding to the background density and a fluctuating part corresponding to the irregularities. The variation of the density will cause the refractive index to fluctuate, which in turn will affect the propagation of the radio waves. In general, for transionospheric propagation, the F2 layer is the most important region.

Group Delay

The presence of charged particles in the ionosphere slows down the propagation of radio signals along the satellite-earth path. The time delay in excess of the propagation time in free space is called the *group delay*. It is an important factor to be considered for digital communication systems. This quantity can be computed from the group velocity derived from (26):

$$\tau = 1.345 N_T / f^2 \qquad (27)$$

where

τ = delay time in nanoseconds in reference to propagation in a vacuum

f = frequency of propagations in gigahertz

N_T = integrations per 10^{16} electrons/m² of electron density along the propagation path

The quantity N_T is called the *total electron content* (TEC) and can be evaluated by

$$N_T = 10^{-16} \int_s N_e(s)\, ds \qquad (28)$$

where s is the ray path in meters. Typically N_T varies from 1 to 200. The exact evaluation of N_T is difficult because N_e has diurnal, seasonal, and solar cycle variations. For a band around 1600 MHz, the signal group delay varies from approximately 0.5 to 100 ns. Fig. 4 is an illustration of worldwide group delay distribution based on a model by Bent et al. [28]. Note from the figure that the regions of maximum time delay occur near the equator at plus and minus 15° to 20° of latitude. These contours move along approximate lines of constant magnetic latitude from east to west as the earth rotates.

The depolarization caused by the Faraday rotation may also cause signal degradation for systems employing linear polarizations.

Scintillation

One of the most severe disruptions along a satellite-earth propagation path for signals from vhf to C-band is caused by ionospheric scintillation. Principally through the mechanisms of forward scattering and diffraction, small-scale irregular structures in the ionization density produce the scintillation phenomenon in which the steady signal at the receiver is replaced by one which is fluctuating in amplitude, phase, and apparent direction of arrival. Depending on the modulation scheme of the system, various aspects of scintillation affect the system performance differently. The most commonly used parameter characterizing the intensity fluctuations is the scintillation index S_4, defined by

$$S_4 = \left(\frac{\langle I^2 \rangle - \langle I \rangle^2}{\langle I \rangle^2} \right)^{1/2} \qquad (29)$$

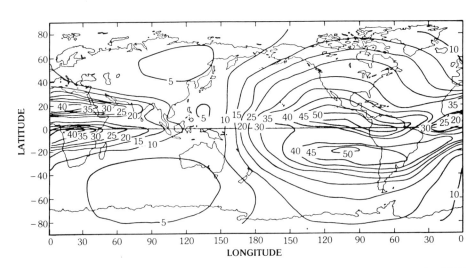

Fig. 4. Worldwide group delay in nanoseconds for a system operating frequency of 1.6 GHz at a universal time of 00 hours derived from the Bent model [28] for conditions appropriate to March 1968.

where I is the intensity of the signal and $\langle \ \rangle$ denotes averaging.

Under weak scintillation conditions which approximately apply when $S_4 \lesssim 0.5$, the scintillation index can be estimated from the following equation [8]:

$$S_4^2 = 8\pi L \lambda^2 r_e^2 \iint_{-\infty}^{+\infty} S_N(\varkappa, 0) \sin^2(\varkappa^2/\varkappa_F^2) \, d^2\varkappa \tag{30}$$

where

L = thickness in meters of the ionosphere layer

r_e = classical radius of electron = 2.818×10^{-15} m

λ = wavelength in meters

$S_N(\varkappa, \varkappa_z)$ = power spectrum of electron density fluctuations in the ionosphere, in meters^{-3}

$\varkappa_F$ = Fresnel wave number = $\sqrt{4\pi/\lambda z}$ in meters^{-1}

z = height of the ionosphere layer in meters

The scintillation index S_4 is related to the peak-to-peak fluctuations of the intensity. The exact relation depends on the distribution of the intensity. Empirically, Chart 1 provides a convenient conversion between S_4 and the approximate peak-to-peak fluctuations P_{fluc} in decibels [29].

Chart 1. Empirical Conversion Table for Scintillation Indices

S_4	P_{fluc}
0.1	1.5
0.2	3.5
0.3	6
0.4	8.5
0.5	11
0.6	14
0.7	17
0.8	20
0.9	24
1.0	27.5

The intensity distribution is best described by the Nakagami distribution [8] for a wide range of S_4 values. As $S_4 \to 1.0$, it approaches the Rayleigh distribution. Occasionally, S_4 may exceed 1, reaching values as high as 1.5. This is due to focusing. For values less than 0.6, S_4 shows a consistent $f^{-1.5}$ frequency dependence for most multifrequency observations in the vhf and uhf bands [30]. Recent equatorial observations at gigahertz frequencies, however, suggested values higher than 1.5 for the spectral index [31]. As the scintillation becomes stronger such that S_4 exceeds 0.6, the spectral index decreases. This is due to the saturation of scintillation for Rayleigh fading under the strong influence of multiple scattering.

Another parameter that is important in system applications is the fade coherence time τ_c, which is defined as the time lag in seconds for the autocorrelation function of the intensity to drop to 50 percent of its maximum value. This coherence time varies from a few tenths of a second to about ten seconds, depending on the signal frequency and the ionospheric conditions [8, 31]. For a given frequency, strong scintillation corresponds to smaller τ_c. Time delays should be greater than τ_c to achieve any diversity gain.

The phase scintillations are in general dominated by slow fluctuations. They become important in systems that are required to maintain phase coherence over long time or spatial intervals.

Geographically, there are two intense zones of scintillations, one at high latitudes and the other centered within ±20° of the magnetic equator as shown in Fig. 5 [32]. Severe scintillations have been observed up to gigahertz frequencies in these two sectors, while in the middle latitudes scintillations mainly affect vhf signals. In all sectors there is a pronounced nighttime maximum of the activity as also indicated in Fig. 5. For equatorial gigahertz scintillation, peak activity around the vernal equinox and high activity at the autumnal equinox have been observed [31, 33].

To model the scintillation phenomenon one needs to understand the physical mechanisms that generate these irregularities. Current knowledge about these irregularities, however, does not provide sufficient basis for the construction of a comprehensive model. It has been suggested that spread-F irregularities are responsible for scintillations [34]. Spread-F irregularities are essentially evening and nighttime events in the F region. Depending on latitudinal locations the

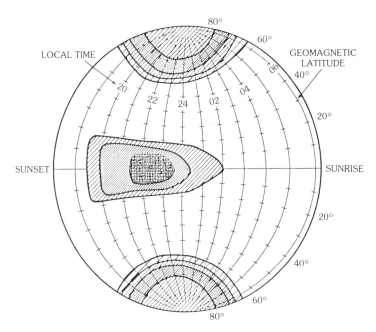

Fig. 5. Depth of scintillation fading (proportional to density of cross-hatching) during low to moderate solar activity. (*After Aarons [32]*, © *1982 IEEE*)

occurrence of spread F has distinctive patterns relating to seasonal variations and magnetic activities. Since spread F has been studied continuously from the early 1930s and enormous amounts of data are available, radio physicists have attempted to devise empirical models for predicting scintillations based on spread-F models. A step-by-step methodology consisting of inputs of morphological data of spread F and more than 50 numerical equations is available for statistical prediction of uhf scintillation [34]. A more recent model has been constructed based on numerous satellite propagation observations, particularly on observations performed in the DNA Wideband Satellite Experiment. Geographical, seasonal, diurnal, and solar activity as well as magnetic activity dependence are included in the model [35, 36, 37]. The models have been shown to be successful in predicting uhf scintillations.

As for scintillations at gigahertz frequencies, however, the available long-term observational data seem to indicate that additional efforts in modeling are needed [33, 35]. An example for an equatorial ionospheric path at 4 GHz is shown in Fig. 6. The annual occurrence statistics of peak-to-peak amplitude fluctuations, P_{fluc} in decibels, are given for two links. In the figure, P curves in solid lines are signal fluctuations as received from a satellite at the east direction of an approximately 20° elevation angle, while the I curves are from a satellite at the west direction at 30° elevation angle. Curves for different years when sunspot numbers (SSN) change from 10 to 165 are labeled. For link budget calculation, P_{fluc} is related to loss L_p given in (4) by

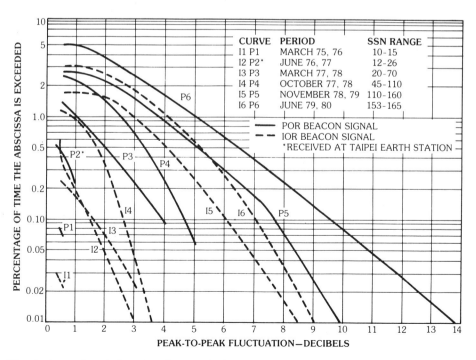

Fig. 6. Annual statistics of peak-to-peak fluctuations observed at Hong Kong earth station (curves I1, P1, I3–I6, P3–P6) and Taipei earth station (curves P2 and I2).

$$L_p = \frac{1}{\sqrt{2}} P_{\text{fluc}} \tag{31}$$

For scintillations at high latitude the CCIR document [35] should be used for further information.

For frequencies above C-band and propagation path at an elevation angle lower than 10° or so, troposphere-induced scintillations become a factor to be considered in computing the path loss. The tropospheric effects are produced by local atmospheric features such as high humidity gradients and temperature inversion layers leading to thin, horizontally stratified layers in which the refractive index is markedly different from the ambient value. Superimposed on these layers are fluctuations caused by internal waves or turbulence.

For a weak scintillation such that $S_4 < 0.6$, the rms fluctuations in log power can be estimated by the following equation assuming the scintillation is caused by Kolmogroff turbulence [6]:

$$\sigma_\chi^2 = 42.25 \, k^{7/6} \int_0^L C_n^2(\zeta) \, \zeta^{5/6} \, d\zeta \tag{32}$$

where

σ_χ = standard deviation of logarithm of received power related to S_4 by $S_4 = 4\sigma_\chi^2$

k = wave number, $2\pi/\lambda$ in meters^{-1}

$C_n(\zeta)$ = refractive index structure constant along the ray path in (meters)$^{-2/3}$

ζ = distance along ray path in meters

L = total length of ray path in meters in the turbulent region

The term C_n^2 varies from 10^{-14} to about 10^{-15} at the ground level to about 10^{-17} to about 10^{-18} at heights of a few kilometers. Observed data indicate that the effects are seasonally dependent and vary geographically. At high elevation angles (greater than 10°), 1-dB peak-to-peak fluctuations can be expected in clear sky and 2- to 6-dB fluctuations in some types of clouds. The fading rates vary from 0.5 Hz to over 10 Hz in general. Aperture averaging is more effective for reducing tropospheric scintillation. To take this effect into account, (32) should be multiplied by an aperture averaging factor. This factor can be evaluated by [6, 38]

$$G(R) = \begin{cases} 1.0 - 1.4(R/\sqrt{\lambda L}), & 0 \leq R/\sqrt{\lambda L} \leq 0.5 \\ 0.5 - 0.4(R/\sqrt{\lambda L}), & 0.5 < R/\sqrt{\lambda_L} \leq 1.0 \\ 0.1, & 1.0 < R/\sqrt{\lambda L} \end{cases} \tag{33}$$

where

R = effective radius in meters of circular antenna aperture, given by $\eta^{1/2}(D/2)$ where η is the antenna efficiency factor and D is the physical diameter of the antenna in meters

Propagation

L = slant distance in meters to height of a horizontal thin turbulent layer
  = $[h^2 + 2R_e h + (R_e \sin\theta)^2]^{1/2} - R_e \sin\theta$

h = height of layer in meters

θ = elevation angle

R_e = effective earth radius = 8.5×10^5 m

λ = operating wavelength in meters

3. Propagation beyond the Horizon via the Ionosphere

While for the satellite-earth propagation path the ionosphere causes additional path loss as discussed in Section 2, it provides the most reliable beyond-the-horizon transmission channels at frequency bands of hf and below and sometimes at the vhf band. When radio signals at hf or lower frequencies are transmitted from the earth surface (or aircraft) to the ionosphere, the refraction due to the ionosphere layers may be sufficient to bend the ray back to earth. In addition, irregular structures of the ionization density in the ionosphere may scatter radio-frequency energy and provide over-the-horizon transmission even at vhf band. In this section the various modes of propagation beyond the horizon via the ionosphere will be discussed.

Ionospheric Propagation at High Frequency

According to magnetoionic theory for radio waves propagating in the ionosphere, two characteristic modes exist. The refractive indices n for the two modes are given by the Appleton-Hartree formula [3, 4, 39]

$$n^2 = 1 - \frac{X}{U - Y_T^2/2(U - X) \pm [Y_T^2/4(U - X)^2 + Y_L^2]^{1/2}} \quad (34)$$

where

$X = 80.6 N_e/f^2$

$Y_T = Y \sin\theta$

$Y_L = Y \cos\theta$

$Y = 2.8 \times 10^{10} B_0/f$

$U = 1 - j(\nu/2\pi f)$

N_e = electron density in electrons per cubic meter

f = frequency in hertz

θ = angle between the propagation direction and the earth's magnetic field, Fig. 7

B_0 = earth magnetic flux density in webers per square meter

ν = collisional frequency in hertz between electrons and molecules

The upper sign in (34) is referred to as ordinary mode, and the lower sign the extraordinary mode.

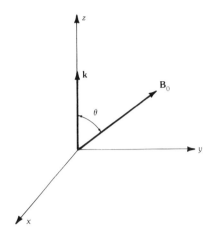

Fig. 7. Propagation geometry of radio waves in a magnetoplasma.

It is interesting to note that for $\theta = 90°$, if one neglects the collisions, the refractive index for the ordinary mode will decrease to zero as $X \to 1$, or as the signal frequency approaches the value

$$f_0 = \sqrt{80.6 N_{max}} \tag{35}$$

where

f_0 = critical frequency in hertz

N_{max} = maximum electron density in electrons per cubic meter in the ionosphere

For a signal frequency f lower than the vertical frequency, a vertically propagated ordinary wave will be turned back by the ionosphere at a height where the electron density N_e is equal to $f^2/80.6$.

For an obliquely incident wave the maximum frequency below which the wave will be bent back to the earth is given by $f_0 \sec \phi_0$, where ϕ_0 is the angle the ray path makes with the vertical. This frequency is known as the theoretical maximum usable frequency, muf. It is this phenomenon that makes long-distance communication using high frequencies via the ionosphere possible. One-hop propagation usually can reach a maximum range of 4000 km. Several hops can be used for even longer distances. The critical frequency for reflection of the extraordinary wave is obtained from (34) by taking the lower sign. For the case $\theta = 90°$,

$$f_x \cong f_0 + f_H/2 \tag{36}$$

where

f_x = critical frequency for extraordinary wave in hertz

f_H = gyrofrequency for electrons = $2.84 \times 10^{10} B_0$ Hz

It is clear that the muf's for the two modes are different.

For a given propagation geometry the muf depends on many parameters, including diurnal and seasonal variation of ionization of the ionosphere, solar flux, magnetic-field disturbances, and propagation configuration with respect to the earth's magnetic field. Different muf's can be assigned to different ionization regions, such as the muf for the E region, muf for the E region irregularity known as "sporadic E," muf for the F region, etc. Fig. 8 shows a typical diurnal variation of critical frequencies. Both f_0 and f_x are included in the figure. In addition, f_z, corresponding to the critical frequency for the occasional z mode, and f_bE_s, corresponding to the critical frequency due to sporadic-E layers [39], are also shown in the figure. These critical frequencies are usually obtained from ionograms. An ionogram is a graph displaying the time delay for a sounding pulse to

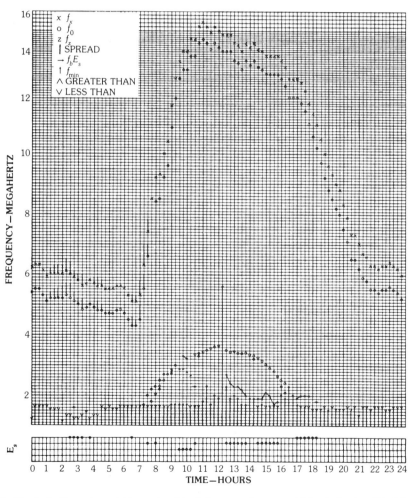

Fig. 8. Typical diurnal variation of critical frequencies at middle latitude: Washington, D.C., December 31, 1958. (*After Davies [39]*)

travel up to the ionosphere and back as a function of frequency. The sounding device is called the *ionosonde*, which essentially is a pulsed radar with a frequency-sweeping capability over a wide range of frequencies from 1 MHz up to 25 MHz in general. The frequency at which the pulse penetrates the ionosphere during vertical sounding is the critical frequency. Modern-day ionosondes can provide digitized output and additional information about the ionosphere.

The main layers in the ionosphere responsible for returning the hf radio signal are the E and F layers. F2 reflections are usually the most important. During the day, one-to-two–hop F-layer reflections are usually dominant. The F1 layer may be important for paths of 2000 km to 3500 km during the day. The E layer provides a useful propagation path during the day for distances up to 2000 km. Sporadic-E (E_s) layers may increase the muf to 150 MHz at times for ranges up to 2400 km. They can also provide a two-hop configuration together with F2-layer reflection with extended range since the D-region absorption can be avoided in the second hop.

Because of the variability of the ionosphere and the propagation geometry, hf propagation is characterized by the need to select operating frequencies as frequently as the ionosphere changes. It is also affected by multipath interference and ionospheric disturbances. Despite all these drawbacks, however, the hf circuit still continues to be one of the most reliable, economical, low- to medium-power long-distance communication systems. Due to a high collisional frequency between electrons and neutral molecules, radio waves suffer most absorptions in the D region. The absorption coefficient defined as absorption per unit length depends on the electron density, the wave frequency, electron collisional frequency, and the earth's magnetic field. For practical applications, an empirical formula applicable in middle latitudes has been derived for the path loss per hop due to absorption [39]:

$$L_a = 430(1 + 0.0035R)(\cos \chi)^{3/4} \sec \phi_D (f \pm f_L)^{-2} \qquad (37)$$

where

L_a = path loss per hop to absorption in decibels
R = sunspot number
χ = solar zenith angle
ϕ_D = angle between the ray and the vertical in the absorbing D region
f = frequency in megahertz
f_L = component in megahertz of electron gyrofrequency along the direction of phase propagation

The positive-and-negative sign in (37) corresponds to the ordinary and extraordinary waves, respectively. The formula shows that absorption increases with sunspot number and with the obliquity of the ray path ϕ_D, which is proportional to the path length. After sunset $\chi > 90°$ and the absorption falls to low values. This is because of the disappearance of the D region after sunset. If all

Propagation

factors remain constant, absorption increases as the wave frequency decreases. The lowest usable frequency for a given link is called the luf.

The formula does not include the effect of the winter anomaly for which exceptionally high absorptions occur. Multiplicative factors must be used to the formula (37) for different months and geographical locations when estimating path loss during the winter anomaly period [39].

In the polar region extremely high D-region absorptions causing long-lasting blackouts of hf propagation are believed to be related closely to the energetic particle precipitation from the solar wind. These events are known as *polar cap absorptions* and are almost always preceded by major solar flare events. Another type of high-latitude absorption is the auroral blackout that occurs near the auroral zones during magnetic storms. Magnetic storms also affect the F region by reducing the critical frequency, thus lowering the muf. Other types of ionospheric disturbances include large-scale traveling ionospheric disturbances (TIDs) and small-scale irregularities. They also affect the propagation of hf waves and often cause deep fades in received signal strength.

Link budget computation for an hf circuit begins in the selection of the *optimum working frequency* (owf or FOT) which is usually chosen above the luf and just below the muf. To calculate the muf for a given path, prediction maps [40] are used. Great-circle paths can be drawn and overlaid on these maps to obtain the muf. These are monthly median values for a given hour. Day-to-day random variability of the muf can be taken into account by considering its random distribution as a function of season, solar activity, etc. With the operating frequency chosen, the path loss due to ionosphere propagation can be calculated as

$$L = L_a + L_g + L_e \tag{38}$$

where

L_a = path loss in decibels due to ionospheric absorption, computed using (37) or empirical curves

L_g = loss in decibels due to ground reflection (value ranges from 4 dB for land to 0.2 dB for ocean per reflection)

L_e = extra path loss in decibels due to defocusing of the rays upon refraction, polarization mismatch, and variability of the loss about monthly median values

Computerized prediction models are available as documented in CCIR publications [40, 41, 42]. These models can be used to estimate hf path loss at a given annual percentage of time for different locations and propagation geometries. The computer program based on [41] can be purchased from National Technical Information Services, US Department of Commerce.

Sky-Wave Propagation at Medium Frequency

Medium-frequency (mf) waves (300 kHz to 3 MHz) are primarily used for am broadcasting. From (37) we note that ionospheric D-region absorption during daytime renders medium frequencies useless for long-distance communication. After sunset, sky-wave propagation at medium frequencies becomes possible. The frequency band from 150 kHz to 1.5 MHz is used for maritime communications and navigation and medium-range broadcasting. For practical engineering evaluations at medium frequencies there are several methods recognized by the CCIR for distinctive regions of the world. The method recommended for northern America is documented here [43].

Let E be the annual median half-hourly field strength in decibels over 1 μV/m at the time of 6 hr after sunset at the receiver. Then

$$E = 10 \log P_t + G_t + (106.6 - 2 \sin \Phi) - 20 \log p - k_R p \times 10^{-3} \quad (39)$$

where

P_t = transmitted power in kilowatts

G_t = transmitting antenna gain in decibels

Φ = geomagnetic latitude in degrees of the path midpoint between transmitter and receiver

p = slant propagation path in kilometers

k_R = loss factor in decibels

It can be shown that

$$\Phi = \sin^{-1}[\sin \phi \sin 78.3° + \cos \phi \cos 78.3° \cos(\lambda - 291°)] \quad (40)$$

with ϕ and λ being the north latitude and east longitude of the midpoint. Furthermore,

$$p = \begin{cases} d, & d < 1000 \text{ km} \\ (d^2 - 40\,000)^{1/2}, & d > 1000 \text{ km} \end{cases} \quad (41)$$

where d is the ground distance, which can be found in the Solar Geographical Data Reports published by NOAA. Also,

$$k_R = bR \times 10^{-2} + K \quad (42)$$

$$b = \begin{cases} 0.4|\Phi| - 16, & \Phi > 45° \\ 0, & \Phi < 45° \end{cases} \quad (43)$$

where R is the smoothed Zurich sunspot number, usually between 20 and 150, which can be found in the Solar Geographical Data Reports published by NOAA. Also,

$$K = [0.0667 \times |\Phi| + 0.2] + 3 \tan^2(\Phi + 3°) \quad (44)$$

Propagation

For local time other than 6 hr after local sunset the value of E can be adjusted by a reduction factor shown in Fig. 9. It has been found that the method provides a good agreement with the well-known FCC curves [44] as well as with measurement data from Europe and other parts of the world [45].

Ionospheric Waveguide-Mode Propagation at Very Low and Extremely Low Frequencies

For signals with frequencies below 30 kHz the wavelength becomes comparable to the scale height of the ionosphere, and the ionosphere behaves as a relatively well defined reflecting layer for these waves. This layer and the highly conducting surface of the earth (at these frequencies) constitute the boundaries for a spherical waveguide. Several special features stand out for the characteristic modes propagating in the ionosphere-earth waveguide. First, natural resonances can exist in this spherical shell structure. Indeed, these resonant oscillations have been observed near 8, 14, and 20 Hz and higher frequency bands. They are believed to be excited by naturally occurring lightning discharges and are known as Schumaun resonance [46]. Secondly, guided modes at vlf and elf bands have relatively low path attenuation, which is quite stable with time. Thirdly, good phase stability exists for the dominant mode. These last two points are in contrast with hf transmissions, which suffer from unreliability during ionospheric disturbances and rapid and deep fading. Therefore vlf and elf systems are attractive candidates for around-the-globe, highly reliable communication, navigation, and timing applications. Indeed, the modern Omega navigation system uses vlf transmissions at 10.2, 11.33, and 13.6 kHz [47]. The disadvantages of systems in these frequency bands are the large antenna installations and limited usable bandwidths. This is especially true due to the necessary high-Q antennas. Typically the bandwidths are limited to 20 to 150 Hz.

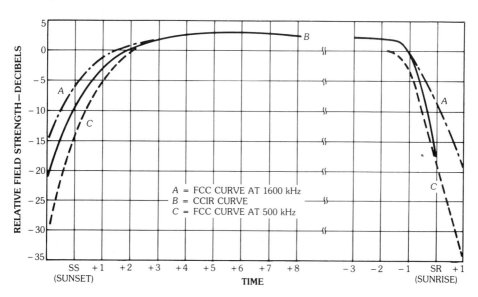

Fig. 9. Adjustment factor for F.

Very low and extremely low frequency propagation in the ionosphere-earth waveguide has been studied using different models [48]. The basic model is a spherical shell structure with a perfectly conducting boundary on the earth surface and an impedance boundary for the ionosphere. Refined models take into account the anisotropy of the ionosphere plasma, the terrain features on the surface, the penetration of the wave into the ionosphere, and the effect of propagation across a sunrise and sunset line. Predicted attenuation coefficients and phase propagation speeds from model computations compare rather favorably with measured data. These results are used in link budget computation for a given path. Since atmospheric noise is quite high in these frequency bands, it has to be included in the computation. A practical procedure for calculating power requirements for a vlf station operating over a specified path in the presence of a given level of background noise can be found in CCIR reports 265-24, 265-5 [46, 49].

Scatter Propagation at Very High and Ultrahigh Frequencies

At vhf and uhf frequencies the radio waves penetrate the ionospheric layers, and the over-the-horizon transmissions via the regular reflection of the ionosphere become impossible. However, irregular structures in the ionosphere are capable of scattering the wave energy back to the earth surface and yield much higher field strength. This is the basic mechanism for scatter propagation. It has been demonstrated that irregularities in the D region can be used to establish a scatter communication link for ranges up to a few thousand kilometers. The prediction of the field strength in such a link is based on the well-known Booker-Gordon formula for a scattering cross section given by [50]

$$\sigma = 2\pi k^4 \sin^2\chi \, S_n(2k \sin \theta/2) \qquad (45)$$

where

σ = scattered power per unit solid angle, per unit incident power density, per unit scattering volume, in meters^{-1}

$k = 2\pi/\lambda$ wave number in meters^{-1}

χ = angle between the incident electric-field vector and the scattering direction

$S_n(K)$ = power spectrum of refractive index irregularities in cubic meters, where K is the wave number

θ = angle between incident and scattered waves

The general geometry is shown in Fig. 10. The strength of the scattered field is determined by a single spectral component of the irregularities corresponding to the scale $2\pi/2k \sin(\theta/2) = \lambda/2 \sin(\theta/2)$. This is the well-known Bragg condition for diffraction from spatial structures. For the ionosphere case the fluctuations in the refractive index are caused by electron density irregularities. The received scattered power is given by

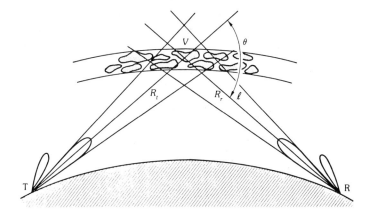

Fig. 10. Geometry of ionospheric propagation based on scattering mechanism.

$$P_r = \frac{\pi P_t G_t G_r \lambda^2 r_e^2}{2 R_t R_r^2} \sin^2\chi \cdot S_N(2k \sin \theta/2) V \qquad (46)$$

where

- P_r = received power in watts
- P_t = transmitted power in watts
- G_t = antenna gain for the transmitting antenna
- G_r = antenna gain for the receiving antenna
- λ = wavelength in meters for the radio wave
- R_t = distance in meters from the transmitter to scattering volume
- R_r = distance in meters from the receiver to scattering volume
- r_e = classical electron radius (2.82×10^{-15} m)
- $S_N(k)$ = power spectrum of electron density fluctuations in meters^{-3}
- V = common volume of the scattering region in cubic meters

Equation 46 can be compared to (2). From (46) it can be shown that the received power is proportional to $\lambda^m/\sin^p(\theta/2)$. The values of m and p depend on the spectral shape of the irregularity spectrum; observations show m varies between 7 and 9.5 while p is found to be in the range of 6 to 8.5. In deriving (45), the irregularities are assumed to be isotropic. This equation has to be modified if field aligned structures are responsible for the scattering.

At the equator, the F-region spread-F type of irregularities can support transequatorial vhf propagation at levels 30 to 40 dB above the level produced by usual scatter propagation. This usually occurs during the hours after local ionospheric sunset. It is closely related to the equatorial irregularities causing scintillations of the transionospheric signal as discussed in Section 2.

Ionization trails produced by meteors in the height range 80 to 120 km also scatter vhf radio signals and may be used to establish scatter links [39].

4. Tropospheric and Surface Propagation

There are three fundamental modes of tropospheric propagation for frequencies above vhf band: line-of-sight propagation, multipath propagation, and diffraction over an obstacle. The surface-wave mode becomes important for frequencies below vhf and may be the dominant mode at mf and lf bands. Numerous research publications are available dealing with theoretical and modeling aspects of each of these modes [2, 51, 52].

Rapid advancement of communication technology in recent years have opened the door for new types of communication service other than the conventional point-to-point communication. It becomes evident that propagation anomalies due to either tropospheric irregularities or surface irregularities affect each type of communication differently simply because of the unique requirement of each individual type. For instance, Doppler effects are much more serious for mobile unit communication than for fixed station communication. Among types of mobile unit communication, land- and sea-mobile units are sensitive to the surface-wave mode of propagation while air-mobile units are not. A broadcasting system involves point-to-area coverage, and so has completely different interference considerations than a point-to-point radio-relay system. A further complication is that even for one specific type of service, propagation degradations are highly region-dependent due to different meteorological conditions. For instance, the CCIR propagation document for a transhorizon radio-relay system [53] outlines the effect of nine climatological regions, including equatorial, continental subtropical, maritime subtropical, desert, Mediterranean, continental temperate, maritime temperate over land, and maritime temperate over sea and polar regions.

Obviously it is impractical to summarize here all engineering models for different communication services and applications at various climatological regions. The objective of this section is to provide a few sets of practical equations which enable one to gain an intuitive physical insight into propagation degradation due to the presence of the troposphere and the earth's surface. Such intuitive insight is always a prerequisite for an engineer to apply judiciously the specific modeling procedure among more than a dozen or so postulated by the CCIR, FCC, and others.

Refractive Index of the Atmosphere

The refractive index n in a medium is a function of ϵ_r and is given by (9). Typical values of ϵ_r and σ (in siemens per meter) relevant to troposphere and surface propagation are shown in Fig. 11. In the atmosphere, $\sigma \cong 0$; therefore

$$n = \sqrt{\epsilon_r} \tag{47}$$

In practice, the radio refractivity N is defined as

$$N = (n - 1) \times 10^6 \tag{48}$$

Propagation

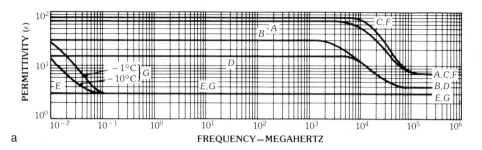

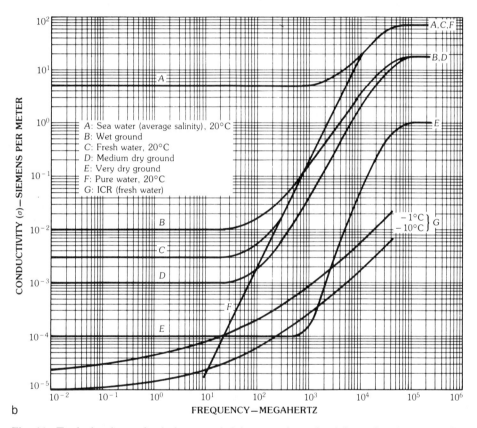

Fig. 11. Typical values of relative permittivity ϵ_r and conductivity σ for the troposphere propagation environment. (*a*) Relative permittivity. (*b*) Conductivity.

It is known that for all frequencies up to 100 GHz, the following approximations apply:

$$N = 77.6 \frac{p}{T} + 3.73 \times 10^5 \frac{e}{T^2} \tag{49}$$

where

p = atmospheric pressure in millibars

T = absolute temperature in kelvins

e = water vapor pressure in millibars

Since p, T, and e are exponential functions of height with different scale heights, one can scale N as a function of height h the same way such that

$$N(h) = N_s e^{-bh} \qquad (50)$$

The CCIR defines a "standard radio atmosphere" for which $N_s = 315$ units and $b = 0.136$ km^{-1}. Worldwide charts of N and b and the gradient of N over the first 1-km thickness from the ground for different months in a year are available in the CCIR document [54].

The refractive index structure in the atmosphere causes continuous refraction of the electromagnetic wave, or bending of a radio ray. The theory for calculation of refraction is well known [1]. The curvature $1/\varrho$ of the radio path at a point is given by

$$\frac{1}{\varrho} = \frac{-\cos\phi}{n}\frac{dn}{dh} \qquad (51)$$

where n is the index of refraction, ϕ the angle of the path with the horizontal, the elevation angle, at the point considered, and dn/dh the vertical gradient of the refractive index. The variability of the gradient determines the variability of the curvature. Since dn/dh is generally negative in the troposphere, the radio ray will usually bend downward due to the atmospheric effect. This downward concave trajectory coupled with the curvature of the earth causes an effective extension of the line-of-sight propagation distance between two antennas on the earth's surface. If the vertical gradient of radio refractivity in a layer of the atmosphere is equal to $-157N$ units/km, the curvature of the paths is equal to the curvature of the earth. The ray would then propagate parallel to the earth surface over a long distance.

Line-of-Sight Propagation

The most common application of line-of-sight (los) propagation for a terrestrial link is for microwave radio-relay transmissions. In this type of transmission, the basic considerations include not only the path loss, as described in (6) and (7), but also the clearance of the path. An illustration is provided in Fig. 12a. An los straight path from transmitter T to receiver R may appear to have enough clearance over a hill with peak P. In reality, due to atmospheric refraction, the los ray simply hits point Q of the hill.

The amount of bending is a function of the vertical gradient dn/dh as given in (51). For practical computations the curvature of the rays can be taken into account by introducing an equivalent earth radius factor K. This factor, K, multiplied by the actual earth radius a_e, is the radius of an equivalent earth. When considering the path profiles, this equivalent earth curvature is such that the ray paths can be represented by straight lines. For the standard radio atmosphere, the effective earth radius factor $K = 4/3$. This concept was used in Fig. 1, where $R_e = 4/3 \times$

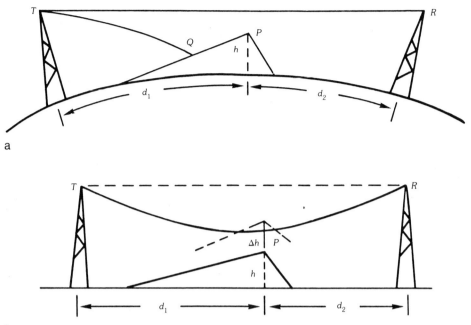

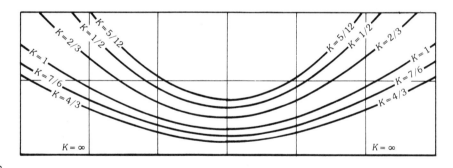

Fig. 12. Geometry of radio-wave propagation. (*a*) Propagation between a transmitter and a receiver of a radio wave in the troposphere. (*b*) Construction of the path profile. (*c*) Standard K curves.

earth's radius. Atmospheric refraction, however, may fluctuate, depending on a variety of factors including time of day, weather condition, the season, and geographical location. The relationship between K and ΔN (in N units), the mean refractivity change from the ground level to a 1-km altitude, is given by

$$K = \frac{1}{1 + 6.37 \Delta N \times 10^{-3}} \qquad (52)$$

Data on the parameter ΔN have been compiled in a CCIR document [54]. Typically K varies from 4/3 to 5/12.

The *radio horizon* is defined as the limiting boundary between a transmitter and a receiver above which line-of-sight propagation is possible, taking into account the earth curvature and atmosphere refraction. The distance between the transmitter and receiver for this grazing radio line-of-sight condition is given approximately by

$$d_h = 3.57\sqrt{K}(\sqrt{h_1} + \sqrt{h_2}) \tag{53}$$

where

d_h = radio horizon distance in kilometers

h_1 = height in meters of transmitting antenna above level ground

h_2 = height in meters of receiving antenna above level ground

The procedure to construct a path profile is illustrated in Fig. 12b. A cross-sectional view of the topological map between the transmitter and receiver with silhouettes of hills, trees, buildings, and other structures that may obstruct the ray is first constructed. To take into account the curvature of the earth and atmospheric refraction, an earth's bulge should be added to the profile. The bulge is given by

$$\Delta h = \frac{d_1 d_2}{12.75\,K} \tag{54}$$

where Δh in meters is the earth's bulge, and d_1 and d_2 are distances in kilometers of a point between transmitter location T and receiver location R, respectively. With the new profile for all combinations of d_1 and d_2, the radio ray can again be plotted as a straight line. The procedure is illustrated as dashed lines in Fig. 12b.

The evaluation of Δh for every point between T and R is tedious. An alternative method is to bend the wave relative to the original topological cross section. This is shown in solid lines in Fig. 12b. As a matter of fact, the bending of waves for different K values can be prepared on a standard sheet as shown in Fig. 12c, which can then be used as an overlay on any cross-sectional view. This is indeed the method used by engineers [55].

The actual los propagation degradation is of course more than bending as mentioned above. The height of the transmitter and receiver towers may not be the same, thus the path may be a slant path. The true ray path can be determined only by a ray-tracing program [56] using a measured atmospheric refractive index profile as the time path length.

The path losses in the line-of-sight propagation situation are mainly due to hydrometeors and gaseous absorptions, as discussed previously in Section 2 for the satellite-earth propagation case. The basic principles in computing the path losses are the same. The main differences are the propagation geometry and the height region where the signal propagates. For gaseous absorption for a terrestrial path, the path loss is given by [57].

$$L_{\text{gas}} = \gamma_a d \tag{55}$$

where

L_{gas} = gaseous attenuation in decibels

γ_a = specific absorption loss in decibels per kilometer

$d = d_1 + d_2$, the path length in kilometers

Fig. 13, which is similar to a corresponding part of Fig. 2, shows γ_a by water vapor and oxygen for a water vapor density of 7.5 g/m³ at an air temperature of 20°C and atmospheric pressure of 1000 mbar.

Similarly, for rain, the path loss is computed by

$$L_R = \gamma_R d\delta \tag{56}$$

where

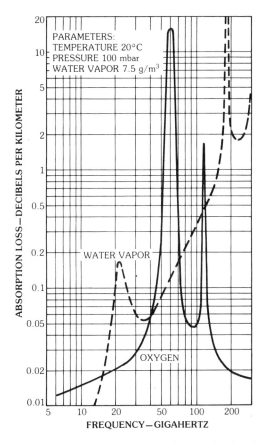

Fig. 13. Absorption by water vapor and oxygen in the standard radio atmosphere. (*After Brodhage and Hormuth [57]*, © *Siemens Aktiengesellschaft, Berlin and Munich*)

L_R = attenuation in decibels due to rain

γ_R = specific attenuation in decibels per kilometer due to rain

d = path length in kilometers

δ = reduction factor for P percent of the time

The term γ_R can be computed by the same equation as in the case of the satellite-earth propagation path.

For the line-of-sight propagation geometry on the ground, the reduction factor is approximately given by

$$\delta = 61[1 - \exp(-d^{0.5}R^{0.7}/61)]d^{0.5}R^{0.7} \tag{57}$$

where

δ = reduction factor for P percent of the time

d = path length in kilometers

R = point rainfall rate in millimeters per hour at a required P percent of the time

In addition to rain and snow, scintillation due to fog and clear air turbulence may also degrade the signal. This effect can be estimated following the procedure discussed in Section 2, under "Scintillation."

All obstacles existing in the space surrounding the direct line-of-sight in the common volume of the main lobes of the transmitting and receiving antennas may affect the propagation process. The free-space line-of-sight propagation condition prevails if the volume of the Fresnel ellipsoid is free from obstacles. The Fresnel ellipsoid is defined as the geometric loci of all points for which the sum of the distances from the two antennas is greater than the direct distance between the antennas by one-half wavelength. The cross section of the Fresnel ellipsoid orthogonal to the direction of propagation is the first Fresnel zone whose linear dimension is given by

$$r_F = 31.6\sqrt{d_1 d_2 \lambda/(d_1 + d_2)} \tag{58}$$

where

r_F = linear dimension of Fresnel zone in meters

d_1, d_2 = distances from the terminals in kilometers

λ = wavelength in meters

Higher-order Fresnel zones are given by $r_F^{(n)} = \sqrt{n}\,r_F$.

Diffraction and reflection effects have to be taken into account if obstacles fall into the Fresnel zone along the line-of-sight path.

Multipath Propagation over an Earth Surface

Multipath propagation is important when a radio-wave transmission can no longer be considered a simple los transmission isolated from the reflecting surfaces in the terrain, buildings, and tropospheric layers. The reflected signals are able to reach the receiving antenna and interact with the los signal. This happens, for example, in the case of low elevation angle transmission and transmission involving wide antenna beams. The principal factor in multipath propagation is the relative phases of the waves that arrive at the receiving antenna via different paths. This is particularly significant at low elevation angles where the reflected wave from the surface may have the same amplitude as the direct wave along an los path. The phase differences depend on the relative positions of the reflectors, whether they are in the first Fresnel zone or higher-order Fresnel zones.

The primary type of multipath propagation is that of a two-path propagation, i.e., a direct wave and a specularly reflected wave from the surface. The geometry is provided in Fig. 1. A general equation for evaluating loss factor F [cf. (3) and (7)] is given by [58, 59]

$$F = 1 + \Gamma g \varrho_s D e^{-j\Delta} \qquad (59)$$

Each of the five parameters Γ, g, ϱ_s, D, and Δ requires a few lines of explanation.

The first parameter Γ is the Fresnel reflection coefficient. It assumes a value either given by Γ_H or Γ_V, depending on whether the wave in consideration is horizontally or vertically polarized, where

$$\Gamma_H = \gamma_H e^{j\theta_H} = \frac{\sin\psi - \sqrt{\epsilon'_r - \cos^2\psi}}{\sin\psi + \sqrt{\epsilon'_r - \cos^2\psi}} \qquad (60)$$

$$\Gamma_V = \gamma_V e^{j\theta_V} = \frac{\epsilon'_r \sin\psi - \sqrt{\epsilon'_r - \cos^2\psi}}{\epsilon'_r \sin\psi + \sqrt{\epsilon'_r - \cos^2\psi}} \qquad (61)$$

where ϵ'_r is given in (9) and ψ is the grazing angle.

For example, according to Fig. 11, $\epsilon_r \cong 80$ and $\sigma \cong 4$ for seawater, the value of ϵ'_r at 1.54 GHz (L-band maritime communication) is $80 - j47$. Since $|\epsilon'_r| \gg 1$ for seawater, (60) yields $\gamma_H \cong 1$, $\theta_H \cong 180°$ for $0 < \psi < 10°$ for a wide frequency range from 100 MHz to at least 30 GHz; values of γ_V and θ_V are shown in Figs. 14a and 14b [52].

Propagation at low elevation angles often involves circular rather than linear polarization. In this case Γ should assume the value of either Γ_R or Γ_L, where

$$\Gamma_{R,L} = \frac{\Gamma_H \pm j\Gamma_V}{\sqrt{2}} \qquad (62)$$

Magnitudes of $\Gamma_{R,L}$ at 1.54 GHz over the sea are shown in Fig. 15 [59].

The parameter g in (59) is the antenna pattern factor between the direct wave and reflected wave. Except for the phased array type of antenna, where g may be complex, the value of g can be considered as a scalar related to the antenna gain

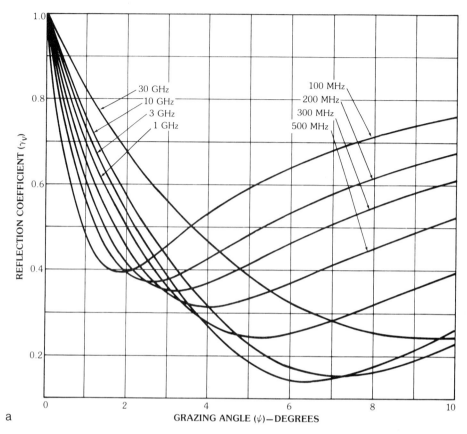

Fig. 14. Reflection coefficient γ_V and phase angle θ_V for vertically polarized wave over the ocean's surface. (*a*) Reflection coefficient. (*b*) Phase angle. (*After Blake [58]*)

function. For example, assume the antenna at point A (Fig. 1) is pointing along R_d toward point D, and the 3-dB beamwidth angle is ϕ_{3dB}. Then

$$g \cong \exp\left[-\left(\frac{\theta_d + \theta_r}{1.7\phi_{3dB}}\right)^2\right] \cong \exp\left[-\left(\frac{\psi}{\phi_{3dB}}\right)^2\right] \quad (63)$$

The last approximation in the above equation is valid for low grazing angles ($\psi \cong 10°$) only. The magnitudes of $\Gamma_{R,L} g$ represented as γ'_R and γ'_L according to (62) with $\phi_{3dB} = 7.22$ (14.4° beamwidth) are also shown in Fig. 15 [59].

Since the reflection coefficient Γ described above concerns only a smooth surface, a further correction factor to take into account surface roughness is required when the height deviations from a smooth surface over the area of the first Fresnel zone exceed the Rayleigh limit $\lambda/16 \sin \psi$. This is provided by ϱ_s in (59). Depending on the nature of surfaces (sea, land, ice, desert, forest, city, etc.) and roughness conditions (sinusoidal, layered, perfectly or partially conducting, hemispherical blobs, etc.), theoreticians have postulated numerous formulations for ϱ_s [5, 52, 58–64]. For the sea surface, a practical first-order equation for ϱ_s is

Propagation

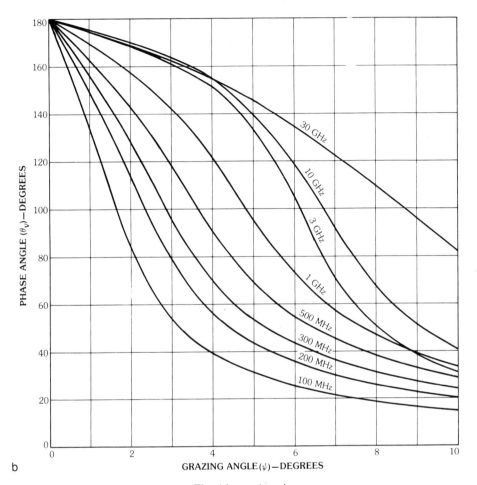

Fig. 14, *continued.*

$$\varrho_s = \exp\left[-2\left(\frac{2\pi\sigma_h \sin \psi}{\lambda}\right)^2\right] \quad (64)$$

where σ_h is the standard deviation of sea wave height, which can be related to the "significant height" $H_{1/3}$ generally used by oceanographers [65, 66]:

$$\sigma_h = \frac{1}{4} H_{1/3} \quad (65)$$

For example, for sea state 3, $H_{1/3}$ is about 4 ft or 1.2? .54 GHz and 5° elevation angle, $\varrho_s \cong 0.23$. Table 3 provides general $t_{1/3}$.

The next factor in (59) is the divergence factor D, w ers the effect of a spherical earth. For most practical applications this fa e approximated by [58]

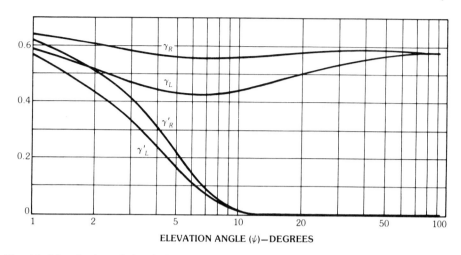

Fig. 15. Magnitudes of circularly polarized γ_R and γ_L over the sea, along with antenna-pattern weighted magnitudes $\gamma'_{R,L} = g\gamma_{R,L}$.

Table 3. Sea State Scale According to Douglas [66] and Wave Height $H_{1/3}$

Sea State	Descriptive Designation	Wave Height $H_{1/3}$ (ft)	Approximate Wind Speeds (kn)
1	Smooth	0–1	0–6
2	Slight	1–3	6–12
3	Moderate	3–5	12–15
4	Rough	5–8	15–20 (whitecaps form)
5	Very rough	8–12	20–25
6	High	12–20	25–30 (gale)
7	Very high	20–40	30–50
8	Precipitous	40	>50 (storm)

$$D \cong \left(1 + \frac{2G_1 G_2}{a_e G \sin \psi}\right)^{-1/2} \qquad (66)$$

where G_1, G_2, and G are shown in Fig. 1.

Factor Δ in (59) is the phase angle difference due to path length difference between the direct wave and the reflected wave, i.e.,

$$\Delta = \frac{2\pi}{\lambda}(R_1 + R_2 - R_d) \cong \frac{4\pi}{\lambda} h_1 \psi \qquad (67)$$

where the last approximation is valid at low grazing angles for a satellite-earth path with h_1 being the height of the transmitting antenna and ψ being the grazing angle in radians [59].

Equation 59 can now be employed for the evaluation of microwave fading. In addition to net attenuation, microwave signals under multipath conditions often fade with increasing peak-to-peak fluctuations as ψ decreases. The magnitude of fades can also be estimated by (59) by taking extreme values of F. Thus the peak-to-peak fluctuations of the signal power, ΔP in decibels, is

$$\Delta P = 20 \log \left(\frac{1 + |\Gamma g \varrho_s D|}{1 - |\Gamma g \varrho_s D|} \right) \tag{68}$$

The temporal variation of the signal is due to the fact that the random reflecting surface is moving or changing with respect to time, or the different paths are varying due to the varying condition of the atmosphere. Indeed, under these conditions, a Doppler spread will appear in the received signal when the input to the channel is a stable single-frequency signal.

In addition to the multipath propagation due to reflections from the earth's surface, tropospheric layers with refractive index structures substantially different from the ambient may provide reflections of the signal, giving rise to multipath propagation conditions. These layers usually are formed over flat ground, especially in humid areas, during calm weather, when the air temperature at higher levels is greater than that near the ground. Such layers can exist both above and below the direct path between the transmitting and receiving antennas. Total reflections, hence multipaths, can occur for small angles of incidence (less than about 0.5°). Sometimes when the conditions are favorable, waveguide modes can be set up between such a layer and the ground, or within the layer itself (elevated duct). Overreach propagation can be produced in such ducts. The maximum wavelength capable of propagating in a duct without loss is given approximately by [67]

$$\lambda_{\max} = 2.5 \, \Delta h \, (\Delta n)^{1/2} \tag{69}$$

where

$\lambda_{\max}$ = maximum wavelength in a duct in meters

Δh = depth of the duct in meters

Δn = difference of refractive index inside and outside the duct

Diffraction

Although, as discussed previously, under "Line-of-Sight Propagation," an los link should be designed for enough clearance by elevated antennas, this is not always possible. Obstacles such as hills, buildings, trees, etc., may be within the first Fresnel zone of the path or may even completely block the los path. The same can of course be said about transmissions under multipath conditions, as described in the preceding subsection, "Multipath Propagation over an Earth Surface." In these situations, wave energy can still be transmitted to the receiver through diffraction. But, in addition to the free-space path loss, the obstacle produces further losses due to the diffraction of waves over the obstacle. For transmissions at vhf and higher frequencies the losses can be evaluated by a simple CCIR model [68] as illustrated below.

For a single, general, blob-shaped obstacle as shown in Fig. 16, the free-space loss L_0 can still be calculated by (6), provided that the path length D is replaced by the actual path length $d_1 + d_2$. The additional medium loss L_p in decibels, due to the blob, can be considered as a sum of three terms:

$$L_p = F(v) + G(\varrho) + E(\chi) \qquad (70)$$

where $F(v)$ is the Fresnel-Kirchhoff knife-edge loss given by Fig. 17 with the dimensionless parameter v given by

$$v = 2\sin(\theta/2)\left[\frac{2(d_1 + R\theta/2)(d_2 + R\theta/2)}{\lambda d}\right]^{1/2} \qquad (71)$$

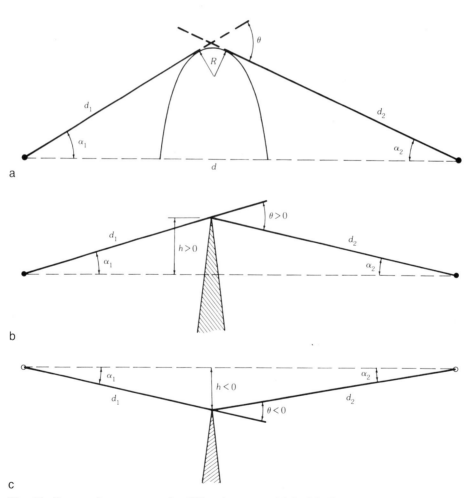

Fig. 16. Propagation geometry for diffraction over a blob. (*a*) Illustrating physical shape of blob. (*b*) Knife-edge diffraction for $h > 0$ and $\theta > 0$. (*c*) Knife-edge diffraction for $h < 0$ and $\theta < 0$.

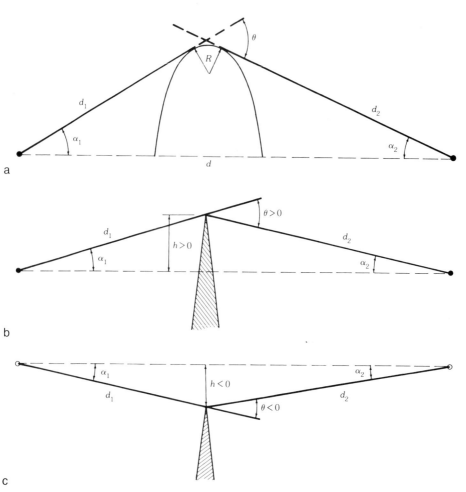

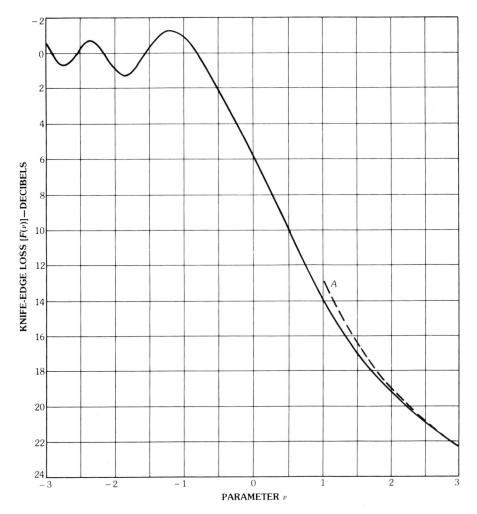

Fig. 17. Knife-edge–diffraction transmission loss relative to free space, where curve $A = 13 + 20 \log v$.

in which λ is the wavelength, d is the straight-line distance, and R is the radius of curvature. The term $G(\varrho)$ refers to a loss due to the curved surface. It can be calculated by

$$G(\varrho) = 7.2\varrho - 2.0\varrho^2 + 3.6\varrho^3 - 0.8\varrho^4 \tag{72}$$

with

$$\varrho^2 = \left(\frac{d_1 + d_2}{d_1 d_2}\right) \bigg/ \left[\left(\frac{\pi R}{\lambda}\right)^{1/3} \frac{1}{R}\right] \tag{73}$$

The last term, $E(\chi)$, refers to additional losses for propagation along the surface away from the blob. It is given by

$$E(\chi) = \begin{cases} [G(\varrho)/\varrho] & \text{for } -\varrho \leq \chi < 0 \\ 12\chi & \text{for } 0 \leq \chi < 4 \\ 17\chi - 6 - 20\log\chi & \text{for } x \geq 4 \end{cases} \quad (74)$$

where

$$\chi = \left(\frac{\pi R}{\lambda}\right)^{1/3} \theta \quad (75)$$

Equation (70) is applicable as long as the antennas are far away from the obstacle. It can also be used for propagation due to diffraction over a spherical earth simply by assigning $R = 8.5 \times 10^6$ m. At the other extreme, the equation can be reduced to only its first term $F(v)$ when $R = 0$. This is the typical case of knife-edge diffraction shown in Fig. 16b. For this case, (71) becomes

$$v = \theta\sqrt{\frac{2d_1 d_2}{\lambda(d_1 + d_2)}} = h\sqrt{\frac{2(d_1 + d_2)}{\lambda d_1 d_2}} \quad (76)$$

For the configuration shown in Fig. 16c, where there is an inadequate clearance for an los path, this equation is also applicable with θ and h given as negative.

Surface-Wave Propagation

Previously in this section, under "Line-of-Sight Propagation," only a direct wave is considered. Following, under "Multipath Propagation over an Earth Surface," the surface-reflected wave is considered together with the direct wave. The next order of business is naturally in the case where wave energy is trapped along the surface. As it turns out, this is the other mode of wave propagation, known as *surface-wave* propagation. This mode arises because of the boundary between the two different media, the air and the earth. The wave is guided along the earth's surface while energy is abstracted from it to supply the losses in the ground.

The intensity of a surface wave decreases exponentially as height increases. The effect has been successfully modeled by an $\exp(-\alpha h/\lambda)$ factor where α is the attenuation factor, and h/λ is the height in wavelengths. Since there is always a practical height for the antenna of a communications system, it follows that the surface wave is important only for low frequencies where the wavelength is at least in the order of meters. This establishes the upper limit of frequencies (300 MHz) for surface waves.

The attenuation of the surface wave depends on the conductivity of the earth and the wave frequency. The attenuation is low at lower frequencies over a good conducting ground. Therefore, in practice, surface waves are most important in the mf and lf bands. Vertical polarization is always used in surface-wave propagation. This is due to the fact that at these frequencies the antenna heights are usually only

Propagation

a fraction of a wavelength. If horizontal polarization were used, the field induced in the earth would virtually cancel out with the field in the antenna. In practice, a surface wave can be conveniently excited by a vertical dipole antenna near the earth's surface. When both transmitting and receiving antennas are located right at the earth's surface, the surface wave is the major component of the wave energy, since the direct line-of-sight and ground-reflected components will nearly cancel each other.

The study of surface-wave propagation with a vertical dipole as a source in the presence of either a plane or a spherical earth is a classical problem. It was originally formulated by Sommerfeld in 1909 and revised by Norton in the late thirties for engineering applications [69], most of which are still being quoted in the current FCC documentations. Over the years, enormous new analytic approaches have been devised, most of which provide clearer ways of bookkeeping. Chapter 2 of Kerr's book [2], a textbook by Jordan and Balmain [70], and others [71, 72] provide excellent coverage on the subject. One aspect which often causes confusion is the definition and the context of the surface waves referred to in the literature. As stated above, under "Diffraction," the diffraction mechanism can also support wave propagation along the surface. This is true even in the case where the surface is perfectly conducting. Blob-shaped conductor and/or sinusoidal corrugations are known to enhance wave diffractions. In theoretical studies the trapped surface wave and the diffracted wave along the surface are difficult to separate mathematically. The term "surface wave" is thus sometimes used loosely for either type of these waves, or both.

Fig. 18 shows the ground-wave attenuation factor F for a vertical dipole over a plane earth in terms of a numerical distance given approximately by [69]

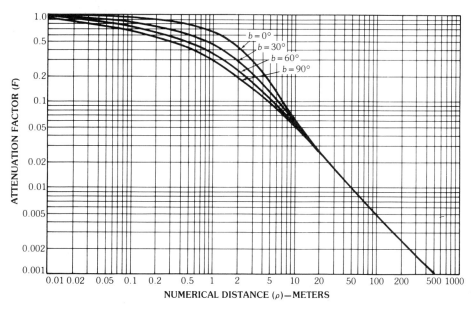

Fig. 18. Ground-wave attenuation factor F. (*After Jordan and Balmain [70]*, © *1968 Prentice-Hall, Inc., Englewood Cliffs, New Jersey; reprinted with permission*)

$$p = \frac{Dk}{2\chi}\cos b \qquad (77)$$

where

$$b = \tan^{-1}\left(\frac{\epsilon_r + 1}{\chi}\right)$$

$$\chi = \frac{\sigma}{\omega\epsilon_0} = \frac{18 \times 10^3 \sigma}{f_{\text{MHz}}}$$

σ = conductivity in siemens per meter of the earth

D = distance in meters between transmitting and receiving antennas

Both the transmitting and receiving antennas are on a plane earth surface.

For the case of spherical earth, the expression of F is very complicated. For hf applications the earth can often be approximated as perfectly conducting. Then again, for transmitter and receiver both on the surface separated by a distance D, the F factor is [73]

$$F \cong 2\left[\pi\left(\frac{k}{2R_e^2}\right)^{1/3} D\right]^{1/2} \exp[-\sqrt{3}/2(k/2R_e^2)^{1/3} D] \qquad (78)$$

where $(k/2R_e^2)^{1/3} D$ is a dimensionless quantity.

Finally, noting the fact that since a surface wave, either trapped or diffracted, depends on the surface condition (example: land, sea, forest, all having different conductivities and permittivities) and surface structure (particularly the sea wave corrugation), the propagation of the surface wave can only be convincingly modeled by experimental data. Indeed, in CCIR documents and other engineering handbooks a variety of highly diversified empirical models are available.

Tropospheric Scatter Propagation

As in the case of the ionosphere, irregular structures in the refractive index in the troposphere are capable of scattering or reflecting electromagnetic energy, thus providing possibilities of propagation over the horizon. The geometry is the same as that shown in Fig. 10 with scattering region in the troposphere. The irregular structures in the troposphere are believed to be either in the form of turbulent eddies or thin turbulent layers [74]. The frequency used usually is in the range of 150 to 4000 MHz. The scatter channel can provide a stable transmission well below the horizon with a field strength much stronger than that predicted by the diffraction theory. Large, rapid fluctuations are usually associated with the signal. These fades are due to the random variations and motions of the scatterers. The bandwidth that can be supported by this mode of propagation depends on the transmission distance, the frequency used, and the antenna beamwidth. If the radiation pattern is broad, multipath signals will narrow the coherence bandwidth. Therefore, to transmit broadband signals, it is necessary to use a narrow-beam antenna. For transmission over large distances (up to 1000 km), only narrow-band applications, such as small-capacity multiplex telephony signals, are practically possible. For shorter-range transmission (200 to 300 km), much broader band applications, including tv signals, can be supported.

The current understanding of the mechanisms which support the tropospheric scatter propagation is not yet complete. Experimental evidence indicates that both the volume scattering as described in Section 3 and partial reflections due to thin layers may be responsible for producing the received field. Because of the complexity of the mechanisms, results from theoretical analysis still do not allow for the quantitative predictions of the received fields and their fluctuations with the accuracy required for practical applications. Empirical procedures have been recommended by the CCIR [53] to calculate the path loss for troposcatter transmission. This procedure takes into account the meteorological condition of the troposphere as well as the altitude of the scattering volume. The attenuation of the medium due to scattering (in addition to the free-space path loss) can be estimated by the following procedure [53]:

Step 1
Construct the path profile by either of the two methods discussed previously in this section, under "Line-of-Sight Propagation," as shown in Fig. 19a.

Step 2
Calculate the scattering angle θ, using the formula in Fig. 19a.

Step 3
The long-term median troposcatter path loss (not exceeded for 50 percent of a year) may be estimated by the formula

$$L(50\%) = 30 \log f - 20 \log D + F(\theta D) - G_e + V \qquad (79)$$

where

L = path loss in decibels

f = frequency in hertz

D = great-circle distance in kilometers

θ = scattering angle in radians

G_e = effective total combined antenna gain in decibels

V = factor for long- and short-term fading in different climate regions, may be up to ± 8 dB

$F(\theta D)$ = the attenuation function in decibels, given in Fig. 19b

NOTE: The different curves in Fig. 19b are for different surface refractivities N_s, which are different for different geographical regions. G_e represents the combined transmitting and receiving antenna gains minus their aperture-to-medium coupling losses.

5. Noise

To conclude this chapter, this section presents a short discussion on noise and its effect on a link. For a given link that is required to maintain a specified signal

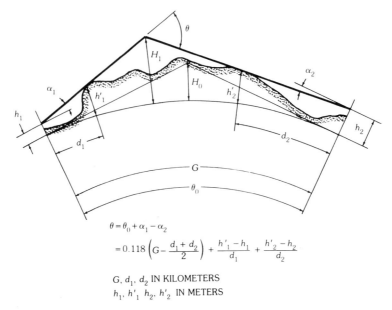

Fig. 19. Path profile and attenuation function for transhorizon propagation. (*a*) Path profile. (*After Picquenard [67], originally published by Macmillan Press, Ltd., London*) (*b*) Graph to obtain $F(\theta D)$. (*Courtesy NTIA, US Dept. of Commerce*)

carrier-to-noise ratio (cnr), the link budget is computed from the equation

$$\text{cnr} = P_r - P_n \tag{80}$$

where

cnr = carrier-to-noise ratio in decibels

P_r = power available at receiving antenna in decibels referred to 1 W, given by (4)

P_n = total received noise power in decibels referred to 1 W

From (4) and (80), the required transmitter power for the link can be calculated. Noise power is generally referred to in terms of temperature, i.e., the noise temperature. Briefly, the noise temperature is related to the noise power P_n in decibels referred to 1 W by

$$P = 10\log(kTB) \tag{81}$$

where

k = Boltzmann's constant = 1.374×10^{-23} J/K

T = noise temperature in kelvins

B = bandwidth in hertz over which the energy is collected in the receiver, i.e., the passband of the receiver

Propagation

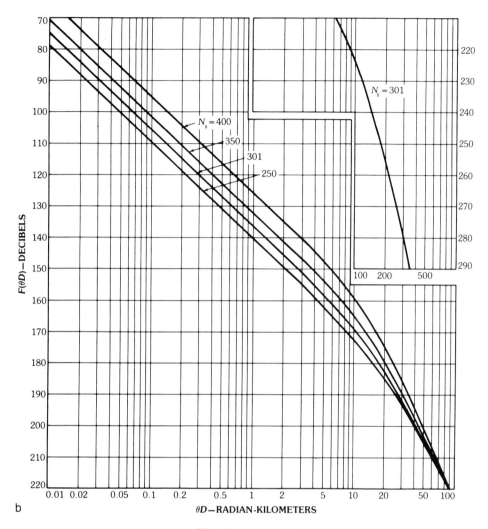

Fig. 19, *continued.*

Equation 81 is also often written as

$$P_n = -204 + 10\log(T/290) + 10\log B \tag{82}$$

It is important to note that noise degradations of the system have to be addressed in a framework of the complete system. For instance, two shf systems can have the same gain over temperature (G/T) performance under clear sky conditions if one employs a large antenna while the other employs a small antenna but with a sensitive low-noise amplifier (lna). Under rainy conditions, however, the noise temperature degradation of the small antenna system is significantly worse than that of the large antenna system. Detailed formulas for the evaluation of overall system temperature of an antenna-receiver chain due to the presence of noise external to the antenna have been presented earlier in the book.

From the viewpoint of propagation it suffices to note that the noise can be characterized as either internal noise or external noise, depending on whether it originates in the receiving system or is generated externally to the antenna. The internal noise includes that due to the antenna and the transmission line, or the receiver itself. It is in general thermal in nature. The external noise includes atmospheric thermal emission from atmospheric constituents and precipitations, extraterrestrial and human-made.

In the following we discuss briefly these various sources of external noise power, which may be received by the receiving antenna.

Atmospheric Noise

Usually atmospheric noise refers to noise caused by disturbances in the atmosphere, such as thunderstorms. Atmospheric noise is erratic, consisting of short pulses occurring randomly in time. It is very strong at the lf band with noise temperature of 2.9×10^{16} K at 10 kHz. It decreases to 2.9×10^9 K at 100 kHz and reaches a minimum of 290 K at 1 MHz, then increases to about 2.9×10^5 K at about 9 MHz and decreases and becomes negligible for frequencies above 15 MHz or so [75]. The atmospheric noise depends on the geographical location of the station. World distributions and characteristics of atmospheric radio noise are available from CCIR publications [76, 77].

Emission by Atmospheric Gases and Precipitation

As discussed in Section 2, under "Gaseous Attenuation," the gaseous constituents of the earth's atmosphere absorb radio-wave energy. Consequently they also emit radio-frequency noise. The emission is strongest at frequencies corresponding to the absorption bands discussed in Section 2. The emitted thermal noise can be computed from radiative transfer considerations [22]. Fig. 20 shows sky noise due to atmospheric gases as a function of frequency and elevation angle of the receiving antenna. The curves are computed for a surface temperature of 20°C, surface pressure of 1 atm (101 325 Pa), and a surface water vapor concentration of 10 g/m^3, corresponding to a relative humidity of 58 percent. Curves for other conditions can also be computed. If one avoids the absorption bands, the sky noise from atmospheric gases for elevation angles greater than 10° is usually less than 100 K in the frequency range of 1 to 100 GHz.

The increase in sky noise due to rain can also be computed in a similar manner. The sky noise temperature increases from about 60 K for a rain attenuation level of 1 dB to about 250 K for a 10-dB rain attenuation, then it saturates to about 270 K for 30-dB attenuation [22] at microwave frequencies.

Extraterrestrial Noise

The sources of extraterrestrial noise are the sun, the galaxy, the cosmic background, and other discrete sources. Fig. 21 gives the sky noise temperature due to the various sources [78]. Also included in the figure are curves for gaseous contributions of dry, clear skies which can be compared with curves shown in Fig. 20.

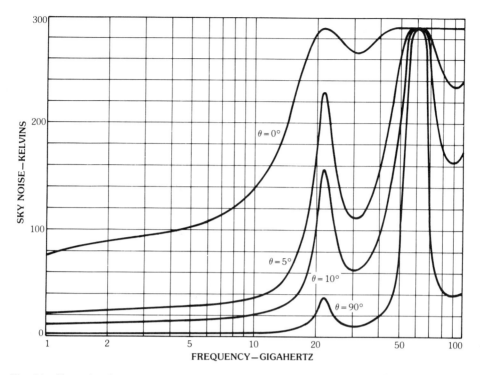

Fig. 20. Sky noise due to atmospheric gases as a function of frequency and elevation angle, for a surface temperature of 20°C, pressure of 1 atm (101 325 Pa), and water vapor concentration of 10 g/m^3. (*After Ippolito [22], © 1981 IEEE*)

Human-Made Noise

Human-made noise can arise from a number of sources, such as power lines, industrial machinery, ignition systems, etc. It varies with location and time and is difficult to estimate under general conditions. Limited observational data indicate that it decreases with frequency. Noise below 2 MHz is generally associated with power lines. At 20 MHz and above, ignition systems are the dominant contributors. An estimate for human-made noise in a business environment is included in Fig. 21. The curve can be extended to low frequencies below 1 MHz [75]. A similar trend is seen for other types of environment, such as residential, rural, and quiet rural areas. The noise levels for those areas are lower, with the quiet rural environment almost 25 dB down from the curve shown in Fig. 21.

From Fig. 21 it is apparent that for a satellite–earth station downlink, the sun constitutes the strongest source of noise and the antenna should be directed away from it. Over most of the uhf range, galactic noise is dominant, but human-made noise in business environments may dominate. Gaseous atmosphere and precipitation are the main contributors to the noise temperature at upper shf band.

For transhorizon communications below uhf frequencies, worldwide maps of radio noise levels were published by the CCIR [79] in 1963. Recently improved maps are available from the NTIA [80]. The improved maps provide corrections, ranging as high as 20 dB, to the 1-MHz noise level values given in the earlier CCIR maps.

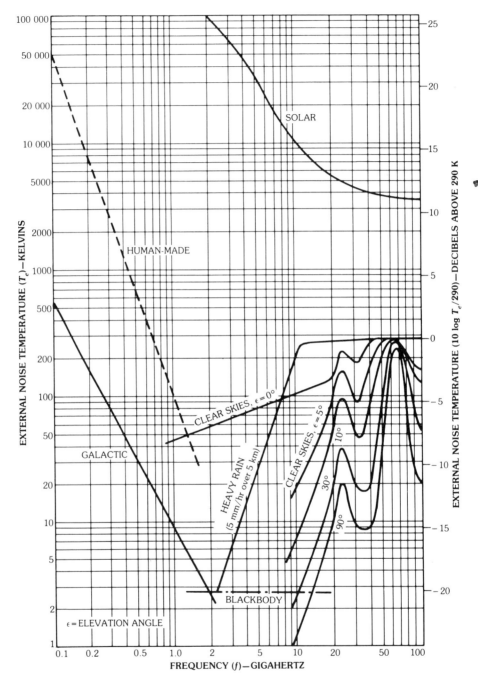

Fig. 21. Apparent temperatures of various external noise sources for downlink slant paths. (*After Dougherty [78]*)

6. References

[1] B. R. Bean and E. J. Dutton, *Radio Meteorology*, New York: Dover Publications, 1968.

[2] D. E. Kerr (ed.), *Propagation of Short Radio Waves*, MIT Radiation Lab Series, vol. 13, New York: McGraw-Hill Book Co., 1951.

[3] K. G. Budden, *Radio Waves in the Ionosphere*, Cambridge: Cambridge University Press, 1961.

[4] K. C. Yeh and C. H. Liu, *Theory of Ionospheric Waves*, New York: Academic Press, 1972.

[5] J. R. Wait, *Electromagnetic Waves in Stratified Media*, 2nd ed., New York: Pergamon Press, 1970.

[6] V. I. Tatarskii, "The effects of the turbulent atmosphere on wave propagation," Springfield, Va.: US Dept. of Commerce TT-68-50464, 1971.

[7] A. Ishimaru, *Wave Propagation and Scattering in Random Media*, vols. I and II, New York: Academic Press, 1978.

[8] K. C. Yeh and C. H. Liu, "Radio wave scintillations in the ionosphere," *Proc. IEEE*, vol. 70, pp. 324–360, 1982.

[9] D. C. Cox, H. W. Arnold, and H. H. Hoffman, "Observations of cloud-produced amplitude scintillation on 19- and 28-GHz earth-space paths," *Radio Sci.*, vol. 16, pp. 885–908, 1981.

[10] H. T. Dougherty and C. M. Rush, "Some propagational aspects of frequency allocation and frequency sharing," AGARD Conference No. 332, Propagation Aspects of Frequency Sharing, Interference and System Diversity, pp. 2.1–2.11, Paris, October 1982.

[11] CCIR Study Group 5 Doc. Report 719-1, Volume V, "Attenuation by atmospheric gases," XVth Plenary Assembly, Geneva, 1982.

[12] CCIR Study Group 5 Doc. Report 564-2, Volume V, "Propagation data required for space telecommunications systems," XVth Plenary Assembly, Geneva, 1982.

[13] D. V. Rogers, "Simple method for estimating 1- to 15-GHz atmospheric absorption," *COMSAT Tech. Rev.*, vol. 13, pp. 157–163, 1983.

[14] T. Oguchi, "Attenuation of electromagnetic waves due to rain with distorted raindrops," *J. Radio Res. Lab.* (Tokyo), vol. 11, pp. 19–37, 1964.

[15] R. K. Crane, "Propagation phenomena affecting satellite communications systems operating in the centimeter and millimeter wavelength bands," *Proc. IEEE*, vol. 59, pp. 173–188, 1971.

[16] D. C. Hogg and T. S. Chu, "The role of rain in satellite communication," *Proc. IEEE*, vol. 63, pp. 1308–1331, 1975.

[17] H. T. Dougherty and E. V. Dutton, "Estimating year-to-year variability of rainfall for microwave applications," *IEEE Trans. Comm.*, vol. COM-26, pp. 1321–1324, 1978.

[18] D. J. Fang and J. M. Harris, "Precipitation attenuation studies based on measurements of ATS-6 20/30 GHz beacon signals at Clarksburg, Md.," *IEEE Trans. Antennas Propag.*, vol. AP-27, pp. 1–11, 1979.

[19] CCIR Study Group 5 Doc. Report 721-1, Volume V, "Attenuation by hydrometeors, in particular, precipitations and other atmospheric particles," XVth Plenary Assembly, Geneva, 1982.

[20] Final Report of the Conference Preparatory Meeting (CPM) for RARC-82, chapter 3, "Radio propagation factors relating to frequencies near 12.5 and 17.5 GHz," Geneva, June–July 1982.

[21] Y. Otsu et al., "Propagation measurements and tv-reception tests with the Japanese broadcasting satellite for experimental purposes," *IEEE Trans. Broadcast Commun.*, vol. BC-25, p. 113, 1979.

[22] L. J. Ippolito, "Radio propagation for space communications systems," *IEEE Proc.*, vol. 69, pp. 697–727, 1981.

[23] P. N. Kumar, "Precipitation fade statistics for 19/29-GHz COMSAT beacon signals and 12-GHz radiometric measurements," *COMSAT Tech. Rev.*, vol. 12, pp. 1–27, 1982.

[24] D. C. Cox, "Depolarization of radio waves by atmospheric hydrometeors in earth-space paths: a review," *Radio Sci.*, vol. 16, pp. 781–812, 1981.

[25] D. J. Fang and C. H. Chen, "Propagation of centimeter/millimeter waves along a slant path through precipitation," *Radio Sci.*, vol. 17, pp. 989–1005, 1982.

[26] CCIR Study Group 5 Doc. Report No. 564, "Propagation data required for space communication system," XIVth Plenary Assembly, Kyoto, Japan, 1978.

[27] K. Rawer, "Intercomparison of different measuring techniques in the upper atmosphere: the international reference ionosphere," *Space Res.*, vol. XV, p. 212, 1975.

[28] R. B. Bent et al., "The development of a highly-successful worldwide empirical ionospheric model and its use in certain aspects of space communications and worldwide total content investigation," *Effects of the Ionosphere on Space Systems and Communications*, ed. by T. M. Goodman, Naval Research Lab, Washington, D.C. 20375, US Government Printing Office Stock No. 008-051-00064-0, 1981.

[29] H. E. Whitney, J. Aarons, R. S. Allen, and D. R. Seemann, "Estimation of the cumulative amplitude probability distribution function of ionospheric scintillations," *Radio Sci.*, vol. 7, pp. 1095–1104, 1972.

[30] R. Umeki, C. H. Liu, and K. C. Yeh, "Multifrequency studies of ionospheric scintillations," *Radio Sci.*, pp. 311–317, 1977.

[31] D. J. Fang and C. H. Liu, "A morphological study of gigahertz equatorial scintillations in the Asian region," *Radio Sci.*, vol. 18, pp. 241–252, 1983.

[32] J. Aarons, "Global morphology of ionospheric scintillations," *Proc. IEEE*, vol. 70, pp. 360–379, 1982.

[33] D. J. Fang and M. S. Pontes, "4/6 GHz ionospheric scintillation measurement at Hong Kong earth station during the peak of sunspot cycle 21," *COMSAT Tech. Rev.*, vol. 11, pp. 293–320, 1981.

[34] D. G. Singleton, "An improved ionospheric irregularity model," *Solar-Terrestrial Predictions Proc.*, US Department of Commerce, Document 003-017-00479-1, vol. 4, pp. D1–D15, 1980.

[35] CCIR Study Group 6 Doc. Report 263-4, Volume VI, "Ionospheric effects upon earth-space propagation," XVth Plenary Assembly, Geneva, 1982.

[36] E. J. Fremouw and C. L. Rino, "A signal-statistical and morphological model of ionospheric scintillation," *Proc. AGARD Conf. on Operational Modeling of Aerospace Propagation Environment*, pap. 18, Ottawa, Canada, 1978.

[37] J. Aarons, E. Mackenzie, and K. Bhavnani, "High latitude analytical formulas for scintillation levels," *Radio Sci.*, vol. 15, pp. 115–127, 1980.

[38] CCIR Study Group 5 Doc. Report 881, Volume V, "Effects of small-scale spatial or temporal variations of refraction on radio wave propagation," XVth Plenary Assembly, Geneva, 1982.

[39] K. Davies, "Ionospheric radio propagation," *National Bureau of Standards Monograph 80*, US Government Printing Office, Washington, D.C., November 1965.

[40] CCIR Study Group 6 Doc. Report 340-4, "CCIR atlas of ionospheric characteristics," XVth Plenary Assembly, Geneva, 1982. (This report is published separately from Volume VI.)

[41] CCIR Study Group 6 Doc. Supplement of Report 252-2, Volume VI, "Second CCIR computer-based interim method for estimating sky-wave field strength and transmission loss at frequencies between 2 and 30 MHz," and CCIR Report 571-1, "Comparisons between observed and predicted sky-wave signal intensities at frequencies between 2 and 30 MHz," XVth Plenary Assembly, Geneva, 1982.

[42] A. F. Barghausen et al., "Predicting long-term operational parameters of high frequency skywave telecommunication systems," *Tech. Rep. ERL 110 ITS 78* (US Environmental Science Services Administration).

[43] CCIR Study Group 6 Doc. Report 575-2, Volume VI, "Method for predicting skywave field-strengths at frequencies between 150 kHz and 1600 kHz," XVth Plenary Assembly, Geneva, 1982.

[44] *FCC Rules and Regulations*, pt. 73, Washington: US Government Printing Office.

[45] C. H. Wang, "Interference and sharing at medium frequency: skywave propagation

considerations," *AGARD Conf. No. 332, Propagation Aspects of Frequency Sharing, Interference and System Diversity*, Paris, October 1982.
[46] A. D. Watt, *VLF Radio Engineering*, New York: Pergamon Press, 1967.
[47] E. R. Swanson, "Omega," *Navigation*, vol. 18, 1972.
[48] J. Galejs, *Terrestrial Propagation of Long Electromagnetic Waves*, New York: Pergamon Press, 1972.
[49] CCIR Study Group 6 Doc. Report 265-5, Volume VI, "Sky-wave propagation and circuit performance at frequencies between about 30 kHz and 500 kHz," XVth Plenary Assembly, Geneva, 1982.
[50] H. G. Booker and W. E. Gordon, "A theory of radio scattering in the troposphere," *Proc. IRE*, vol. 38, p. 401, 1950.
[51] V. A. Fock, *Electromagnetic Diffraction and Propagation Problems*, New York: Pergamon Press, 1965.
[52] P. Beckman and A. Spizzichino, *The Scattering of Electromagnetic Waves from Rough Surfaces*, New York: Macmillan, 1963.
[53] CCIR Study Group 5 Doc. Report 239-4, Volume V, "Propagation data required for trans-horizon radio-relay system," XVth Plenary Assembly, Geneva, 1982.
[54] CCIR Study Group 5 Doc., Volume V, "Radiometeorological data," XVth Plenary Assembly, Geneva, 1982.
[55] GTE Lenkurt Publications, *Engineering Considerations for Microwave Communication Systems*, San Carlos, Calif.: GTE Lenkurt Incorporated, 1970.
[56] R. H. Ott, "Scattering, diffraction, propagation and refraction of acoustic and EM waves," *Remote Sensing of the Troposphere*,, Chapter 8, Boulder: Univ. of Colorado, and NOAA, June 13–30, 1972.
[57] H. Brodhage and W. Hormuth, *Planning and Engineering of Radio Relay Links*, New York: John Wiley & Sons, 1977, p. 87.
[58] L. V. Blake, *Radar Range-Performance Analysis*, Norwood, Massachusetts: Artech House, 1986.
[59] D. J. Fang and R. H. Ott, "A low elevation angle *L*-band maritime propagation measurement and modeling," *IEE Conf. Pub. 222*, pp. 45–50, Third International Conference on Satellite Systems for Mobile Communications and Navigations, London, 1983.
[60] S. O. Rice, "Reflection of electromagnetic waves from slightly rough surfaces," *Theory of Electromagnetic Waves*, ed. by M. Kline, New York: Interscience, pp. 351–378, 1951.
[61] H. Davis, "The reflection of electromagnetic waves from a rough surface," *Proc. IEEE*, vol. 101, pp. 209–214, 1954.
[62] D. E. Barrick, "Remote sensing of sea state by radar," *Remote Sensing of the Troposphere*, ed. by V. Derr, chapter 12, pp. 12-1 to 12-46, Washington: US Government Printing Office, 1972.
[63] D. J. Fang, "Scattering from a perfectly conducting sinusoidal surface," *IEEE Trans. Antennas Propag.*, vol. AP-20, pp. 388–390, 1972.
[64] R. H. Ott, "An alternative integral equation for propagation over irregular terrain, pt. II," *Radio Sci.*, vol. 6, no. 4, April 1971.
[65] B. Kinsman, *Wind Waves, Their Generation and Propagation on the Ocean Surface*, New York: Prentice-Hall, 1965.
[66] M. W. Long, *Reflectivity of Land and Sea*, Lexington, Massachusetts: D.C. Heath & Co., 1975.
[67] A. Picquenard, *Radio Wave Propagation*, New York: John Wiley & Sons, 1974.
[68] CCIR Study Group 5 Doc. Report 715-1, Volume V, "Propagation by diffraction," XVth Plenary Assembly, Geneva, 1982.
[69] K. A. Norton, "The propagation of radio waves over the surface of the earth and in the upper atmosphere," *Proc. IRE*, vol. 24, p. 1367, 1936; *Proc. IRE*, vol. 25, p. 1203, 1937; and *Proc. IRE*, vol. 25, p. 1192, 1937.
[70] E. C. Jordan and K. G. Balmain, *Electromagnetic Waves and Radiating Systems*, 2nd ed., New York: Prentice-Hall, 1958.

[71] J. R. Wait, "Electromagnetic Surface Waves," in *Advances in Radio Research*, ed. by J. A. Saxton, London and New York: Academic Press, 1964.

[72] H. M. Barlows and J. Brown, *Radio Surface Waves*, Oxford: Clarendon Press, 1962.

[73] R. H. Ott, "Fock currents for concave surface," *IEEE Trans. Antennas Propag.*, vol. AP-22, pp. 357–360, 1974; also see J. R. Wait, "EM surface waves," in *Advances in Radio Research*, ed. by J. Saxton, New York: Academic Press, 1964.

[74] F. DuCastel, *Tropospheric Radio Wave Propagation Beyond the Horizon*, New York: Pergamon Press, 1966.

[75] CCIR Study Group 1 Doc. Report 670, "Worldwide minimum external noise levels, 0.1 Hz to 100 GHz," XIVth Plenary Assembly, Kyoto, Japan, 1978.

[76] CCIR Study Group 6 Doc., Volume VI, "Characteristics and applications of atmospheric radio noise data," XVth Plenary Assembly, Geneva, 1982.

[77] CCIR Study Group 6 Doc., Volume VI, "Measurement of atmospheric radio noise from lightning," XVth Plenary Assembly, Geneva, 1982.

[78] H. T. Dougherty, "A consolidated model for uhf/shf telecommunication links between earth and synchronous satellites," *NTIA Rep. 80-45*, US Department of Commerce, August 1980.

[79] CCIR Study Group 6 Doc. Report 322-2, "World distribution and characteristics of atmospheric radio noise," Xth Plenary Assembly, Geneva, 1963.

[80] A. P. Spaulding and J. S. Washburn, "Atmospheric radio noise: worldwide levels and other characteristics," *NTIA Rep. 85-173*, Boulder, April 1985.

Chapter 30

Antenna Response to Electromagnetic Pulses

K. S. H. Lee
Dikewood, Division of Kaman Sciences Corporation

CONTENTS

1. High-Altitude Electromagnetic Pulses	30-3
2. The Effects of Antenna Mounting Structures	30-4
3. Principal Antenna Elements	30-7
4. Techniques for Analyzing Antenna Response to Broadband Electromagnetic Pulses	30-8
Low-Frequency Techniques 30-9	
The Babinet Principle 30-14	
5. EMP Responses of Aircraft Antennas	30-17
L-Band Blade Antennas 30-17	
UHF Communication Antennas 30-17	
VHF Communication Antennas 30-17	
VHF Localizer Antennas 30-23	
VHF Marker Beacon Antennas 30-23	
HF Fixed-Wire Antennas 30-23	
VLF/LF Trailing-Wire Antennas 30-27	
6. References	30-31

Kelvin S. H. Lee received the BS, MS, and PhD degrees in engineering and applied science, in 1960, 1961, and 1963, respectively, from the California Institute of Technology.

Dr. Lee, Vice President of Kaman Sciences Corporation, Dikewood Division, is the Manager of Dikewood's California Research Branch in Santa Monica and is in charge of its R&D programs in electromagnetics. He is the editor of the newly published book *EMP Interaction: Principles, Techniques, and Reference Data* (Hemisphere Publishing Corporation, New York, 1986).

He has been actively involved with system response to EMP, lighting, and high-power microwaves (HPM) in both the development and the experimental verification of analytical coupling models, in system-level testing, and in hardening design and hardness assessment of many military and commercial systems.

Dr. Lee has made numerous contributions to the electromagnetic theory involving the electrodynamics of moving media, the complex Doppler effect, electromagnetic shielding, reversible and irreversible power, dipole antennas, transient electromagnetic phenomena, and low-frequency approximations in electromagnetic theory.

1. High-Altitude Electromagnetic Pulses

Of the many different types of electromagnetic pulses (EMP) from a nuclear detonation the high-altitude EMP (HEMP) has received the most attention mainly because it covers a large area. HEMP is generally assumed to be a plane wave with the electric-field amplitude given by [1]

$$E_1(t) = E_0(e^{-\alpha t} - e^{-\beta t})u(t) \tag{1}$$

or

$$E_2(t) = E_0(e^{\alpha t} + e^{-\beta t})^{-1} \tag{2}$$

where u is the unit-step function, $1/\alpha \cong 250$ ns, $1/\beta \cong 2$ ns, and E_0 is tens of kilovolts per meter. The waveforms are plotted in Fig. 1 and their time- and frequency-

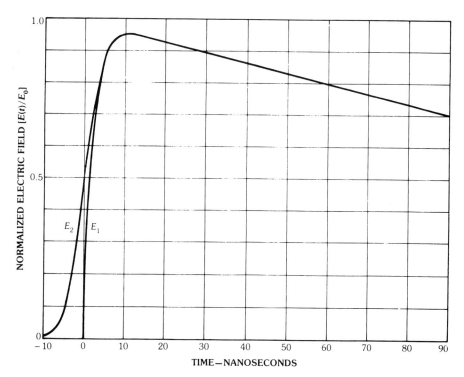

Fig. 1. Two canonical high-altitude EMP waveforms for which $\alpha = 4 \times 10^6$ s^{-1} and $\beta = 5 \times 10^8$ s^{-1}. (*After Lee [1]*)

domain characteristics are given in Tables 1 and 2. Note that E_1 has a discontinuous time derivative at $t = 0$, whereas the time derivative of E_2 at $t = 0$ is continuous. Both spectra, as shown in Fig. 2, have the first breakpoint at $\alpha/2\pi$ and roll off as $1/f$ until the second breakpoint at $\beta/2\pi$ is reached, from which point onward $E_1(\omega)$ decreases as $1/f^2$ whereas $E_2(\omega)$ drops off more rapidly. The 20-dB point of the half-width of both spectra is about 6 MHz.

2. The Effects of Antenna Mounting Structures

In most cases antennas are not directly exposed to the EMP plane-wave field described in the preceding section. The structures on which the antennas are mounted will modify not only the incident field but also the input impedances of the antennas. If the dimensions of the antenna are small compared with the local radii of curvature of the mounting structure, the effect of the mounting stucture can be approximated by that of an infinite ground plane so far as the antenna's input impedance is concerned. On the other hand, the *external* field to which the antenna is exposed contains the resonances of the mounting structure which can easily be excited by EMP. Figs. 3 and 4 are the scale-model measurements of the axial current density at the top and bottom midforward fuselages of the 747 and 707 aircraft [2]. The incident time-harmonic plane-wave field is striking from the top with **E** parallel to the fuselage. Fig. 5 shows the total current at three locations of a cylinder [3]. The resonances of the structures are obvious from these figures, which

Table 1. Time-Domain Characteristics of Two Canonical EMP Waveforms (*After Lee [1]*)

Parameter	$E_1(t)/E_0$	$E_2(t)/E_0$
Rise times (10%–90%)	$2.2/\beta$	$4.4/\beta$
Peak values	$1 + (\alpha/\beta)[\ln(\alpha/\beta) - 1]$	$1 + (\alpha/\beta)[\ln(\alpha/\beta) - 1]$
Duration of pulse (for decay to 10% of peak)	$2.3/\alpha$	$2.3/\alpha$
$\int_{-\infty}^{\infty} \cdots dt$	$1/\alpha$	$1/\alpha$

Table 2. Frequency-Domain Characteristics of Two Canonical EMP Waveforms (*After Lee [1]*)

Parameter	$|E_1(\omega)|/E_0$	$|E_2(\omega)|/E_0$
Low-frequency approx. ($\omega \ll \alpha$)	$1/\alpha$	$1/\alpha$
Midfrequency approx. ($\alpha \ll \omega \ll \beta$)	$1/\omega$	$1/\omega$
High-frequency approx. ($\omega \gg \beta$)	β/ω^2	$(2\pi/\beta)e^{-\omega\pi/\beta}$

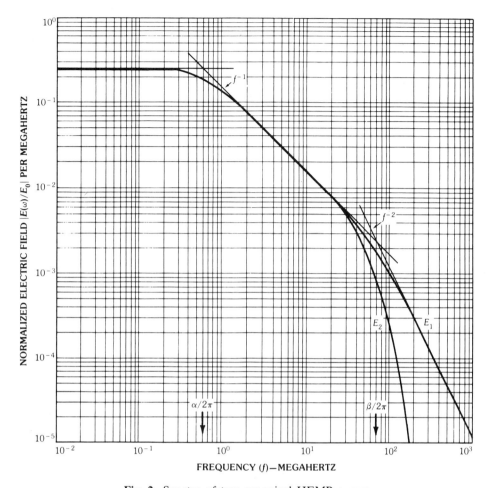

Fig. 2. Spectra of two canonical HEMP curves.

need be scaled by the EMP spectra given in Section 1 to obtain the excitation field for the mounted antenna. For example, take E_0 to be 50 kV/m in (1), and make use of Figs. 2 through 5 and the formula

$$I_{\text{peak}} = |\Delta\omega| |I(\omega_0)| \tag{3}$$

where $|I(\omega)|$ is the spectrum of $I(t)$ and I_{peak} is the peak amplitude of the damped sinusoid defined by

$$I(t) = I_{\text{peak}} e^{-|\Delta\omega|t/2} \sin \omega_0 t \tag{4}$$

with ω_0 being the resonance frequency and $|\Delta\omega|$ the 3-dB bandwidth. One obtains Table 3 for the first-resonance induced current or current density.

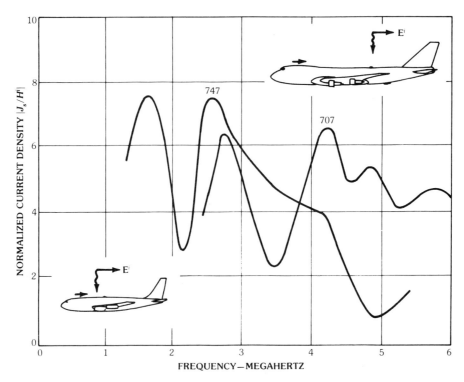

Fig. 3. Normalized current density at the top of midforward fuselages of 747 and 707 aircraft.

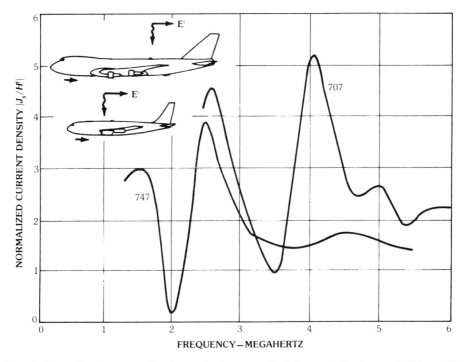

Fig. 4. Normalized current density at the bottom of midforward fuselages of 747 and 707 aircraft.

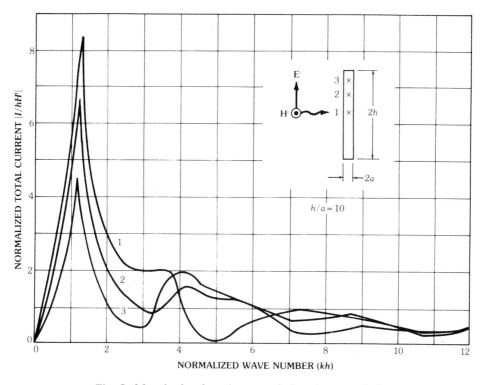

Fig. 5. Magnitude of total current induced on a cylinder.

Table 3. Characteristics of the First Damped Sinusoid Induced by a 50-kV/m High-Altitude EMP

Characteristic	747		707		Cylinder ($2h = 10$ m)		
	Top	Bottom	Top	Bottom	1	2	3
First resonance frequency (MHz)		1.60		2.65	11.20		
Bandwidth (MHz)		0.65		0.55	3.15		
Peak current density (A/m)	365	146	176	126			
Peak current (kA)					1.47	1.23	0.83

3. Principal Antenna Elements

The high-altitude EMP, as described in Section 1, has not only a high field strength but also a broad frequency spectrum. These two characteristics of EMP make it necessary to analyze the response of an antenna not only within its in-band frequencies but also in out-of-band frequencies in which the antenna's behavior is usually of no concern. There are many antennas that have the same in-band but completely different out-of-band characteristics. This arises from the competition for the same antenna market among many manufacturers who arrive at different antenna designs for the same in-band specifications. It is important to bear this in mind in analyzing the EMP response of an antenna.

Between the connector and the antenna gap usually referred to in antenna theory there exist many different kinds of matching or compensating networks to tune the antenna impedance to a desirable level at the connector. The matching network, often referred to as the *internal network*, may be transmission lines, baluns, hybrid circuits, transformers, or just lumped elements. Beyond the antenna gap is the radiating element. The equivalent circuit looking outward from this gap is sometimes called the *external network* of the antenna (Fig. 6). The aim in analyzing antenna response to EMP is to calculate V_{oc} (I_{sc}) and Z_{in} (Y_{in}) across the connector terminals A, B, from the knowledge of which one can calculate the voltage and current at any linear or nonlinear load attached to the connector.

4. Techniques for Analyzing Antenna Response to Broadband Electromagnetic Pulses

The response of an antenna is completely characterized by its Thevenin or Norton equivalent circuit at the antenna's input connector, such as the terminals A and B in Fig. 6. The circuit parameters in the Thevenin or Norton equivalent network are V_{oc}, Z_{in} or I_{sc}, Y_{in}. The source terms V_{oc} and I_{sc} can be traced back to the *induced* open-circuit voltage V_{ind} and short-circuit current I_{ind} across the antenna gap (Fig. 6). These latter source quantities can be factored into a product of effective height or area and the local electric or magnetic field on the mounting

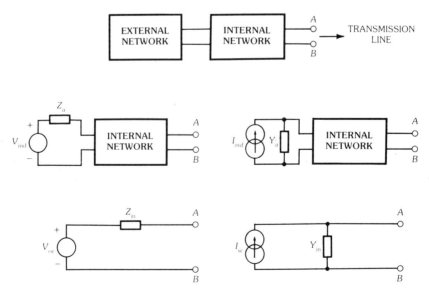

Fig. 6. Principal antenna components.

Antenna Response to Electromagnetic Pulses

structure. Examples of the local field are given in Section 2. In this section techniques are given for calculating the circuit parameters of the external network shown in Fig. 6.

Low-Frequency Techniques

As far as their responses to EMP are concerned, most antennas can be considered electrically small. In Chart 1, formulas are given for the fundamental quantities of two basic dipole antennas. The definitions of these quantities are as follows:

$\mathbf{A}_e, \mathbf{A}_m$ = equivalent area of electric, magnetic dipole antenna [4]

$\mathbf{h}_e, \mathbf{h}_m$ = equivalent height of electric, magnetic dipole antenna [4]

V_e, V_m = equivalent volume of electric, magnetic dipole antenna [4]

$\mathbf{P}, \mathbf{M}$ = electric, magnetic polarizability tensor

$\mathbf{p}, \mathbf{m}$ = electric, magnetic dipole moment

$V_{\text{ind}}, I_{\text{ind}}$ = induced open-circuit voltage, short-circuit current

C_a, L_a = antenna input capacitance, inductance

Note that

$$\begin{aligned}
\mathbf{A}_e \mathbf{h}_e &= \mathbf{h}_e \mathbf{A}_e \\
\mathbf{A}_m \mathbf{h}_m &= \mathbf{h}_m \mathbf{A}_m \\
V_{\text{ind}}^* I_{\text{ind}} &= j\omega \mathbf{p} \cdot \mathbf{E}^* - j\omega\mu_0 \mathbf{m} \cdot \mathbf{H}^* \\
&= j\omega(\epsilon_0 \mathbf{E} \cdot \mathbf{h}_e \mathbf{A}_e \cdot \mathbf{E}^* - \mu_0 \mathbf{H} \cdot \mathbf{h}_m \mathbf{A}_m \cdot \mathbf{H}^*)
\end{aligned} \tag{5}$$

Chart 1. Formulas for Electrically Small Electric and Magnetic Dipole Antennas

Rod	Loop
$I_{\text{ind}} = j\omega\epsilon_0 \mathbf{A}_e \cdot \mathbf{E} = j\omega C_a V_{\text{ind}}$	$V_{\text{ind}} = j\omega\mu_0 \mathbf{A}_m \cdot \mathbf{H} = j\omega L_a I_{\text{ind}}$
$V_{\text{ind}} = \mathbf{h}_e \cdot \mathbf{E}$	$I_{\text{ind}} = \mathbf{h}_m \cdot \mathbf{H}$
$\epsilon_0 \mathbf{A}_e = C_a \mathbf{h}_e$	$\mu_0 \mathbf{A}_m = L_a \mathbf{h}_m$
$V_e = \mathbf{A}_e \cdot \mathbf{h}_e$	$V_m = \mathbf{A}_m \cdot \mathbf{h}_m$
$\mathbf{p} = \epsilon_0 \mathbf{P} \cdot \mathbf{E} = \epsilon_0 \mathbf{h}_e \mathbf{A}_e \cdot \mathbf{E}$	$\mathbf{m} = \mathbf{M} \cdot \mathbf{H} = \mathbf{A}_m \mathbf{h}_m \cdot \mathbf{H}$

Since $V_e = \mathbf{h}_e \cdot \mathbf{A}_e$ and $V_m = \mathbf{h}_m \cdot \mathbf{A}_m$, it can be seen that V_e and V_m are, respectively, the measure of (maximum) electric and magnetic energy extracted from the external field that is stored in the antenna.

A useful quantity to measure the real power absorbed by an antenna is the effective receiving area, whose maximum value for an electrically small antenna is given by [5,6]

$$A_r = \frac{3\lambda^2}{8\pi} \quad \text{for a rod or slot antenna}$$
$$= \frac{3\lambda^2}{4\pi} \quad \text{for transmission through a slot antenna} \tag{6}$$

where λ is the wavelength of the incident wave. When A_r is multiplied by the time-average incident Poynting vector $\overline{S}_0$, one has the maximum time-average real power absorbed by the antenna. Combining (5) and (6), one has the following useful formula:

maximum time-average complex power absorption of an electrically

small antenna $= A_r \overline{S}_0 + 2j\omega(V_e \overline{W}_e - V_m \overline{W}_m)$ \hfill (7)

where $\overline{W}_e$ and $\overline{W}_m$ are the time-average electric and magnetic energy of the external wave.

For an electrically small ellipsoidal center-fed antenna as shown in Fig. 7, the normalized equivalent area is given by the formula [7]

$$\frac{A_e}{\pi bc} = \frac{(a^2 - b^2)\sqrt{a^2 - c^2}}{abc} \frac{1}{F(\phi/\alpha) - E(\phi/\alpha)} \tag{8}$$

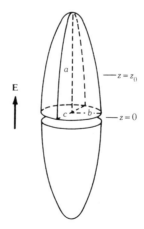

Fig. 7. Center-driven ellipsoidal receiving antenna.

Antenna Response to Electromagnetic Pulses

where $a > b > c$, F and E are respectively the incomplete elliptic integrals of the first and second kind, and

$$\phi = \sin^{-1}\sqrt{1 - c^2/a^2}$$
$$\alpha = \sin^{-1}\sqrt{(a^2 - b^2)/(a^2 - c^2)}$$

Equation 8 is plotted in Fig. 8 for various b/a and c/a values.

When $c \ll b$, the ellipsoid degenerates into an elliptic disk or blade, which can be used to approximate most aircraft blade antennas. In this case, (8) becomes

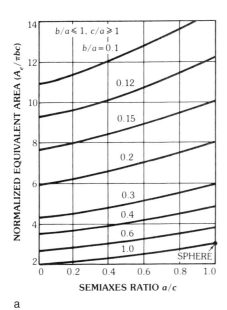

a

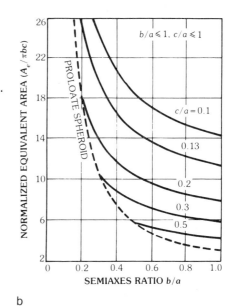

b

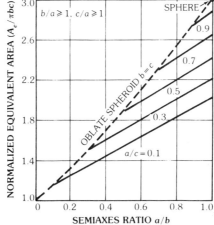

c

Fig. 8. Normalized equivalent area. (*a*) For various b/a. (*b*) For various c/a. (*c*) For various a/c.

Table 4. Normalized Equivalent Area of an Elliptic Disk (Blade Antenna) (*After Lee [1]*)

b/a	$A_e/\pi ab$	$A_c/\pi ab$	a/b
10^{-3}	137.0980	1.0000	10^{-3}
10^{-2}	20.0328	1.0002	10^{-2}
0.1	3.6945	1.0112	0.1
0.2	2.4420	1.0324	0.2
0.3	1.9890	1.0581	0.3
0.4	1.7376	1.0864	0.4
0.5	1.5865	1.1162	0.5
0.6	1.4836	1.1469	0.6
0.7	1.4091	1.1782	0.7
0.8	1.3527	1.2100	0.8
0.9	1.3087	1.2413	0.9
1.0	1.2732	1.2732	1.0

$$A_e = \frac{\pi ab}{\sqrt{1-m}} \frac{m}{K(m) - E(m)} \qquad (9)$$

where K and E are complete elliptic integrals of the first and second kind, and $m = 1 - b^2/a^2$. Table 4 gives the values of the normalized equivalent area of a blade for various b/a ratios.

Most aircraft blade antennas are fed not at the center ($z = 0$) but rather at $z = z_0$ (Fig. 7). In this case the equivalent area is

$$A_e(z_0) = A_e(1 - z_0^2/a^2) \qquad (10)$$

where A_e is given by (9). Equation 10 suggests that an ideal transformer with $n = 1 - z_0^2/a^2$ can be introduced for reference to the center-fed quantities, as shown in Fig. 9.

There are certain useful relationships among the electric and magnetic polarizabilities, **P** and **M**, defined in Chart 1. Let V devote the geometric volume of the ellipsoid and let the coordinate axes be along the principal axes (1, 2, 3) of the ellipsoid. Then [7]

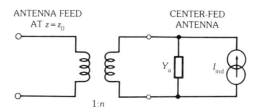

Fig. 9. Use of ideal transformer for referring to center-fed antenna parameters Y_a and I_{ind}.

$$P_{ij} = M_{ij} = 0, \quad i \neq j$$

$$\frac{1}{P_{ii}} - \frac{1}{M_{ii}} = \frac{1}{V} \tag{11}$$

$$\sum_{i=1}^{3} \frac{1}{P_{ii}} = \frac{1}{V}$$

from which one can deduce

$$P_{11} = -2M_{33} \tag{12}$$

for a body rotationally symmetric about, say, the 3 axis, and

$$P_{11} = P_{22} = P_{33} = 3V$$
$$M_{11} = M_{22} = M_{33} = -\frac{3}{2}V \tag{13}$$

for a sphere.

Figs. 10 and 11 show, respectively, the normalized equivalent area and capacitance of an annular slot and center-fed spherical dipole antenna [8,9]. For the annular slot antenna one has the following limiting forms:

$$\frac{A_e}{\pi ab} \cong \frac{8}{\pi^2} + \frac{32}{\pi^4}\frac{a}{b}, \qquad a \ll b$$

$$\frac{C}{8\epsilon a} \cong \frac{1}{2}\ln\left(\frac{2\pi a}{b-a}\right) - 0.19, \qquad a \to b \tag{14}$$

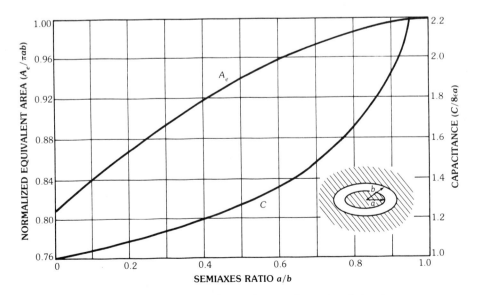

Fig. 10. Equivalent area and capacitance of annular slot antenna.

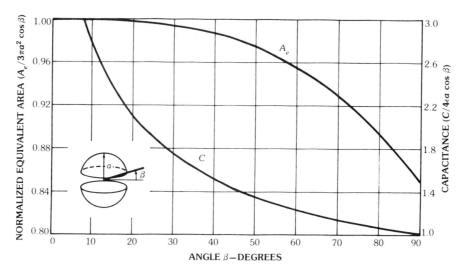

Fig. 11. Equivalent area and capacitance of spherical dipole antenna versus the gap half-angle.

For the spherical dipole antenna the limiting form is

$$\frac{C}{4\epsilon a} \cong 0.9872 - \ln \beta, \quad \beta \leq 10° \tag{15}$$

The Babinet Principle

The Babinet principle is an indispensable tool for analyzing slot antennas. Fig. 12a shows an inductively loaded, folded, slot antenna, and Fig. 12c is its complementary counterpart, a capacitively loaded, folded, strip antenna. According to the Babinet principle, their electrical quantities are related as follows [1]:

$$Z_a Z'_a = Z_0^2/4$$
$$\hat{\mathbf{k}} \times \mathbf{h}_e = \frac{1}{2} \frac{Z_0}{Z'_a} \mathbf{h}'_e \tag{16}$$

where the unprimed quantities are referred to the slot antenna and the primed to the strip antenna. Moreover, $\hat{\mathbf{k}}$ is the unit vector of the propagation vector, $\mathbf{E}_n$ and $\mathbf{H}_{sc}$ are the short-circuit fields normal and tangential to the metallic screen, and $\mathbf{E}^i$ and $\mathbf{H}^i$ are the incident fields. Figs. 12b and 12d are the equivalent circuits corresponding respectively to Figs. 12a and 12c. With (16), one can easily deduce the induced current source I_{ind} in Fig. 12b, viz.,

$$I_{\text{ind}} = \frac{1}{Z_a} V_{\text{ind}} = \frac{1}{Z_a} \mathbf{h}_e \cdot \mathbf{E}^i = \frac{Z_0}{2 Z_a Z'_a} (\hat{\mathbf{k}} \times \mathbf{E}^i) \cdot \mathbf{h}'_e$$
$$= 2 \mathbf{H}^i \cdot \mathbf{h}'_e = \mathbf{h}'_e \cdot \mathbf{H}_{sc} \tag{17}$$

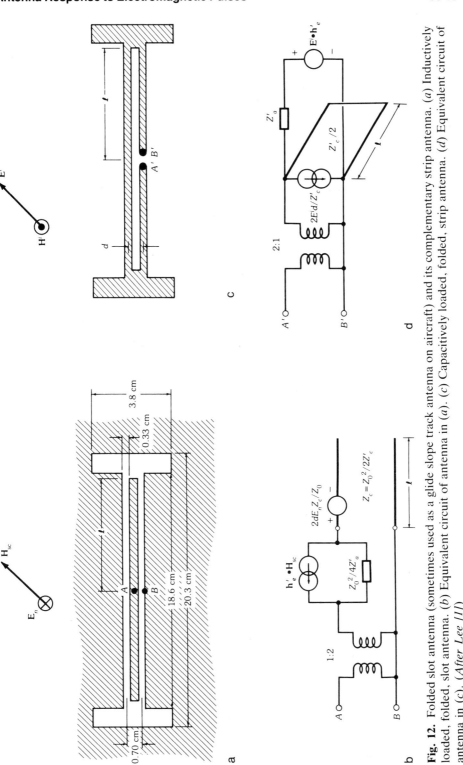

Fig. 12. Folded slot antenna (sometimes used as a glide slope track antenna on aircraft) and its complementary strip antenna. (*a*) Inductively loaded, folded, slot antenna. (*b*) Equivalent circuit of antenna in (*a*). (*c*) Capacitively loaded, folded, strip antenna. (*d*) Equivalent circuit of antenna in (*c*). (*After Lee [11]*)

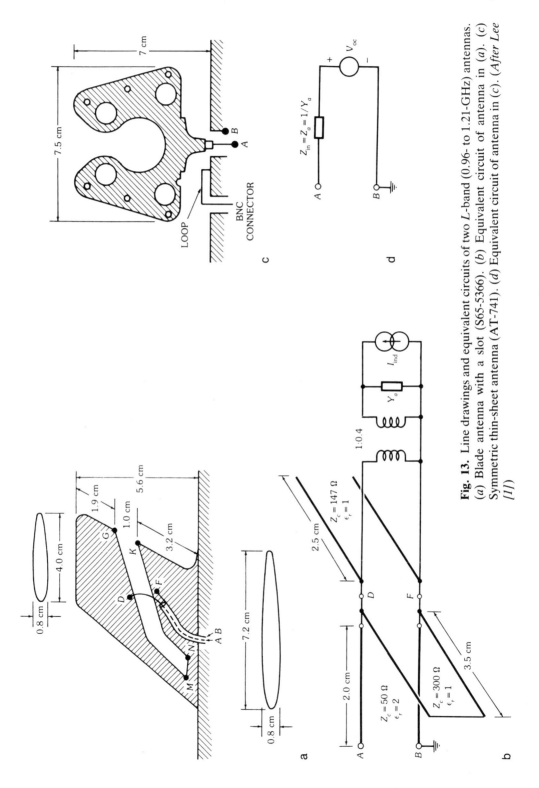

Fig. 13. Line drawings and equivalent circuits of two L-band (0.96- to 1.21-GHz) antennas. (a) Blade antenna with a slot (S65-5366). (b) Equivalent circuit of antenna in (a). (c) Symmetric thin-sheet antenna (AT-741). (d) Equivalent circuit of antenna in (c). (*After Lee [11]*)

5. EMP Responses of Aircraft Antennas

The EMP responses of many different types of aircraft antennas have been analyzed. Typical examples follow [1, 10–12].

L-Band Blade Antennas

Figs. 13a and 13c show two different antennas that both operate between 0.96 GHz and 1.21 GHz. Although their overall dimensions and in-band properties are similar, their out-of-band properties are quite different.

Fig. 13a is a blade antenna with a slot (the slot often being referred to as a notch). The equivalent circuit is given in Fig. 13b. The induced current I_{ind} and antenna admittance Y_a in the equivalent circuit of Fig. 13b are the quantities referred to the same blade antenna having a feeding gap across the entire blade at D, F. The two transmission lines with characteristic impedances 147 Ω and 300 Ω are due to the slot, whereas the transmission line between D, F and A, B represents the coaxial cable with a characteristic impedance of 50 Ω. The frequency variation of the antenna's input impedance is shown in Fig. 14a.

The interior structure of another L-band antenna is shown in Fig. 13c. The actual antenna element is a thin metal sheet cut out in a regular symmetric pattern. The holes cut out in the sheet have the effect of increasing the antenna inductance, since the current has to flow around the holes, thereby lengthening the current path. The input impedance is displayed in Fig. 14b, while the effective heights of both antennas are given in Fig. 15.

Both L-band antennas have roughly the same input impedance and effective height at in-band frequencies. The notch antenna has an inductive input impedance and a very small effective height for low frequencies, whereas the other blade antenna has a capacitance input impedance and considerable effective height for low frequencies.

UHF Communication Antennas

A very common uhf antenna is shown in Fig. 16a, and Fig. 16b is the corresponding equivalent circuit in which L is the inductance of the short stub. This stub not only provides a dc path for lightning protection between the upper part of the antenna and the aircraft skin but also mechanically fastens the two parts of the antenna. The different sections of the transmission lines and the end capacitance C_e are for tuning purposes, so that within the operating frequency band (225 to 400 MHz) the vswr is about 2:1 at the input connector. The capacitance C_p accounts for the two ends of the gap. Figs. 17a and 17b show the frequency variations of Z_{in} and h_e together with measured data for Z_{in}.

VHF Communication Antennas

A schematic diagram of a typical vhf communication antenna is shown in Figs. 18a and 18b. A careful examination of these figures reveals that Fig. 18c is the appropriate equivalent circuit for this antenna. The transmission lines with impedance $Z_c = 176$ Ω represent the empty portion of the slot, whereas the transmission line with impedance $Z_c/\sqrt{\epsilon_r} = 119$ Ω represents the dielectric-filled portion of the slot. The impedance Z_c can be estimated from the analysis of two

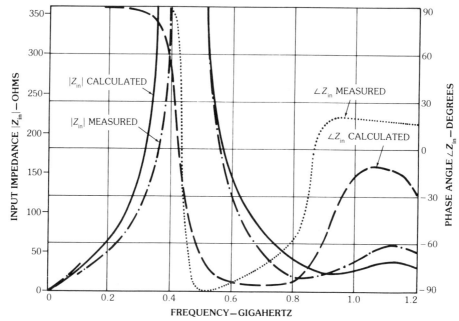

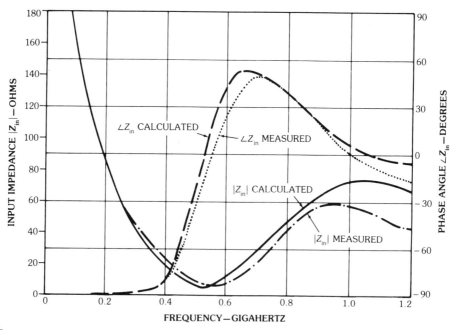

Fig. 14. Input impedance of two *L*-band (0.96- to 1.21-GHz) antennas. (*a*) Of blade antenna with a slot (S65-5366). (*b*) Of symmetric thin-sheet antenna (AT-741). (*After Lee [1]*)

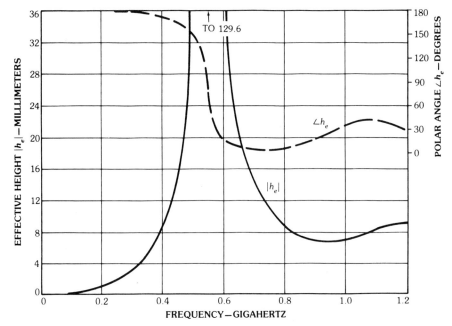

a

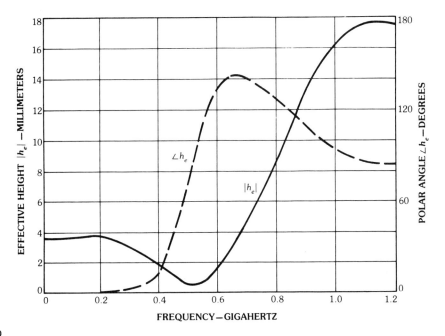

b

Fig. 15. Effective heights of two *L*-band (0.96- to 1.21-GHz) antennas. (*a*) Of blade antenna with a slot (S65-5366). (*b*) Of symmetric thin-sheet antenna (AT-741). (*After Lee [1]*)

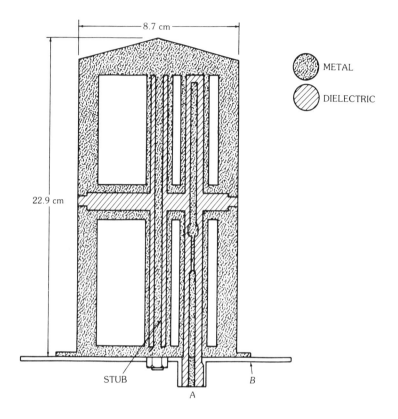

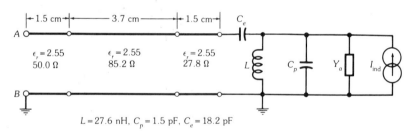

Fig. 16. Line drawing and equivalent circuit of the uhf communication antenna AT-1076. (*a*) Line drawing of antenna. (*b*) Equivalent circuit of (*a*). (*After Lee [1]; after Lee, Liu, and Marin [12]*, © *1978 IEEE*)

coplanar strips of unequal width by the method of conformal mapping. The positions of the tap points (D, F), the length of the dielectric portion of the slot, and the three added coaxial cables act as a matching network, so that the vswr is less than 1.75:1 at in-band frequencies (116 to 156 MHz). The ideal transformer in the equivalent circuit accounts for the fact that the short-circuit current between D and F is just a fraction of I_{ind}, which is the total induced current through a cross section of the antenna (located at D or F and being parallel to the antenna base)

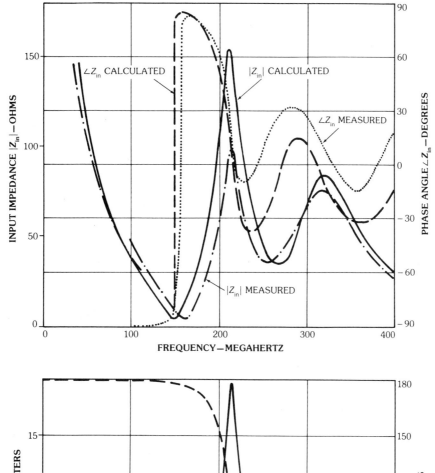

Fig. 17. Input impedance and effective height of the uhf communication antenna AT-1076 (225 to 400 MHz). (*a*) Input impedance. (*b*) Effective height. (*After Lee [1]; after Lee, Liu, and Marin [12],* © *1978 IEEE*)

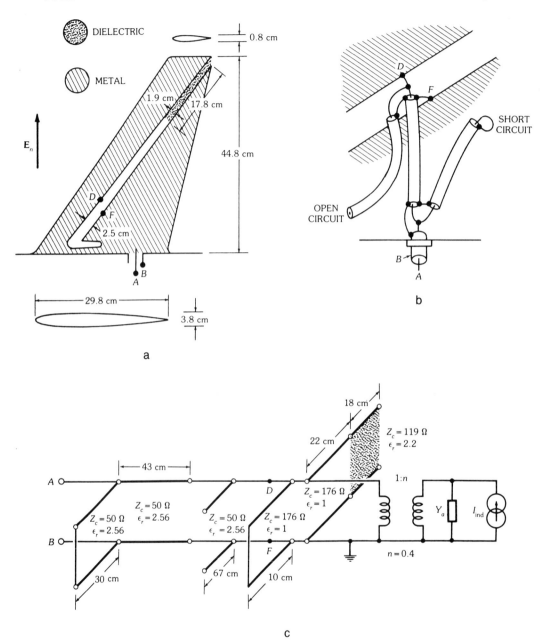

Fig. 18. Line drawings and equivalent circuit of the vhf communication antenna S65-8262-2. (*a*) Construction of antenna. (*b*) Antenna feed circuit. (*c*) Equivalent circuit of (*a*). (*After Lee [1]*)

when the slot is absent. The ideal transformer also accounts for the effects of the locations of D, F on the antenna's input admittance, since Y_a in Fig. 18c is the admittance across a gap bisecting the entire antenna at the location of the cross section discussed above. The transformer ratio n can be estimated by the method described in Section 4, under "Low-Frequency Techniques." Fig. 19 shows the calculated input impedance together with some measured data.

VHF Localizer Antennas

Figs. 20a and 20b show all the electrical elements of a localizer antenna and Fig. 20c is the equivalent circuit. The external elements Z_a and V_{ind} can be calculated from the theory of a circular loop with due account for factors of two resulting from the presence of the ground plane. The tuning capacitors at the ends of the loops (Fig. 20a) are represented by C in Fig. 20c. The coaxial cable wound around the center rods serves as a balun, converting a balanced signal into an unbalanced one for the terminals A, B and A', B within the in-band frequencies (108 to 112 MHz). The wires wrapped around the 100-Ω resistor can be identified as a hybrid circuit which splits an incoming signal into two equal parts for the two terminals A, B and A', B. Because of the close coupling between the wires, M can be taken equal to L. Fig. 21a the input impedance Z_{in} referred to A, B with A', B open circuited, or vice versa. Also shown in the figure are some measured data. Fig. 21b gives the frequency variation of V_{oc} across A, B or A', B for an incident plane wave with the electric-field vector perpendicular to the ground plane and the magnetic-field component B_0 perpendicular to the plane of the loop.

VHF Marker Beacon Antennas

The marker beacon antenna shown in Fig. 22a is flush-mounted at the bottom of the fuselage and operates around 75 MHz. The bowl can be approximated by a hemispherical indentation. The equivalent circuit is given by Fig. 22b in which L_1 and L_2 are respectively the inductances of the small and large loops formed by the metal rod and the feed wire. Figs. 23a and 23b give the frequency variations of Z_{in} and V_{oc}/B_0, where B_0 is the magnetic field tangential to the aircraft skin and perpendicular to the plane of the large loop, and V_{oc} and Z_{in} are the open-circuit voltage and input impedance across A, B.

HF Fixed-Wire Antennas

Two hf fixed-wire antennas on an aircraft are shown in Fig. 24a. A schematic drawing of the stick-model aircraft used in the calculations is shown in Fig. 24b. To find the input impedance between A, B when the port A', B is terminated by the impedance Z_L, let the antenna be driven with the voltage V between A, B. The induced currents on the antenna wires and the aircraft can be decomposed into (*a*) radiating currents on the aircraft, (*b*) a TEM mode such that the wire currents are of equal magnitude and in the same direction with the "return current" along the fuselage, and (*c*) a TEM mode such that the wire currents are of equal magnitudes but in opposite directions with no net current along the fuselage. The details of this decomposition can be found in reference 13. An equivalent circuit of the antenna is shown in Fig. 24c. The admittance Y_a is defined at an imaginary gap across the fuselage at the location of the antenna feed point, and can be calculated using a

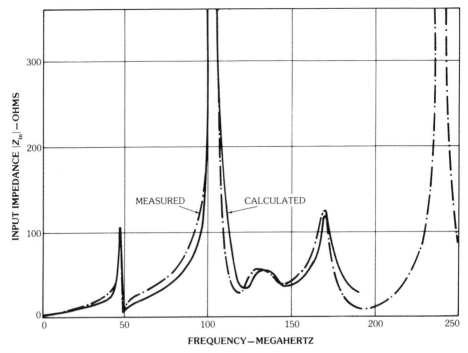

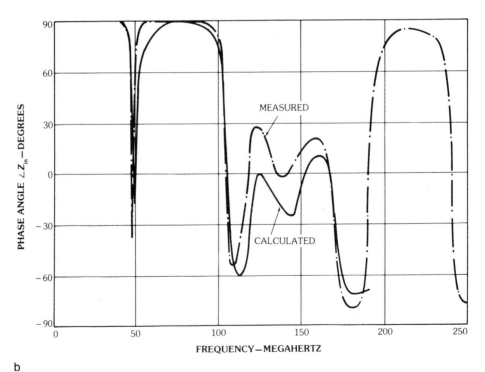

Fig. 19. Input impedance of the vhf communication antenna S65-8262-2 (116 to 156 MHz). (*a*) Magnitude of input impedance. (*b*) Phase angle of input impedance. (*After Lee [1]*)

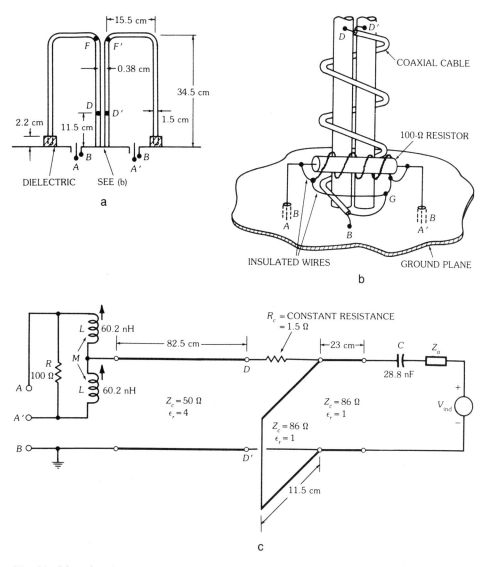

Fig. 20. Line drawings and equivalent circuit of the localizer antenna S65-147. (*a*) Loop structures. (*b*) Signal-splitting circuit. (*c*) Equivalent circuit of antenna in (*a*). (*After Lee [1]; after Lee, Liu, and Marin [12],* © *1978 IEEE*)

stick-model aircraft. The transmission line with the characteristic impedance Z'_c represents the first TEM mode described above, whereas the transmission line with impedance Z''_c represents the second TEM mode. The ideal transformers account for the coupling between the TEM modes and the radiating currents. The current I_{ind} is induced on the fuselage by a plane wave at the location of the antenna feed gap and in the absence of the antenna wires (see Fig. 24b). The other current generators are found from the transmission-line theory [14].

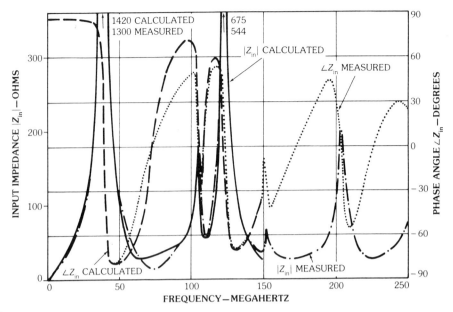

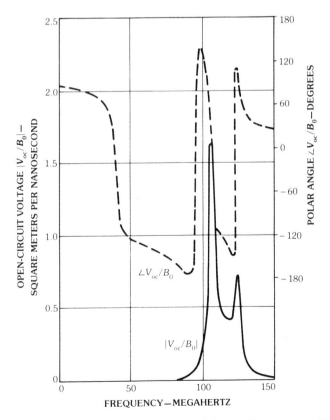

Fig. 21. Input impedance and open-circuit voltage of the localizer antenna S65-147 (108 to 112 MHz). (*a*) Input impedance. (*b*) Open-circuit voltage. (*After Lee [1]; after Lee, Liu, and Marin [12]*, © *1978 IEEE*)

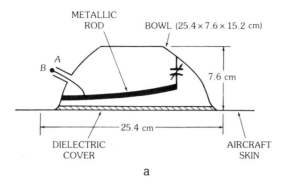

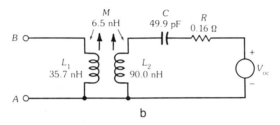

Fig. 22. Line drawing and equivalent circuit of the AT-536/ARN marker beacon antenna. (*a*) Antenna structure. (*b*) Equivalent circuit of antenna in (*a*). (*After Lee [1]; after Lee, Liu, and Marin [12]*, © *1978 IEEE*)

Fig. 25 gives the frequency variations of the input impedance and effective height. In this case the effective height relates the open-circuit voltage to the incident electric field, and is a function of the angle of incidence and polarization of the incident plane wave. In Fig. 25 the values $Z_L = 0$, $Z_L = \infty$, and $\theta = 120°$ have been used. Quantities with superscript "sc" correspond to $Z_L = 0$, and quantities with superscript "oc" correspond to $Z_L = \infty$. The variation of h_e with θ can be found in [13]. The transient response of this antenna to the high-altitude EMP waveform given by (1) can be found in reference 15.

VLF/LF Trailing-Wire Antennas

A vlf/lf (17- to 60-kHz) transmitting dual-wire antenna is shown in Fig. 26a. The response of this type of antenna to EMP can be obtained using two different steps. In the first step, which is valid in the frequency range where the aircraft is electrically small, one may model the aircraft by a capacitance to characterize its effect on the antenna response. The interaction between the two wires can be calculated by decomposing the wire currents into common-mode (antenna-proper) currents and differential-mode (transmission-line) currents. In the second step, which is valid in the frequency region around the first few aircraft resonances, the aircraft can be modeled by intersecting sticks. Fig. 26b is the equivalent circuit appropriate for the situation depicted in Fig. 26a. Explicit approximate formulas for I'_{ind}, I''_{ind}, Z'_a, h, Z''_a, Z_c, α, and β can be found in reference 16.

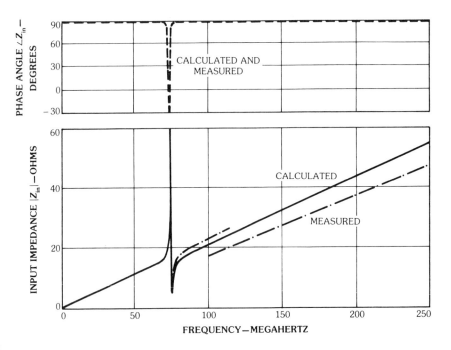

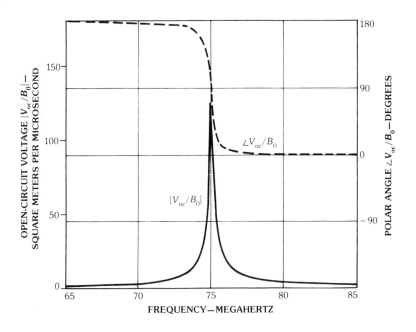

Fig. 23. Input impedance and relative open-circuit voltage of the AT-536/ARN marker beacon antenna (75 MHz). (*a*) Input impedance. (*b*) Open-circuit voltage. (*After Lee [1]; after Lee, Liu, and Marin [12]*, © *1978 IEEE*)

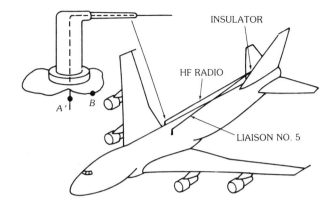
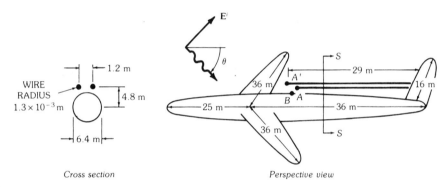
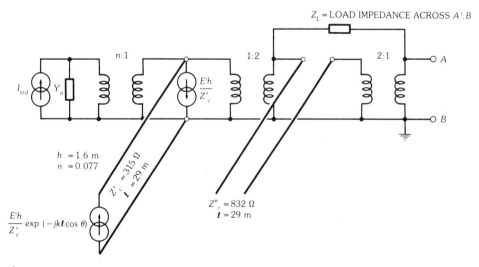

Fig. 24. Line drawings and equivalent circuit of the two hf fixed-wire antennas on the E-4 aircraft. (*a*) Aircraft with two fixed-wire hf antennas. (*b*) Wire position and incident electric field. (*c*) Equivalent circuit of antenna. (*After Lee [1]*)

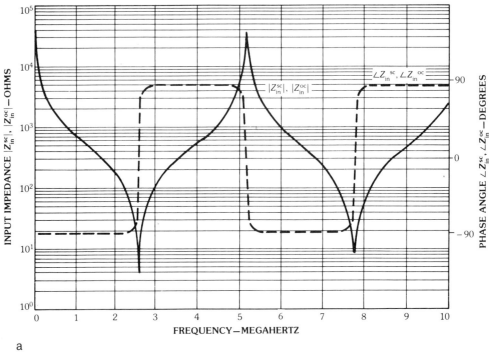

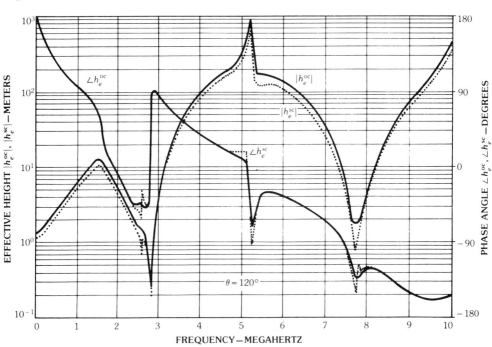

Fig. 25. Input impedance and effective height of the hf (2- to 30-MHz) fixed-wire antennas on the E-4 aircraft. (*a*) Input impedance. (*b*) Effective height. (*After Lee [1]*)

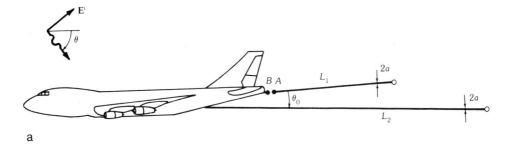

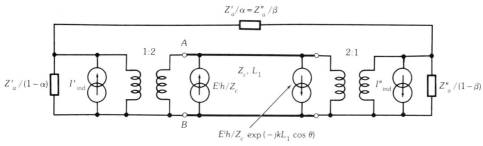

Fig. 26. Line drawing and equivalent circuit for the dual-wire antenna. (*a*) Aircraft with trailing-wire transmitting antenna. (*b*) Equivalent circuit of antenna in (*a*). (*After Lee [1]; after Marin, Lee, and Castillo [16]*, © *1978 IEEE*)

6. References

[1] K. S. H. Lee, *EMP Interaction: Principles, Techniques, and Reference Data*, chapter 2.1, New York: Hemisphere Publishing Corp., 1986.

[2] V. V. Liepa, "Surface field measurements on scale model EC-135 and E-4 aircraft," *Interaction Application Memos 15 and 17*, Air Force Weapons Lab, New Mexico, March and August 1977.

[3] R. W. Sassman, "The current induced on a finite, perfectly conducting, solid cylinder in free space by an electromagnetic pulse," *EMP Interaction Notes, Note 11*, Air Force Weapons Lab, New Mexico, July 1967.

[4] C. E. Baum, "Parameters for some electrically-small electromagnetic sensors," *EMP Sensor and Simulation Notes, Note 38*, Air Force Weapons Lab, New Mexico, March 1967.

[5] J. D. Kraus, *Antennas*, chapter 3, New York: McGraw-Hill Book Co., 1950.

[6] K. S. H. Lee, "Maximum power penetration through an electrically small aperture," *IEEE Trans. Antennas Propag.*, vol. AP-31, no. 3, pp. 518–519, May 1983.

[7] K. S. H. Lee, "Electrically small ellipsoidal antennas," *EMP Sensor and Simulation Notes, Note 193*, Air Force Weapons Lab, New Mexico, February 1974.

[8] R. W. Latham and K. S. H. Lee, "Capacitance and equivalent area of a disk in a circular aperture," *EMP Sensor and Simulation Notes, Note 106*, Air Force Weapons Lab, New Mexico, May 1970.

[9] R. W. Latham and K. S. H. Lee, "Capacitance and equivalent area of a spherical dipole sensor," *EMP Sensor and Simulation Notes, Note 113*, Air Force Weapons Lab, New Mexico, July 1970.

[10] T. K. Liu, K. S. H. Lee, and L. Marin, "Broadband responses of deliberate aircraft

antennas, part I," *EMP Interaction Notes*, *Notes 228*, Air Force Weapons Lab, New Mexico, May 1975.

[11] K. S. H. Lee and L. Marin, "Deliberate aircraft antenna model development," *AFWL-TR-76-218* (2 volumes), Air Force Weapons Lab, New Mexico, May 1977.

[12] K. S. H. Lee, T. K. Liu, and L. Marin, "EMP response of aircraft antennas," *IEEE Trans. Antennas Propag.*, vol. AP-26, no. 1, pp. 94–99, January 1978; also *IEEE Trans. Electromagn. Compat.*, vol. EMC-20, no. 1, pp. 94–99, February 1978.

[13] L. Marin, "Response of the hf fixed-wire antenna on the E-4," *AFWL-TR-78-42*, Air Force Weapons Lab, New Mexico, February 1979.

[14] K. S. H. Lee, "Two parallel terminated conductors in external fields," *IEEE Trans. Electromagn. Compat.*, vol. EMC-20, no. 2, pp. 288–296, May 1978.

[15] G. Bedrosian and L. Marin, "Transient response of the hf fixed-wire antennas on the E-4," *AFWL-TR-78-43*, Air Force Weapons Lab, New Mexico, February 1979.

[16] L. Marin, K. S. H. Lee, and J. P. Castillo, "Broadband analysis of vlf/lf aircraft wire antennas," *IEEE Trans. Antennas Propag.*, vol. AP-26, no. 1, pp. 141–145, January 1978.

Chapter 31

Radome Electromagnetic Design

G. P. Tricoles
General Dynamics

CONTENTS

1. Introduction — 31-3
   Radome Performance Parameters 31-3
   Variables in the Radome Performance Parameters 31-4
2. Physical Description of Radome Effects — 31-5
   Fields in Radome-Bounded Regions; Constituent Waves 31-5
   Field Distribution Measurement 31-6
   Wavefront Aberrations and Boresight Error 31-7
   Diffraction Methods for Radome Design 31-8
3. Plane Wave Propagation through Flat Dielectric Sheets — 31-10
   Incidence on a Plane Boundary between Two Homogeneous Dielectrics 31-10
   Incidence on Flat, Homogeneous, Dielectric Sheets 31-13
   Incidence on Multilayer Dielectric Sheets 31-14
4. General Aspects of Theoretical Design — 31-18
5. Plane Wave Propagation through Dielectric Sheets — 31-20
   Direct-Ray Method 31-20
   Surface Integration 31-24
6. Antenna Patterns — 31-25
7. Examples of Boresight Error — 31-27
8. The Moment Method — 31-27
9. Comments on Materials — 31-29
10. References — 31-30

Gus Tricoles received a BA in physics from the University of California, Los Angeles, in 1955, an MS in applied mathematics from San Diego State College in 1958, and an MS in physics in 1962 and a PhD in applied physics in 1970, both from the University of California, San Diego.

He was employed by General Dynamics from 1955 to 1959 and by Smyth Research Associates from 1960 to 1962; in 1962 he returned to General Dynamics, where he is now Engineering Manager of Electromagnetics in the Electronics Division in San Diego.

Dr. Tricoles' main work has been in propagation and microwave holography with applications to radomes, direction finding, and geological prospecting. He developed analytical methods for radome design and experimental, diagnostic methods for radomes. He has designed radomes for several aircraft and missiles, developed anisotropic radomes and guided-wave antennas, and modeled retinal receptors. He holds 14 patents and has published 20 papers and written many reports. He is a member of US Commissions A, B, and C of URSI, the IEEE Antenna Standards Committee, a Fellow of the Optical Society of America, and a Fellow of the Institute of Electrical and Electronic Engineers.

1. Introduction

Radomes are enclosures for antennas. Most radomes are hollow dielectric shells although some contain perforated metallic layers or metallic reinforcing structures. Radomes are used with large antennas on the earth's surface to reduce wind loading and to prevent accumulation of ice or snow; these radomes usually have spherical contours. Many aircraft and missiles have radomes; some are blunt, but a nose radome may be pointed to reduce aerodynamic drag. See Fig. 1. Although a radome may perform a useful function, it can degrade an electronic system connected to the enclosed antenna. Radomes reduce radar range because they reflect and attenuate incident waves. They also cause errors in determining the direction of a reflecting object or radiating source; in missiles, this directional error significantly affects the accuracy of guidance. Radomes also can increase radar clutter by reflections that are analogous to multipath. These effects on systems constitute the electromagnetic performance of a radome. A radome's electromagnetic performance is described by four parameters. They are as follows:

Radome Performance Parameters

Far-Field Radiation Patterns—Far-field patterns describe the angular distribution of the intensity radiated or received by an antenna. Radomes can change beamwidths, side lobe levels, directions of pattern maxima, values of maxima or minima, and axial ratio. Patterns of antennas in radomes are measured with conventional apparatus except for special fixtures to support the antenna and radome.

Power Transmittance—Power transmittance of a radome describes the intensity at the peak of the antenna pattern's main beam. It is the ratio of peak intensity received with the radome present to the intensity received when the radome is absent. Although transmittance can be measured by comparing intensities at the peaks of antenna patterns measured with and without the radome, accuracy is greater if the measurements are made with the enclosed antenna held fixed and receiving energy from a distant transmitter. The radome is pivoted about the enclosed antenna, and data are recorded continuously during the radome motion.

Boresight Error—Boresight error is an angular shift; it depends on the nature of the enclosed antenna. For monopulse antennas, it is the shift that the radome causes in the direction of the difference mode pattern's deepest minimum or the shift that is obtained by comparing the phases in a pair of antennas. For antennas with patterns that determine direction from the main beam peak, boresight error is the shift in direction of the main beam. Boresight error is usually measured by pivoting the radome about the fixed, enclosed receiving antenna. Several kinds of detection are used [1].

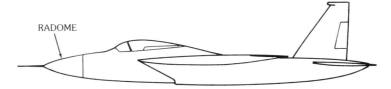

Fig. 1. Aircraft with nose radome.

Boresight Error Slope—Boresight error slope is the rate of change of boresight error with respect to the angle between the axes of the radome and the antenna. The antenna axis is some preferred direction, such as the monopulse null direction. The radome axis is an axis of revolution, if one exists, or some arbitrary direction. Boresight error slope is significant for guided missiles that are steered to maintain a constant target bearing.

Variables in the Radome Performance Parameters

The radome performance parameters depend on several variables, which may be classified as either constrained or design variables.

Constrained Variables—The parameters depend on some variables that are constrained by the system that uses the radome. The frequencies and polarization of the radiation have strong influence; so do the enclosed antenna's size, polarization, aperture distribution, position, and range of orientations. Radomes must obviously be big enough to provide antenna clearance, and many aircraft and missile radomes are pointed to reduce drag. In addition, the radome's environment is significant. Ground-based radomes are subjected to wind loads and weather. Aircraft and missile radomes must withstand aerodynamic loads and the high temperatures generated during flight. The temperatures and loads restrict the materials that can be used for fabrication. All these variables influence radome performance, but the constraints put them, to varying degrees, beyond a designer's control.

Design Variables—Although some variables are tightly constrained, values of several additional variables can be chosen in a design to achieve (or approach) desired performance. We call the latter class design variables. Some design variables describe a radome's composition; others, its configuration. The radome wall configuration is one of the most important design variables. Radome walls can consist of one homogeneous layer or of several; the dielectric constants and thicknesses can be chosen within restrictions imposed by material strength and temperatures. In addition, small local thickness variations or gradual tapers improve performance. Recently, materials with configurational anisotropy have been used to increase frequency bandwidths and to reduce effects produced by changing wave polarization [2,3]. Metallic layers also have been included in radomes to give protection from lightning or static electricity [4].

2. Physical Description of Radome Effects

This section describes what radomes do to electromagnetic waves. When a wave illuminates a radome, several mechanisms occur; these are refraction, reflection, attenuation, guided-wave excitation, and scattering by the tip or base. Because the mechanisms are diverse, occur simultaneously, and depend on frequency and polarization, radome analysis tends to be complicated.

Fields in Radome-Bounded Regions; Constituent Waves

Fig. 2 shows the kinds of waves that occur when a plane wave illuminates the outside of a pointed radome. The sketch is two-dimensional, so it does not show the effect of transverse radome curvature and wave polarization. Experiments, theory, and computation show that the field in a radome-bounded region consists of several constituent fields [5,6]. These constituents interfere, so the antenna receives a complicated, nonplanar wave, even for specific polarization, frequency, and direction of the incident wave. One of the constituents is directly propagated through the illuminated side of the radome; it includes multiple internal reflections. A second constituent is a wave that propagates first through the illuminated side and is then reflected by the opposite side. Guided waves are excited; their propagation constants differ from those of the direct and reflected waves. A vertex may scatter a field. The base of the radome and the adjacent airframe also scatter, but absorbing material can suppress these fields, so they are omitted.

Knowledge about the constituent waves is useful. It helps in formulating and

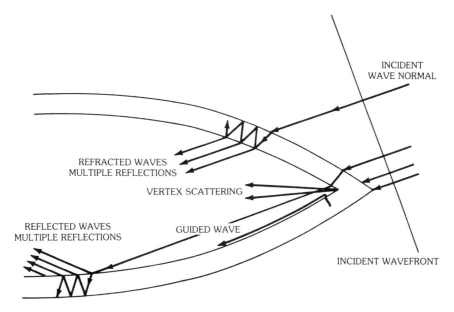

Fig. 2. A wave incident on a pointed radome produces several kinds of waves in the radome-bounded region.

evaluating numerical design methods, which usually are approximate because radomes often have shapes not defined by any separable coordinate surface. Knowing each wave's properties strengthens empirical design approaches, which may involve fabricating and testing several different radomes to connect configuration and measured performance. For example, identifying reflected waves helps locate reflecting regions, which can be modified to reduce antenna side lobe levels.

Field Distribution Measurement

Fig. 3 shows an arrangement of apparatus for measuring the field distribution in a radome-bounded region. A small probe scans the region and receives waves from a transmitting antenna which is connected to a signal generator. A coherent reference field also is supplied to the receiver, a network analyzer that measures phase and intensity.

Very nonuniform distributions can be produced. For example, Fig. 4 shows intensity measured with a small probe antenna for β, the angle between the radome axis and propagation direction of the incident wave, equal to 36°. (The angle β is called the *gimbal angle*.) The radome was pointed, with base diameter 33 cm and length 66 cm; the thickness was 0.49 cm. The dielectric constant was 4.0. The spacings of the fringes show they result from interference of direct (refracted) and reflected waves that are suggested in Fig. 2. Notice that the reflected-field magnitude is large, approximately that of the incident field. Large reflections can occur despite the high transmittance in the reflection-free region because incidence on the lower side of the radome in Fig. 2 is more nearly grazing. The dependence of reflection on incidence direction is described in Section 5. Radome reflections can increase the side lobe levels of an enclosed aperture antenna.

In addition to intensity, phase distributions are deformed. Fig. 5 shows phase measured for β equal to 4° at 9 GHz in a 0.15-cm-thick axially symmetric radome with length 66 cm and base diameter 33 cm. The dielectric constant was 3.2. The

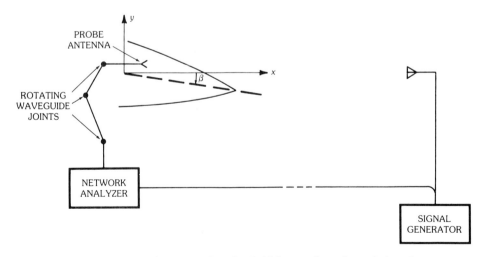

Fig. 3. Apparatus for measuring the field in a radome-bounded region.

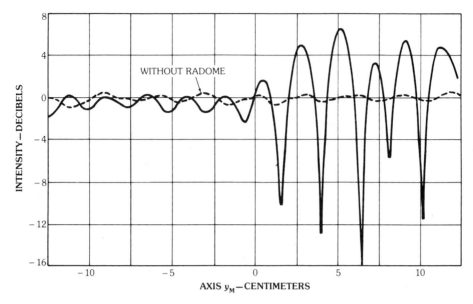

Fig. 4. Power transmittance in an axially symmetric shell measured at 16.5 GHz, with the probe on a plane perpendicular to the radome axis for $\beta = 36°$.

minimum occurs at the shadow of the tip. The distorted wavefronts cause boresight error.

Fig. 6 shows boresight error typical of pointed radomes. In this case the radome had a dielectric of constant of 5.5; length and base diameter respectively were 24.1 and 11.6 wavelengths at midband frequency. The *E*-plane is the plane that contains the electric field and the normal to the antenna. Note that boresight error depends on the orientation of polarization relative to the scan direction of the antenna.

Wavefront Aberrations and Boresight Error

Approximate diffraction analysis shows that radome boresight error is related to wavefront deformations that are asymmetric, like tilts. Although most practical radomes are three-dimensional, let us for the present consider Fig. 7, where the *z* axis points out of the page. For a fixed direction of an incident wave let the distorted wave phase Φ be expanded as a polynomial in the coordinates of Fig. 7 so that

$$\Phi = a + by + cy^2 + dy^3 + ez^2 y \tag{1}$$

The phase Φ is measured over a plane; it is a variation from the incident wave's phase over that plane, and it has units of radians. The coefficient *a* has units of phase, *b* has dimension of reciprocal length, and so on. Some typical values of Φ are given in Section 5.

When (1) is used in a scalar diffraction integral for far-field patterns, the boresight error $\delta\beta$ is [7]

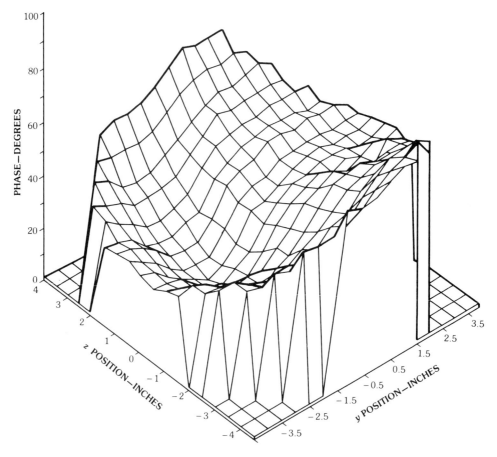

Fig. 5. Measured phase distribution in a thin radome.

$$\delta\beta = (\lambda/2\pi)[b + (e + 3d)R^2/6] \qquad (2)$$

where R is the radius of the aperture and λ is wavelength. Equation 2 was derived for conical scan antennas. It assumes no intensity variations accompany Φ and uniform antenna aperture distribution. It shows that error arises from linear tilts given by b and e, as well as cubic tilts given by d. The antenna radius also appears, suggesting that boresight error depends on antenna size for a specific radome; in fact, experience shows antenna size is significant.

Diffraction Methods for Radome Design

Although (2) gives a physical interpretation of boresight error, quantitative performance estimates of specific radome designs require computing the diffraction patterns of the enclosed antenna. Computers can do the calculations quickly and economically in most cases; however, accuracy depends on the theoretical basis of algorithms, especially for small antennas and radomes, so empirical design is common. Selection of a general design approach seems to depend on many fac-

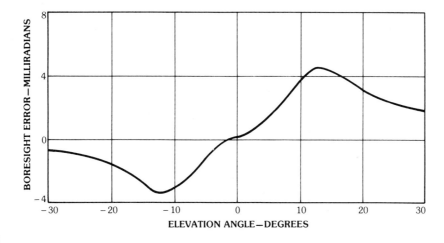

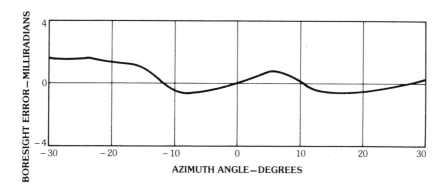

Fig. 6. Measured boresight error for a ceramic radome with thickness approximately 0.5λ, base diameter 11.6λ, and length 24.1λ, enclosing an antenna of diameter 10λ. (*a*) *E*-plane. (*b*) *H*-plane.

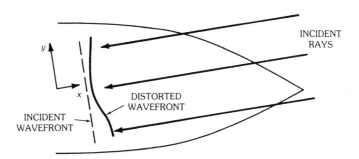

Fig. 7. Incident and distorted wavefronts in a two-dimensional view.

tors, which include costs, schedules, the availability of antennas, and even the background of the designer. Although analysis and computation may be tedious, they seem useful and economical if accurate. Empirical approaches have limited range, and the cost of tooling makes mistakes expensive.

Sections 5 and 6 describe theoretical design methods that vary in degree of approximation. Experimental tests of the methods are included to indicate accuracy. The following section analyzes propagation through flat dielectric sheets as a preliminary to methods for analyzing curved dielectric shells.

3. Plane Wave Propagation through Flat Dielectric Sheets

The analysis of plane wave propagation through flat, multilayer dielectrics of infinite breadth is basic to radome design. The analysis gives transmittances for use in approximate computations that analyze curved radomes by treating them as locally plane. The flat-sheet transmittances, combined with determinations of incidence angle, estimate power transmittance of curved radomes. The local flat-sheet approximation is suggested in Fig. 8; note that the radome is approximated by a number of sheets. The incidence angle is the angle α in Fig. 9.

This section summarizes propagation through flat sheets; it also describes some common radome wall configurations that show how propagation depends on dielectric constants and thicknesses of the layers, wavelength, polarization, and incidence angle. The section starts with formulas for the reflection and transmission at an infinite planar interface between two homogeneous regions; formulas for propagation through a flat homogenous sheet are then described.

Incidence on a Plane Boundary between Two Homogeneous Dielectrics

At a plane infinite interface, as in Fig. 10, an incident plane wave excites reflected and refracted plane waves. The amplitudes of these waves are given by well-known Fresnel formulas (see [8] and [9], pp. 38–41). The dielectrics are

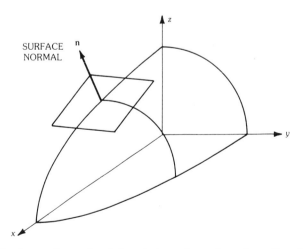

Fig. 8. Approximate local description of a radome by a flat sheet.

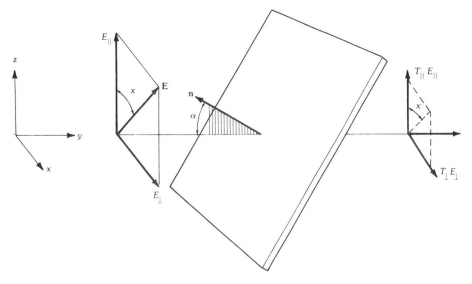

Fig. 9. Plane wave incident at angle α on a flat panel.

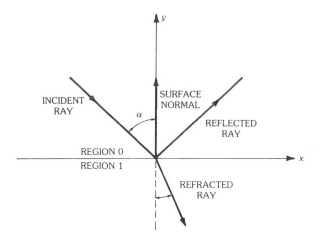

Fig. 10. Reflection and refraction at a plane interface between two homogeneous dielectric regions.

assumed homogeneous, isotropic, linear, and nonmagnetic. A rectangular electric-field component of an incident, linearly polarized wave is resolved into components parallel and perpendicular to the plane of incidence, which is defined with reference to Fig. 9 as the plane containing the surface normal **n** and the incident wave normal. Transmittances and reflectances for the two components differ because distinct boundary conditions apply. The name *parallel polarization* denotes the component parallel to the incidence plane; *perpendicular polarization* refers to the component orthogonal to the incidence plane.

Interface Reflectance and Transmittance for Perpendicular Polarization—In the coordinate system of Fig. 10, the z component of the incident wave is

$$E_z^i = \exp[-jk_0(x \sin \alpha - y \cos \alpha)] \qquad (3)$$

The reflected wave is

$$E_z^r = r_{01} \exp[-jk_0(x \sin \alpha + y \cos \alpha)] \qquad (4)$$

and the refracted, or transmitted, wave in region 1 is

$$E_z^t = t_{01} \exp[-jk_1(x \sin r - y \cos r)] \qquad (5)$$

The factors r_{01} and t_{01} are the Fresnel interface reflectance and transmittance, respectively, and k_0 and k_1 are the propagation constants in regions 0 and 1, respectively. Formulas for r_{01} and t_{01} are derived by applying the boundary conditions, namely, continuity of E_z at the interface ($y = 0$) and continuity of the tangential component of the magnetic field **H**. This component is found by applying the Maxwell equation curl $\mathbf{E} = -j\omega\mu\mathbf{H}$ and taking the projection onto the interface. Because the derivation is standard, only the results are cited. For dielectrics that have loss tangents less than about 0.02, the reflectance between regions labeled j and k is

$$r_{jk} = [(P_j - P_k) - j(L_j P_j - L_k P_k)]/[(P_j + P_k) - j(L_j P_j + L_k P_k)] \qquad (6)$$

where P_j is $(\varkappa_j - \sin^2\alpha)^{1/2}$ with $\varkappa_j$ the dielectric constant of the jth layer, i.e., $k_j = k_0\sqrt{\varkappa_j}$; L_j is $\varkappa_j(\tan \delta)_j/2P_j^2$; M_j is $(\tan \delta)_j - L_j$; and $(\tan \delta)_j$ is the loss tangent of the jth layer. The value of k_0 is that for free space. The transmittance is

$$t_{jk} = 1 + r_{jk} \qquad (7)$$

Interface Reflectance and Transmittance for Parallel Polarization—For the electric-field component parallel to the plane of incidence, the reflectance for low-loss dielectric regions j and k is

$$\hat{r}_{jk} = [(\varkappa_j P_k - \varkappa_k P_j) - j(\varkappa_j M_k P_k - \varkappa_k M_k P_j)]/[(\varkappa_j P_k + \varkappa_k P_j) - j(\varkappa_j M_j P_k + \varkappa_k M_k P_j)]^{-1} \qquad (8)$$

The transmittance is

$$\hat{t}_{jk} = 1 + \hat{r}_{jk} \qquad (9)$$

An important distinction exists between r_{01} and $\hat{r}_{01}$; for parallel polarization, $\hat{r}_{01}$ can vanish but r_{01}, for perpendicular polarization, cannot. The vanishing of $\hat{r}_{01}$ is well known; it is the Brewster phenomenon. From (8) and Snell's law, $\hat{r}_{01}$ is zero when the incidence angle is $\alpha_\beta = \tan^{-1}\sqrt{\varkappa}$, where $\varkappa$ is k_1/k_0; this angle is Brewster's angle. The Brewster phenomenon influences radome design because the parallel and perpendicular components are affected differently by radomes, as will be described in Section 5.

Incidence on Flat, Homogeneous, Dielectric Sheets

For plane wave incidence on a flat, homogeneous, isotropic, and nonmagnetic dielectric sheet, the transmitted field is the sum of multiple reflections as suggested by Fig. 11. For now, the polarization is assumed perpendicular or parallel. To be definite, consider first perpendicular polarization. The incident wave, with unit magnitude, produces a wave with magnitude $t_{12}t_{23}$; the Fresnel transmittances describe propagation through the interfaces. The incident wave also produces a second wave that is transmitted into the sheet, reflected twice, and then transmitted out of the slab; this wave has magnitude $t_{12}r_{23}r_{21}t_{23}$. This process continues, giving a series of terms. Initially assume the dielectric is lossless.

The phase of the various terms is evaluated at a plane at distance r from the origin. Let $E^i(0, d_2)$ be the incident field at $(0, d_2)$. With p_2 the path length in the sheet, the term with amplitude $t_{12}t_{23}$ changes phase by $k_2 p_2 + k_3(r_3 - p_2 \sin r_2 \sin r_3)$. Use the abbreviation Φ_2 for $k_2 p_2 - k_3 p_2 \sin r_2 \sin r_3$ so the phase of the first term is rewritten as $\Phi_2 + k_3 r_3$. For the second term the phase change is $3\Phi_2 + k_3 r_3$. Higher-order reflections have additional even multiples of Φ_2. The field propagated through the sheet is given by a geometric series which sums as

$$E_{13} = E^i(0, d_2) t_{12} t_{23} \exp(-j\Phi_2 - L_2 \Phi_2) \exp(-jk_3 r) D_2^{-1} \qquad (10)$$

where

$$D_2 = 1 - r_{21} r_{23} \exp(-j\Phi_2 - L_2 \Phi_2) \qquad (11)$$

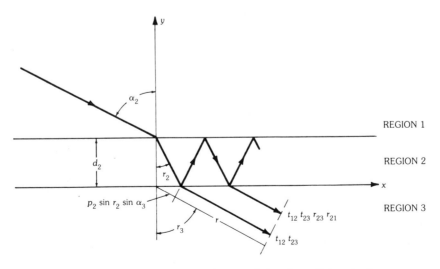

Fig. 11. Multiple reflection and transmission in a dielectric sheet.

It is convenient to express the formulas in terms of the angle of incidence. In particular, by Snell's law,

$$\Phi_2 = k_2 p_2 - k_3 p_2 \sin r_2 \sin r_3$$
$$= k_0 d_2 (\varkappa_2 - \varkappa_1 \sin^2 \alpha_2)^{1/2} \quad (12)$$

The transmittance is found from (10) normalized to (dividing by) the incident field at the reference plane. Thus the complex amplitude transmittance of the flat sheet is

$$T_{13} = t_{12} t_{23} D_2^{-1} \exp(-j\Phi_2 - L_2 \Phi_2) \exp(jk_2 d_2 \cos \alpha_2) \quad (13)$$

The power transmittance (for $\tan \delta = 0$) is

$$|T_{13}|^2 = |t_{12} t_{23} D_2^{-1} \exp(-L_2 \Phi_2)|^2 \quad (14)$$

The Fresnel reflectances and transmittances are given by (6) through (9).

The *insertion phase delay* is the phase change produced by the sheet; it is

$$\Delta \Phi = \arg(t_{12} t_{23} D_2^{-1}) - k_0 d_2 (\varkappa_2 - \varkappa_1 \sin^2 \alpha_2)^{1/2} + jk_2 d_2 \cos \alpha_2 \quad (15)$$

It can be measured with a microwave interferometer.

For parallel polarization, replace r_{21}, r_{32}, t_{12}, and t_{23} by $\hat{r}_{21}$, $\hat{r}_{32}$, $\hat{t}_{12}$, and $\hat{t}_{23}$. The reflectance is (for perpendicular polarization)

$$R_{13} = [r_{12} + r_{21} \exp(-j_2 \Phi_2)] D_2^{-1} \quad (16)$$

Incidence on Multilayer Dielectric Sheets

Multilayer radome walls are sometimes used to obtain strength or broadband performance. The general formulas for analyzing them are complicated ([9], pp. 55–59); therefore, numerical examples will be given.

Computation for a Homogeneous Sheet—Fig. 12 shows computed values of power transmittance for a flat sheet with dielectric constant 3.2 and loss tangent 0.015. The graphs show values of $|T_\perp|^2$ for perpendicular polarization and $|T_\parallel|^2$ for parallel. These values are typical for laminated composites made of quartz fabric and an epoxy resin. The graphs show that $|T|^2$ depends on thickness, incidence angle, and polarization. Fig. 13 shows transmittance for a sheet with a dielectric constant of 5.5.

For normal incidence (incidence angle zero), peaks occur at thicknesses that are half the wavelength in the dielectric; this wavelength is $\lambda_0/\sqrt{\varkappa}$, where λ_0 is the free-space wavelength. At 60° incidence, for parallel polarization, $|T_\parallel|^2$ is high for all thicknesses. However, at 60° for perpendicular polarization, $|T_\perp|^2$ is low between maxima, which occur at thicknesses greater than those for normal incidence. The low transmittances restrict bandwidths of radomes.

Although graphs like those in Figs. 12 and 13 describe $|T|^2$ for flat sheets, the

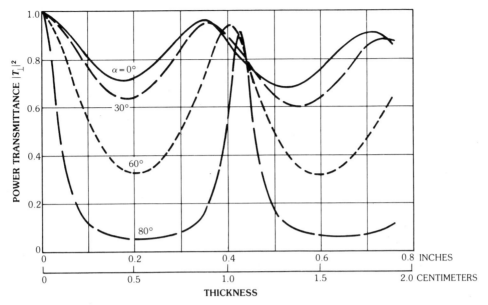

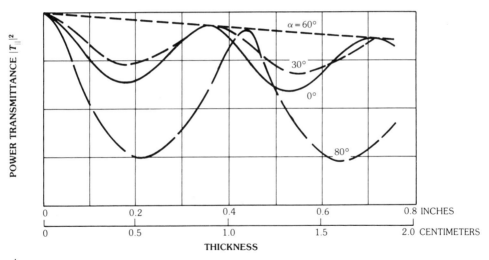

Fig. 12. Power transmittance of a flat sheet with dielectric constant 3.2 and loss tangent 0.015, at 9.375 GHz with incidence angle α. (*a*) For perpendicular polarization. (*b*) For parallel polarization.

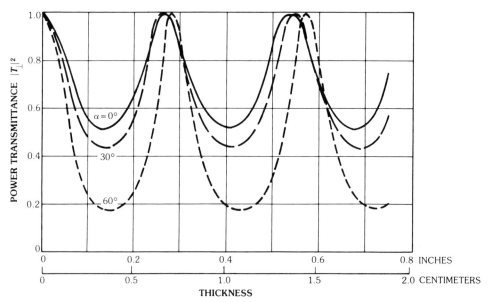

a

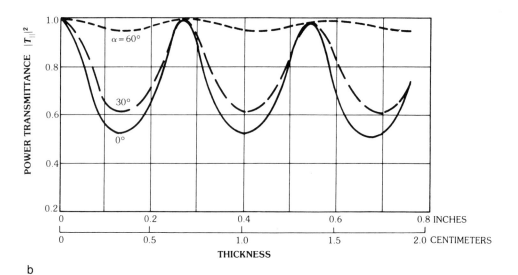

b

Fig. 13. Power transmittance of a flat sheet with dielectric constant 5.5 and loss tangent 0.0003, at 9.375 GHz with incidence angle α. (*a*) For perpendicular polarization. (*b*) For parallel polarization.

Radome Electromagnetic Design

dependence on polarization and incidence angle will be shown, in Section 5, to be significant in the design of curved radomes.

An Approximation for Reflectances of a Single-Layer Sheet—An approximation to the reflectance in (16) provides some useful insight. For simplicity consider $\tan \delta = 0$. Suppose the flat sheet is in free space, so region 3 has the same properties as region 1. Thus $r_{23} = r_{21}$. Now electromagnetic theory shows that $r_{21} = -r_{12}$. If r_{12} has small absolute value relative to unit value, then the product $r_{21}r_{23}$ can be omitted in D_2, so from (16) the reflectance is approximately

$$R^a_{13} = r_{12}[1 - \exp(-j2\Phi_2)] \quad (17)$$

The power reflectance is approximately

$$|R^a_{13}|^2 = 2r_{12}^2(1 - \cos 2\Phi_2) \quad (18)$$

The power reflectance is least, and therefore transmittance is maximized, when 2Φ is $2\pi m$, where m is an integer.

Special Cases—From (18), $|R^a_{13}|^2$ vanishes for $\Phi_2 = 0$, and, from (12), when $d_2 = 0$. In fact $|R_{13}|^2$ is small when d_2 is small enough. A sheet or radome with small thickness is commonly called a *thin-wall design*. Fig. 12 shows the high transmittance for small d/λ_d.

From (18), $|R^a_{13}|^2$ vanishes for $\Phi_2 = \pi$, or, from (12), when

$$d'_2 = \lambda/2\sqrt{\varkappa_2 - \varkappa_1 \sin^2 \alpha_2} \quad (19)$$

For thicknesses near d'_2 the reflectance is low. Sheets that satisfy this condition are called *half-wave radomes*. Fig. 12 shows the high transmittance for thickness approximately half the wavelength in the dielectric.

A-Sandwich—A common design (called an A-sandwich) consists of two thin layers, or skins, having high dielectric constant spaced by a third low dielectric constant layer called a *core*. The core thickness is approximately $\lambda_d/4$. The reflections from each sheet are identical, except for small propagation losses. Because of the quarter-wave spacing the field reflected from one thin sheet travels two quarter-wavelengths farther than the wave reflected from the other sheet; therefore, the two reflections are out of phase. In the single-layer, half-wave case, the negative sign of r_{21} (equal to $-r_{12}$) requires a path difference of two half-wavelengths for the reflections to cancel.

A-sandwich wall designs have some advantages over half-wave walls. They are light and have more broadband transmittance, for moderate incidence angles.

Fig. 14 shows computed transmittance of an A-sandwich as a function of frequency. In this figure the thicknesses are: skins, 0.035 in (0.89 μm); core, 0.25 in (6.35 mm). The dielectric constants are: skins, 3.2; core, 1.1. The loss tangents are: skins, 0.015; core, 0.005. Fig. 15 compares the transmittance of two A-sandwich flat sheets. In one case (Fig. 15a), the skins have the same thickness. In

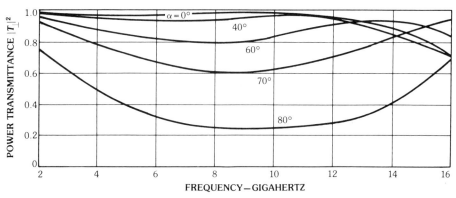

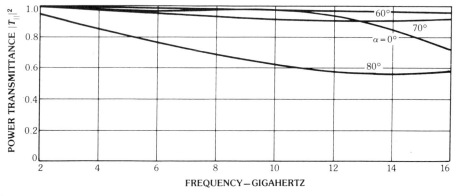

Fig. 14. Power transmittance of A-sandwich wall. The incidence angle is α. (*a*) For perpendicular polarization. (*b*) For parallel polarization.

the other case (Fig. 15b), the A-sandwich is asymmetric; the inner skin is thinner than the outer. The asymmetric design has more broadband performance.

4. General Aspects of Theoretical Design

Radome design is an analytic rather than synthetic process because antennas and radomes are complicated. They involve more variables than does synthesis of filters or one-dimensional antennas. Radome analysis starts with a candidate design, which is a statement of configuration and composition, and proceeds to evaluate performance (boresight error, transmittance, side lobe levels). The analysis may require iteration (sometimes called trial and error). The evaluation may be experimental, theoretical, or numerical; all three can be considered analytical.

Theoretical and numerical design methods evaluate performance parameters by computing the diffraction patterns of the enclosed antenna. The computations

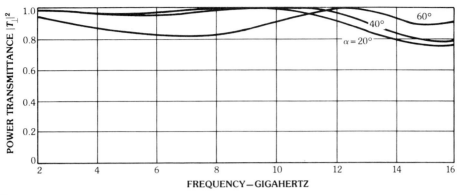

a

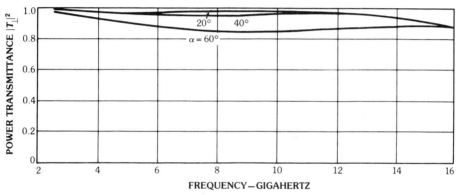

b

Fig. 15. Power transmittance of two A-sandwich panels for perpendicular polarization, with dielectric constants 2.85 (skins) and 1.15 (core), and loss tangents 0.003 (skins) and 0.004 (core). (*a*) Symmetric case: outer layers 0.04 in (1.016 mm) thick; central layer 0.30 in (7.62 mm) thick. (*b*) Asymmetric case: outer layers 0.04 in (1.016 mm) and 0.01 in (0.245 mm) thick; central layer 0.30 in (7.62 mm) thick.

are extensive because a pattern is necessary for each frequency, polarization direction, and orientation of the antenna relative to the radome. The antenna aperture distribution also is a variable. Monopulse antennas require three patterns for each frequency, polarization, and direction to describe the sum and two difference modes.

Receiving rather than transmitting operation is assumed in the following sections. Because of reciprocity, intensity patterns for reception and transmission are identical. Receiving operation justifies assuming a plane wave incident on the outside of a radome. This view seems simpler than analyzing the complicated field produced near a radome by a transmitting antenna.

Antenna pattern calculations involve two steps. The first determines the field produced in a radome-bounded region by a plane wave incident on the outside.

This step describes propagation through a dielectric shell. The second step is to couple the internal field, which is an array of Huygens sources, to the antenna [10, 11].

5. Plane Wave Propagation through Dielectric Sheets

Field distributions in radome-bounded regions are useful. Computed values are used in algorithms for computing boresight error or patterns of enclosed antennas. Comparison of computed and measured field values enables one to evaluate approximations in theories and algorithms.

The field in a radome-bounded region is evaluated on a surface. An appropriate surface for patterns calculations is a plane near and parallel to the receiving aperture; see Fig. 16. At a typical point P_R, the amplitude and phase of each rectangular field component are described by the complex-valued transmittance

$$T(P_0, P_R) = E(P_0, P_R)/E^i(P_R) \qquad (20)$$

where $E^i(P_R)$ is the incident (no radome) field component at P_R and $E(P_0, P_R)$ is the component with the radome present. The notation suggests that $T(P_0, P_R)$ depends on P_0 and P_R; it also depends on the direction of the incident wave.

This section describes two approximate methods for computing $T(P_0, P_R)$. Both methods approximate the radome locally by a flat dielectric sheet; many sheets, with varying orientations, are used to approximate the entire radome. One of the methods is simpler than the other. As might be expected, the simpler method is less accurate.

Direct-Ray Method

This method [12] is the simpler. The simplification results from assuming that field propagation in the radome-bounded region is along rays parallel to the incident ray or wave normal. We call this model the *direct-ray method*.* Fig. 16

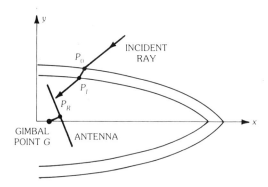

Fig. 16. A sketch of a two-dimensional direct-ray model.

---

*This name emphasizes the parallelism of the incident ray and propagated ray.

shows a direct ray. It intersects the radome at P_0, is refracted, leaves the radome at P_I, and from P_I to P_R is parallel to the incident ray. The parallelism depends on approximating a radome locally by a flat sheet as in Fig. 17. The flat sheet approximation implies multiple internal reflections, which also appear in Fig. 17. The incident ray that is refracted at P_0', reflected at P_I' and P_0, and emergent at P_I is also a direct ray. The similarity between Figs. 17 and 11 is apparent; however, many radomes have both axial and circumferential curvature, so the approximating sheet depends on P_0 and P_R.

The thickness of the approximating sheet is defined to have the same ray path length as the distance $\overline{P_I P_0}$. This thickness is $\overline{P_I P_0} \cos r$ where r is the angle of refraction calculated by Snell's law with the radome's refractive index and the incidence angle at P_0. The difference between the actual thickness and $\overline{P_I P_0} \cos r$ is negligible in many cases where curvature is not too great. For example, boresight error calculations made with the thickness values showed differences of 5 percent at peak values of 3 mrad; in general, differences were less than 2 percent.

To determine the field at P_R, start with the incident field

$$\mathbf{E}^i(P_R) = E_0^i \exp[-j\mathbf{k} \cdot \mathbf{r}(P_R)] \mathbf{e} \qquad (21)$$

where E_0^i is the scalar amplitude of the incident wave, which is assumed linearly polarized; j is $\sqrt{-1}$, k is the propagation constant; $\mathbf{k}$ is a vector with magnitude k and in the direction of the incident plane wave; $\mathbf{e}$ is a rectangular unit vector in the direction of the electric field; and $\mathbf{r}(P_R)$ is the position vector to P_R. At P_0 the incident field is resolved into components parallel and perpendicular to the plane of incidence, as suggested in Fig. 9. Let $\mathbf{1}_\perp$ and $\mathbf{1}_\parallel$ be unit vectors perpendicular and parallel, respectively, to the plane of incidence, which contains $\mathbf{k}$ and $\mathbf{n}$ and which is normal to the surface at P_0. Let $\mathbf{e} \cdot \mathbf{1}_\perp$ be α_p and $\mathbf{e} \cdot \mathbf{1}_\parallel$ be β_p. The incident field component is

$$E^i(P_0) = E_0^i[\alpha_p \mathbf{1}_\perp + \beta_p \mathbf{1}_\parallel] \qquad (22)$$

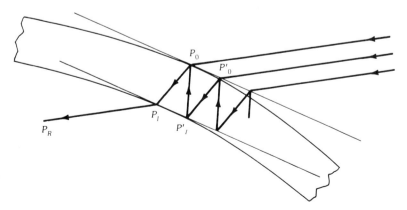

Fig. 17. An approximating flat sheet and multiple internal reflections.

(Note that α_p is not the incidence angle α.) The approximating thickness $\overline{P_I P_0} \cos r$ and the normal direction of the shell are used to compute transmittance and phase delays in (13) and (15) (for both parallel and perpendicular polarization). The field at P_I is then

$$\mathbf{E}'(P_I) = [\alpha_p E^i(P_0) |T_\perp| e^{-j\Delta_\perp}]\mathbf{1}_\perp + [\beta_p E^i(P_0) |T_\parallel| e^{-j\Delta_\parallel}]\mathbf{1}_\parallel \tag{23}$$

Assume that the receiving antenna responds to a linear field component that is parallel to the direction of a unit vector $\mathbf{1}_d$. The rectangular component of the electric field transmitted through the radome and in the direction of $\mathbf{1}_d$ is, from (23),

$$E'_d(P_I) = |T_e| E^i(P_0) \exp[-j \arg(T_e)] \tag{24}$$

where

$$|T_e|^2 = (|T_\perp|\alpha_p \delta)^2 + (T_\parallel \beta_p \delta')^2 + 2|T_\perp| \alpha_p \beta_p \delta \delta' \cos \Delta \tag{25}$$

and δ is $\mathbf{1}_\perp \cdot \mathbf{1}_d$, δ^1 is $\mathbf{1}_\parallel \cdot \mathbf{1}_d$, and Δ is $\Delta \Phi_\perp - \Delta \Phi_\parallel$. In addition,

$$\arg(T_e) = \Delta_\parallel + \tan^{-1}\{|T_\perp \alpha_p \delta|[|T_e|^2 - (|T_\perp| \delta_p \sin \Delta)]\}^{-1/2} \sin \Delta \tag{26}$$

At P_R the rectangular field component propagated through the radome and parallel to $\mathbf{1}_d$ is

$$E'_d(P_R) = T_e E^i(P_R) \tag{27}$$

The direct-ray method is adequate for radomes with antennas that have diameters bigger than approximately ten wavelengths and lengths approximately twice the diameter. However, it is inaccurate for antennas with diameter only about five wavelengths in radomes and with length about ten wavelengths. The source of error appears to be in representing Huygens' principle by a set of parallel, direct rays and using only one ray ($\overline{P_0 P_R}$) to P_R. This approximation is questionable for rays that intersect a pointed radome near its tip. The large circumferential curvature near the tip means that the direction of the surface normal changes significantly for small lateral displacements of rays near the tip. According to the direct-ray method, $T(P_0, P_R)$ would change rapidly for small lateral displacements of P_0 near the tip. To see this point, consider axial incidence, as in Fig. 18 for vertical polarization. For the ray labeled 1, the wave polarization is parallel to the plane of incidence, but for ray 2 the polarization is perpendicular. Computed values of $T(P_0, P_R)$ would vary between those for parallel and perpendicular polarization.

Measurements with apparatus like that in Fig. 3, however, show $T_e(P_0, P_R)$ varies gradually and has values between those for parallel and perpendicular polarization. These measurements were done in a systematic study that involved three radomes of distinct sizes but with similar shapes and identical dielectric constants. Fig. 19 shows the radome profiles and gives dimensions. Fig. 20 shows

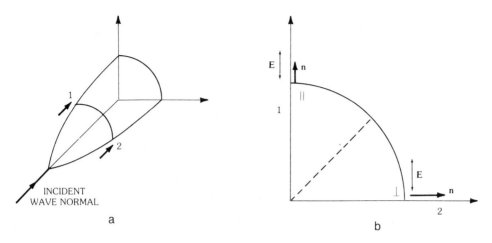

Fig. 18. Axially incident rays have distinct phase delays because the surface normal direction varies. (*a*) With axially incident rays 1 and 2. (*b*) Polarization of rays 1 and 2 with respect to surface normal.

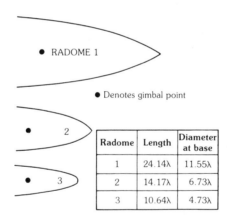

Radome	Length	Diameter at base
1	24.14λ	11.55λ
2	14.17λ	6.73λ
3	10.64λ	4.73λ

Fig. 19. Three radome configurations; all radomes had dielectric constant 5.55, and all were approximately 0.5λ thick.

insertion phase delay computed by the direct-ray method [$\arg(T_c)$ of (26)] for radome 2 of Fig. 19. The calculations differ appreciably from values measured with a small probe with aperture 1.6 by 1.0 wavelengths. The minimum in the values computed by the direct-ray method is the shadow of the tip.

For a given radome the deficiencies in the direct-ray method seem more significant for smaller antennas because the errors, in the shadow of the tip, extend over a larger portion of a small antenna.

The discrepancies between theory and measurement can be considered as caused by undersampling of the radome by the direct-ray method. The following subsection describes a method that more densely samples the tip region.

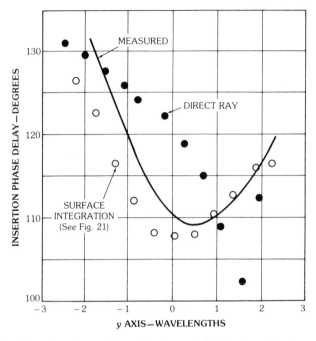

Fig. 20. Insertion phase delay of radome 2 for horizontal polarization.

Surface Integration

This method models Huygens' principle by representing the field at a typical point P_R as resulting from a set of converging rays that originate on a wavefront surface outside the radome. Fig. 21 suggests the difference between the direct-ray method and surface integration; the figure is two-dimensional but the converging rays arise from an area, not a line.

The transmittance for surface integration uses the flat sheet approximation at each of a set of points P_0; these are designated by the subscript m. The transmittance for surface integration is a sum corresponding to converging rays; it is

$$T_{si}(P_{0m}; P_R) = \sum_{M=1}^{M} T(P_{0m}, P_R) \tag{28}$$

The sum computationally represents an integral over the wavefront. The number of rays M and their distribution influence the transmittance. Determination of M and the distribution of rays is a truncation and sampling problem.

Comparison of measured and computed values showed that good accuracy was obtained by bounding the integration area to the first Fresnel zone centered about P_R. The field is then

$$E_d^t(P_R) = T_{si} E^i(P_R) \tag{29}$$

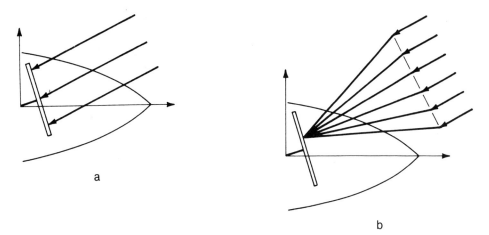

Fig. 21. Two methods for computing transmittance. (*a*) Direct-ray method. (*b*) Surface integration.

Fig. 20 shows phase delay values computed by surface integration. In this case and in many others, surface integration is more accurate than the direct-ray method. However, for sufficiently large antennas and radomes the direct-ray method gives useful predictions of boresight error. The direct-ray method has the advantage of lower computer cost than surface integration.

6. Antenna Patterns

Although Huygens' principle was generally described in Chapters 1 and 2, this section utilizes a special form suitable for computing patterns of antennas in a radome [13].

Pattern computations are based on considering the wavefront in the radome-bounded region as an array of secondary sources. Schelkunoff's induction theorem states that each secondary source is a combination of electric and magnetic current elements with moments that are proportional, respectively, to the magnetic and electric intensities tangential to the wavefront. For a wave in a homogeneous medium, the induction theorem becomes the equivalence theorem, and the densities of the magnetic current **M** and electric current **J** are

$$\mathbf{M} = \mathbf{E}^i \times \mathbf{1}_s \tag{30}$$

and

$$\mathbf{J} = \mathbf{1}_s \times \mathbf{H}^i \tag{31}$$

where $\mathbf{E}^i$ and $\mathbf{H}^i$ are, respectively, the electric and magnetic intensities of the incident plane wave, and $\mathbf{1}_s$ is a unit vector normal to the wavefront. For a linearly polarized wave with rectangular components E_x and H_y, each element of area dA

on the wavefront is an elementary source, a Huygens' source. If $E_x = \eta H_y$, with η being the intrinsic impedance of free space, then in a spherical polar coordinate system with the polar axis along the z axis, and ϕ the azimuthal angle measured in the xy plane, the ϕ component of the distant radiated field is

$$E_\phi = f(r, \phi) \, E \times dA \qquad (32)$$

where r is the distance from the element to the observation point. Thus the radiated field is expressed in terms of a product. One factor describes the source alone, and the other describes the relative positions of the source and observation points and the orientation of the source.

For a wavefront in the volume bounded by the radome, each element of area is an elementary source and, thus, is equivalent to an electric current $\mathbf{J}$ and a magnetic current $\mathbf{J}_m$, where $\mathbf{J} \propto \mathbf{H}$, and $\mathbf{J}_m \propto \mathbf{E}$. Here $\mathbf{E}$ and $\mathbf{H}$ are, respectively, the electric and magnetic intensities tangential to the wavefront.

Now consider an aperture antenna receiving the wavefront. The antenna is connected to a detector by a waveguide. Let the field in the waveguide at P', due to an element of unit strength at a point P_R, be

$$dE_g = F(P_R, P') E'_d(P_R) \, dx \, dy \qquad (33)$$

where E'_d is given by (27) or (29). The function F is the receiving aperture distribution, which can be measured, at least approximately [13].

The field due to the entire incident wave is

$$E_g = \iint F(P_R, P') E'(P_R) \, dx \, dy \qquad (34)$$

The integral extends over that portion of the incident wavefront such that F has appreciable values. This surface is flat. The phase variation over this surface is given by that phase of T_{si} in (29) or T_e in (27).

Power patterns are obtained from $|E_g|^2$. Transmittance is obtained from peak values of the main beam, by comparing values with and without the radome.

Boresight error calculation depends on the antenna. For conical scan antennas, error is obtained from an intensity minimum determined from overlapping patterns for distinct beam direction. For amplitude comparison monopulse, boresight error is obtained from angular shifts of power pattern minima. For phase comparison monopulse, the shift is in the direction giving a quadrature relation of sum and difference patterns.

It is assumed that a reasonable approximation to F can be measured with a suitable probe for any particular receiving antenna. Call this approximation F_m.

In practice,

$$E_g = \sum_{j,k} F_m(P_R, P') \, E'_d(P) \qquad (35)$$

The sum in (35) is the evaluation of the integral for E_g. Dimensions are preserved on omitting the differentials by factoring the distance between equally spaced points P and P'. This distance is omitted for convenience. The preceding formulation is largely physical. It is suggested by and based on the induction theorem. Its practical use depends on knowing the function F_m. Some assumptions have been made on the nature of the secondary sources. It is assumed that for the wavefront propagated through the dielectric shell, $E = \eta H$ at any point; it is not assumed that E is a constant over some plane.

Another factor in actual cases is reflected energy. Fields reflected from the receiving antenna would change the field on the aperture plane, changing the strength of a secondary source. In addition, fields which enter the antenna after reflection from the radome must be considered; for example, if side lobe estimates are being made. Fig. 2 shows the path of a possible reflected ray which represents energy above that received from a collimated ray bundle entering the antenna after passing through the radome without reflection. Reflected rays may be very significant in some cases.

7. Examples of Boresight Error

For radome 1 of Fig. 19, Fig. 22 compares measured boresight error and values computed by the direct-ray method for the antenna axis rotated parallel and perpendicular to the plane of the electric field. Fig. 22 shows calculations made with an assumed uniform amplitude distribution that had a phase reversal on opposite sides of the aperture. Fig. 22 also shows calculations made with an aperture distribution that was measured with a small probe at a distance of four wavelengths and backward propagated to the aperture. The direct-ray method gives reasonable accuracy, and aperture distribution affects accuracy [14].

Fig. 23 shows boresight error for radome 3, the smallest in Fig. 19. The direct-ray method failed, but surface integration gave accurate predictions.

Surface integration also was applied to radome 1; it was more accurate than the direct-ray method.

Some alternative methods have been described [15–17].

8. The Moment Method

The analytical methods of Sections 5 and 6 are approximate. Propagation is described by direct rays or surface integration; both methods approximate the radome as locally plane. Guided waves are omitted. Coupling of the wavefront to the antenna is approximate.

A more accurate description is given by the moment method [18]. This method computes the field in a radome for a given incident field from the solution of an integral equation. The external fields, near or far, are computed from the internal field values.

In Richmond's formulation the total field $\mathbf{E}^T$ is defined as the sum of the incident (known) field and the scattered field $\mathbf{E}^s$; that is,

$$\mathbf{E}^T = \mathbf{E}^i + \mathbf{E}^s \tag{36}$$

The scattered field follows from vector and scalar potentials which are integrals over polarization currents. Thus

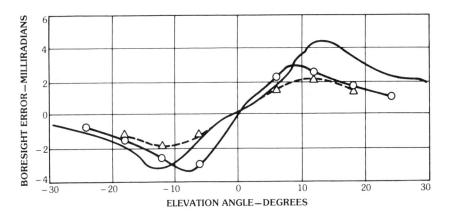

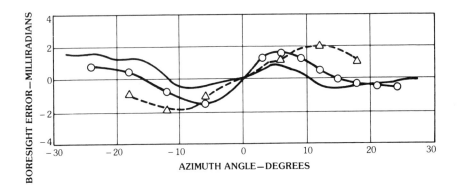

Fig. 22. In-plane boresight error of radome 1: measured (—); computed, uniform aperture amplitude (△); computed, measured aperture distribution (○).

$$\mathbf{E}^s = -j\omega\mathbf{A} - \operatorname{grad}\phi_s \tag{37}$$

$$\mathbf{A} = (\mu_0/4\pi)\int g\mathbf{J}\,dV \tag{38}$$

and

$$\phi_s = (4\pi\epsilon_0)^{-1}\int g\varrho\,dV \tag{39}$$

where ϱ is $-\epsilon_0\nabla\cdot(\varkappa - 1)\mathbf{E}$, $\mathbf{J}$ is $j\omega\epsilon_0(\varkappa - 1)\mathbf{E}$, and g is $r^{-1}\exp(-jkr)$. Notice that the unknown $\mathbf{E}^T$ appears in the potentials and thus in $\mathbf{E}^s$ by (37). When $\mathbf{E}^s$ from (37) is used in (36), the integral equation results.

To solve the integral equation the radome is subdivided into small volumes, or cells, and the field in each cell is assumed constant.

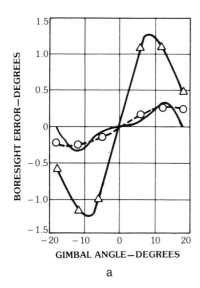 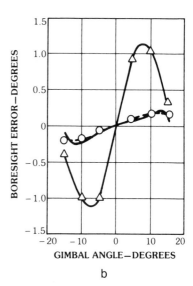

Fig. 23. Boresight error of radome 3: measured (—); computed direct-ray (△); computed surface integration (○). (*a*) *E*-plane. (*b*) *H*-plane.

Richmond gave explicit calculations for dielectric cylinders of arbitrary shape for both transverse electric and transverse magnetic fields. The arbitrarily shaped cylinders were represented by cells that were circular cylinders. The cells should be small relative to the wavelength to justify assuming that the field is constant in any cell. The integrals for **A** and ϕ_s reduce to sums of integrals; each integral is a sum that extends through a cell. The unknown $\mathbf{E}^T$ can be factored from under the integrals and appears as a factor. The integral equation is evaluated for each cell to reduce it to a set of simultaneous algebraic equations. Computations based on Richmond's method give good agreement with measured near fields; of course, cells must be small enough [19].

Recently the moment method has been applied to hollow cones [20, 21]. The cells are angular sectors of annuli, so ogival radomes can be analyzed. Measured near fields agreed well with calculations.

Although the moment method is accurate, current computers limit matrix inversion, so analysis is feasible for small cones; lengths are a few wavelengths. Iterative methods are being studied. Despite some limitations the moment method is powerful; it has estimated strengths of guided waves on dielectric slabs.

9. Comments on Materials

Although the selection of a material is significant for radome design and development, this chapter omits materials except to list some standard materials in Table 1. The dielectric constant values are approximate. The values depend on temperature. Descriptions of electromagnetic, thermal, and mechanical properties are available [22].

Table 1. Common Radome Materials

Material	Dielectric Constant
Laminate: epoxy resin, E-glass	5 ± 0.3
Laminate: polyamide resin, quartz fabric	3.3 ± 0.2
Ceramic: Pyroceram (Corning Glass Works)	5.5 ± 0.1
Ceramic: slip cast fused silica	3.2 ± 0.1

Acknowledgments

The results and methods described in this chapter were obtained with considerable help from the author's colleagues, R. A. Hayward and E. L. Rope. Helpful discussions were held with T. E. Fiscus of our laboratory, L. E. Weckesser of the Johns Hopkins University's Applied Physics Laboratory, G. Tatnall and K. Foulke of the US Naval Air Development Center, and D. Paris and K. Huddleston of the Georgia Institute of Technology; many others helped. The Naval Air Systems Command and James Willis of the Command supported much of the work.

10. References

[1] T. J. Lyon, "Operational radomes," chapter 7 in *Radome Engineering Handbook*, ed. by J. D. Walton, New York: Marcel Dekker, 1970.

[2] G. Tricoles and E. L. Rope, "Circular polarization radome technique," *Tech. Rep. AFAL-TR-69-164*, US Air Force Avionics Lab, Wright-Patterson AFB, Ohio 45433, August 1969.

[3] D. G. Bodnar and H. L. Bassett, "Analysis of an anisotropic dielectric radome," *IEEE Trans. Antennas Propag.*, vol. AP-23, pp. 841–846, 1975.

[4] E. A. Pelton and B. A. Munk, "A streamlined metallic radome," *IEEE Trans. Antennas Propag.*, vol. AP-22, pp. 799–803, 1974.

[5] G. Tricoles and E. L. Rope, "Guided waves in a dielectric slab, hollow wedge, and hollow cone," *J. Opt. Soc. Am.*, vol. 55, pp. 328–330, 1965.

[6] G. Tricoles and N. L. Farhat, "Microwave holography: application and techniques," *Proc. IEEE*, vol. 65, pp. 108–121, 1977.

[7] R. A. Hayward, E. L. Rope, and G. Tricoles, "Radome boresight error and its relation to wavefront distortion," *Proc. 13th Symp. on Electromagnetic Windows*, ed. by H. L. Bassett and J. M. Newton, pp. 87–89, Georgia Institute of Technology, Atlanta, 1976.

[8] J. A. Stratton, *Electromagnetic Theory*, New York: McGraw-Hill Book Co., 1941, pp. 492–497.

[9] M. Born and E. Wolf, *Principles of Optics*, New York: Pergamon Press, 1964.

[10] B. B. Baker and E. T. Copson, *The Mathematical Theory of Huygens' Principle*, 2nd ed., Oxford: Oxford University Press, 1953.

[11] S. A. Schelkunoff, "Kirchoff's formula, its vector analogue, and other field equivalence theorems," in *Theory of Electromagnetic Waves*, ed. by M. Kline, New York: Dover Publications, 1951, pp. 43–59.

[12] G. Tricoles, "Radiation patterns and boresight error of a microwave antenna enclosed in an axially symmetric dielectric shell," *J. Opt. Soc. Am.*, vol. 54, pp. 1094–1101, 1964.

[13] G. Tricoles, "Radiation patterns of a microwave antenna enclosed by a hollow dielectric wedge," *J. Opt. Soc. Am.*, vol. 53, pp. 545–557, 1963.

[14] R. A. Hayward, E. L. Rope, and G. Tricoles, "Accuracy of two methods for radome analysis," *Dig. 1979 IEEE Antennas Propag. Symp.*, pp. 598–601. IEEE cat. no. 79CH1456-3 AD.

[15] K. Siwiak et al., "Boresight error induced by missile radomes," *IEEE Trans. Antennas Propag.*, vol. AP-27, pp. 832–841, 1979.

[16] D. T. Paris, "Computer aided radome analysis," *IEEE Trans. Antennas Propag.*, vol. AP-18, pp. 7–15, 1970.

[17] D. G. Burks, E. R. Graf, and M. D. Fahey, "A high frequency analysis of radome-induced radar pointing error," *IEEE Trans. Antennas Propag.*, vol. AP-30, pp. 947–955, 1982.

[18] J. H. Richmond, "Scattering by a dielectric cylinder of arbitrary cross section shape," *IEEE Trans. Antennas Propag.*, vol. AP-13, pp. 334–341, 1965; "TE wave scattering by a dielectric cylinder of arbitrary cross section shape," *IEEE Trans. Antenna Propag.*, vol. AP-14, pp. 460–464, 1966.

[19] G. Tricoles, E. L. Rope, and R. A. Hayward, "Wave propagation through axially symmetric dielectric shells," *General Dynamics Electronics Division Report R-81-125*, final report for Contract N00019-79-C-0638, San Diego, June 1981.

[20] G. Tricoles, E. L. Rope, and R. A. Hayward, "Electromagnetic waves near dielectric structures," *General Dynamics Electronics Division Report R-83-047*, final report for Contract N00019-81-C-0389, San Diego, February 1983.

[21] G. Tricoles, E. L. Rope, and R. A. Hayward, "Electromagnetic analysis of radomes by the moment method," *Proc. Seventeenth Symp. Electromagnetic Windows*, ed. by H. Bussell, pp. 1–8, Georgia Institute of Technology, Atlanta, Georgia 30332.

[22] J. D. Walton, ed., *Radome Engineering Handbook*, New York: Marcel Dekker, 1970.

Chapter 32

Measurement of Antenna Radiation Characteristics on Far-Field Ranges

E. S. Gillespie
California State University, Northridge

CONTENTS

1. Introduction — 32-3
2. Radiation Patterns and Pattern Cuts — 32-6
3. Antenna Ranges — 32-8
   Positioners and Coordinate Systems   32-8
   Antenna Range Design Criteria   32-14
   Outdoor Ranges   32-19
   Indoor Ranges   32-25
   Instrumentation   32-30
   Range Evaluation   32-33
4. Measurement of Radiation Characteristics — 32-39
   Amplitude Patterns and Directivity Measurements   32-39
   Antenna Gain Measurements   32-42
   Polarization Measurements   32-51
5. Modeling and Model Measurements — 32-63
   Theory of Electromagnetic Models   32-65
   Materials for Electromagnetic Models   32-69
   Scale-Model Construction, Instrumentation, and Measurement   32-74
6. References — 32-87

Edmond S. Gillespie received the BEE degree from Auburn University in 1951 and the MS and PhD degrees from UCLA in 1961 and 1967, respectively. He is a registered professional engineer in the State of California.

From 1951 through 1955 Dr. Gillespie was a member of the staff of the Sandia Corporation, Albuquerque, New Mexico, principally engaged in the development of airborne antennas. In 1955 he joined the Lockheed-California Company, where he was primarily concerned with the design of aircraft antennas and also directed the research activities of the antenna laboratory. During parts of 1961–1963, he was on leave from Lockheed attending UCLA, where he was appointed an associate in engineering. He taught courses in solid-state electronics, electromagnetics, and microwave engineering. In 1963 he left Lockheed, where he had held the position of Senior Research Specialist, and joined the UCLA Institute of Geophysics and Planetary Physics as a senior engineer. He was engaged in a study of the scattering of surface waves by obstacles. In 1965 he joined the faculty of the School of Engineering and Computer Science at the California State University, Northridge. Currently he is a professor of electrical engineering and Director of the Microwave and Antenna Engineering Program.

Dr. Gillespie is a Fellow of the IEEE, and he is a member of many technical societies and committees. He was responsible for the development of ANSI/IEEE Standard 149-1979, *Test Procedures for Antennas*, and IEEE Standard 145-1982, *Definitions of Terms for Antennas*.

1. Introduction

To assess the performance of a communication, radar, or telemetry system, it is necessary to know the characteristics of the antennas employed to within specified confidence limits. This usually requires a measurement program, the extent of which depends on the severity of the system requirements as well as economic factors. Chart 1 is a tabulation of *antenna** characteristics that are often measured.

For a simple line-of-sight communications system one might be required to measure only a limited number of amplitude patterns and the peak gain of the antennas. For a spacecraft antenna or an airborne radar antenna, however, an extensive measurement program encompassing many of the entries of Chart 1 may be required. Antennas which are designed to operate at frequencies above about 100 MHz usually can be tested at antenna test facilities away from their operational environment. There are exceptions such as large earth station and radioastronomy antennas which must be tested *in situ* [2]. On-site measurements of this type are beyond the scope of this chapter, although much that is presented is applicable. The characteristics of some types of antennas are greatly affected by their immediate environment. For example, a common situation where this occurs is when antennas are located on large supporting structures such as aircraft, spacecraft, or ships, the physical size of which may preclude their being tested at an antenna test facility. In those cases, scale-modeling techniques can be employed and the modeled structure can be treated as an integral part of the antenna being measured ([2], pp. 56–58). In this way the properties of the reduced-size model system can be measured at an antenna test facility.

The various measurements indicated by the entries in Chart 1 require specialized equipment, facilities, and personnel to perform them. This is particularly obvious for the major headings of the tabulation; however, it is also the case for the measurements listed under them. For example, consider those labeled as radiation characteristics. *Noise temperature*, *EIRP*, and *G/T* measurements are generally performed on-site and are, in fact, system measurements since they usually include amplifiers and transmitters along with the antennas being tested. Typically, there are separate, dedicated facilities, referred to as antenna ranges, for *boresight* and *radome* measurements, for *radar cross-section* measurements, and for amplitude and phase *radiation pattern* measurements. These ranges include a source antenna to illuminate the test antenna, positioners to support and orient them, the electronic instrumentation, and the physical space between and around the antennas.

---

*Throughout this chapter, the definitions of terms employed will conform to Institute of Electrical and Electronics Engineers' (IEEE) standards [1, 2], and those appearing in reference 1 will be italicized the first time they appear in the text.

Chart 1. Commonly Performed Antenna Measurements

1. Radiation Characteristics
    Amplitude and power radiation patterns
    Phase
    Directivity
    Gain
    Polarization
    Radiation efficiency
    Boresight
    Radome
    Equivalent isotropically radiated power (EIRP)
    Figure of merit (G/T)
    Noise temperature
    Passive intermodulation
    Scattering/radar cross section

2. Circuit Characteristics
    Self-impedance/reflection coefficient
    Mutual impedance

3. Power-Handling Capability
    Metallic and dielectric heating
    Voltage breakdown/corona
    Multipacting

4. Structural Integrity
    Vibration
    Surface deformation

5. Electromechanical
    Positioning accuracy
    Agility
    Et cetera

6. Environmental
    Wind
    Rain
    Salt spray
    Temperature
    Altitude
    Et cetera

This chapter is primarily concerned with *far-field* radiation pattern ranges and the measurements that are performed on them, namely, amplitude and *phase pattern*, *directivity*, *gain*, and *polarization* measurements. Far-field antenna ranges are those which are designed to produce a plane or quasi-plane wave over the *aperture* of the antenna being tested, thus allowing one to measure the far field of that antenna directly. This definition not only includes those ranges whose lengths or spacings between the source and test antenna are great enough so that far-field conditions are met, but also compact ranges. These latter ranges are ones in which the quasi-plane wave is achieved by collimating the field of a small source antenna by use of a large lens or reflector. Since the planar field exists in the aperture region of the *lens* or *reflector*, the required range length is reduced to relatively short distances [2, 3].

The success of any measurement is critically dependent on the design of the measurement system employed. For antenna measurements this includes not only the instrumentation but the space between and around the source and test antennas. Since the range length can be, in some cases, hundreds of meters in length, a substantial investment in land may be required in addition to that in instrumentation. Furthermore, the most serious sources of measurement error are introduced in the link between the source and test antennas. The causes of these errors include the finite range length, unwanted specular reflections from surfaces and objects, multipath between antennas, adverse weather conditions, and extraneous interfering signals. These effects cannot be completely eliminated; but, rather, the ranges should be designed in such a way as to reduce them to acceptable levels. Outdoor ranges pose particular difficulties because of weather conditions such as wind, rain, snow, and humidity, which cannot be controlled. However, detection systems can be installed to determine wind and moisture levels to alert the operator when measurements should be suspended. All other effects can be controlled by good antenna range design and management.

The design criteria for antenna ranges were fairly well established during World War II after the introduction of microwave radar systems. This work was summarized by Cutler, King, and Kock [4]. Hollis and his colleagues have refined and extended these criteria as well as antenna measurement techniques in general. Their work, including the work of other individuals, is presented in their landmark book, *Microwave Antenna Measurements* [5]. Johnson, Ecker, and Hollis [6], and Kummer and Gillespie [7], have published significant review papers on antenna measurements. IEEE Standard 149-1979, *Test Procedures for Antennas* [2], contains further extensions of these works and reference 8 is an excellent presentation on antenna measurements. The reader is referred to these sources for details that may be omitted here.

In this presentation the basic criteria for the design of antenna ranges are given. These form a guide for the range designer. No attempt has been made to present engineering details of range design, which are too varied and specialized to be included. The quality of an antenna range is dependent on the designer's knowledge and attention to operational details. When designing a test facility, one must literally visualize, step by step, the measurement procedures to be employed on a proposed range and provide a design that will allow the experimenter to perform measurements in an accurate and, where possible, convenient manner. Designing the range for anticipated future use is most important because of the large capital outlay that is required and the difficulty one has in making later modifications to the range.

The measurement procedures described are presented in the same manner, that is, only the basic principles are presented. Using them the experimenter can design the measurement procedure for the particular facility where the measurement is to be performed. The first step is usually the construction of an error budget based on a knowledge of the instrumentation employed, the physical layout of the antenna range, and the limits of uncertainty required for the measurement. This will guide the experimenter in the detail design of the experiment.

Because of the extreme measurement accuracy required by modern systems employing antennas and the resulting sophistication of the instrumentation required to achieve such results, the antenna measurements engineer must possess a broad knowledge of many fields. These include electromagnetics, electronics, digital and analog control, metrology, and data processing, to name a few. Like all engineering work the art is as important as the scientific aspects and it can only be obtained by experience.

2. Radiation Patterns and Pattern Cuts

The radiation pattern of an antenna is the spatial distribution of any of several quantities which characterize the field excited by the antenna. These quantities include *power flux density*, *field strength*, *radiation intensity*, *directivity*, gain, phase, and *polarization*. It is not practical to measure these quantities at every point in space, but, rather, one must sample them over a defined measurement surface. The most common surfaces are planar, cylindrical, and spherical. Any path over a measurement surface is called a *pattern cut* and the distribution of any of the above-mentioned quantities over a pattern cut is also called a radiation pattern. Practical considerations usually dictate that the cuts coincide with the natural coordinates of the measurement surface.

Planar and cylindrical measurement surfaces find their greatest use for measurements in the *near-field region* of an antenna (see Chapter 33), whereas the spherical surface can be employed for either near-field or far-field measurements. The spherical surface that is appropriate for far-field measurements is the *radiation sphere* of an antenna. The location of points of the radiation sphere are given in terms of the θ and ϕ coordinates of a standard spherical-coordinate system whose origin is located at the center of the sphere.

There are two standard types of pattern cuts over the radiation sphere. The path formed by the locus of points for which θ is a specified constant and ϕ is a variable is called a conical cut or ϕ-cut. The path formed by the locus of points for which ϕ is a specified constant and θ is the variable is called a great-circle cut or θ-cut (see Fig. 1).

In a practical measurement situation there must be a probe antenna to sample the field as depicted in Fig. 2. The probe's receiving coordinate system is the $x'y'z'$ Cartesian coordinate system as shown. Note that the z' axis is pointing in the radial direction which is determined by the values of θ and ϕ of the radiation sphere's coordinate system. The rotation of the probe about the z' axis is measured with respect to the unit vector $\hat{\theta}$ with the resulting angle designated as χ. The angles θ, ϕ, and χ are simply the Euler angles that relate the two coordinate systems. Continuous variations of θ or ϕ will yield radiation patterns, whereas a continuous variation of χ will result in a *polarization pattern*.

Among the decisions that an experimenter must make are the polarization of the probe and its path over the radiation sphere. In addition, the orientation of the antenna being tested with respect to the spherical-coordinate system of the radiation sphere must be specified. There are two common choices: the beam axis can be oriented toward one of the poles (usually the north pole) or the equator (usually either the x or the y direction) as shown in Fig. 3.

For directional antennas which are *linearly polarized*, the choice depends on the two orthogonally polarized field components to be measured over the radiation

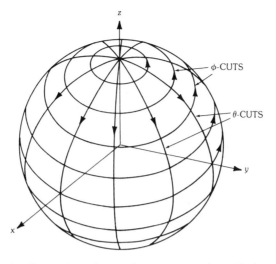

Fig. 1. An illustration of θ- and ϕ-cuts over the radiation sphere.

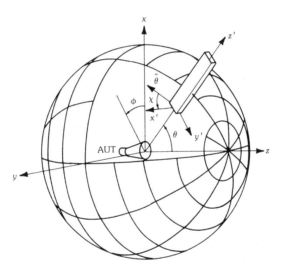

Fig. 2. An illustration of the coordinate systems of the antenna under test (AUT) and the probe used to sample the field over the radiation sphere.

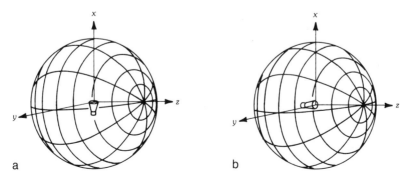

Fig. 3. Standard orientations of the antenna under test with respect to the spherical-coordinate system of the radiation sphere. (*a*) Equatorial orientation. (*b*) Polar orientation.

sphere. For example, if the two components that are to be measured are E_θ and E_ϕ, then the orientation of the antenna's *beam axis* should be toward the equator. If it were oriented toward the pole, then for conical cuts in the vicinity of the pole, one would simply obtain polarization patterns, and for great-circle cuts there would be drastic changes in the direction of polarization from cut to cut.

On the other hand, if Ludwig's third definition of *copolarization* and *cross polarization* are chosen [9], one would orient the beam axis toward the north pole of the radiation sphere. A linearly polarized probe would be employed and orientated for a *polarization match* at the pole. Note that all θ-cuts which are specified by the value of ϕ pass through the pole and each forms a different angle χ_0 with respect to the probe. The procedure is to keep the angle χ_0 constant over each θ-cut (great-circle cut) as well as equal to that angle obtained at the pole with the probe oriented for a polarization match. In this way, for each great-circle cut, there is a polarization match at $\theta = 0$. This procedure will result in a measurement of the *copolar radiation pattern* over the radiation sphere as specified by Ludwig's third definition. To obtain the *cross-polar radiation pattern*, one rotates the probe 90° at the pole and repeats the procedure with the probe oriented to $\chi_0 + 90°$ with respect to the axis where χ_0 is the angle for the copolar pattern.

It should be kept in mind that the copolarization of an antenna is a defined quantity. Systems considerations usually dictate its choice. With the availability of modern computing equipment, one can measure any two orthogonal far-field components of the antenna's field and their relative phases and then compute the components that correspond to any specified copolarization and cross polarization.

3. Antenna Ranges

An antenna range is a laboratory facility which consists of the instrumentation and physical space required to measure the radiation patterns of antennas away from their operational environments. For far-field measurements it is usual for the antenna being tested to be operated in the receive mode. The probe antenna discussed in the previous section is then replaced by a source (transmitting) antenna whose radiation pattern and polarization are defined by the particular measurement problem for which it is employed. Unless otherwise specified, the antenna under test will be considered to be operated in the receive mode for the remainder of this chapter.

Positioners and Coordinate Systems

Two orthogonal rotational axes are required in order to provide the required relative motion of the source antenna with respect to the antenna being tested. Fig. 4 illustrates these axes and their relation to the measurement coordinate system. The antenna under test is located at the origin and is fixed relative to the coordinate system. The source antenna is located at point S, with OS being the line of sight between the source antenna and the antenna under test. The line OA is the θ rotational axis and OZ is the ϕ rotational axis. Note that both OS and OZ are always perpendicular to OA.

There are two basic means of achieving the required motions about the θ and ϕ rotational axes. One approach is to move the source antenna physically over a

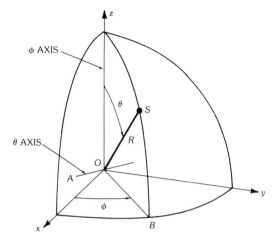

Fig. 4. The two orthogonal rotational axes required for a spherical-coordinate positioning system.

prescribed path. Ranges that employ this approach are called movable line-of-sight ranges. The second approach is one where the source antenna is fixed and the antenna under test is provided with a means of rotation about the two orthogonal axes, OA and OZ. Ranges employing this approach are called fixed line-of-sight ranges and they are the more common of the two.

The movable line-of-sight ranges can be realized by the use of a vertical arch that is either fixed or mounted on an azimuthal positioner. The antenna under test is usually mounted on an azimuthal positioner or a turntable so that its motion can be independent of that of the source antenna, as shown in Fig. 5a. There are two modes of operation: (1) The antenna under test is rotated incrementally in ϕ and the source antenna moves continuously over the arch for each θ-cut. (2) The source antenna is moved incrementally in θ and the antenna under test is rotated continuously for each ϕ-cut. If the arch is mounted on an azimuthal positioner then, with the antenna under test remaining fixed, the source antenna is moved incrementally in θ and the arch is rotated continuously for each ϕ-cut as depicted in Fig. 5b. An alternate approach is to use a gantry arrangement or simply a rotating arm to position the source antenna as shown in Fig. 5c. Note that the rotating arm could have been oriented so that the motion of the source antenna is in the horizontal plane. Also, not shown in Fig. 5 is the microwave-absorbing material that would be required to cover the reflecting surfaces of the positioners.

The movable line-of-sight ranges find their greatest use for testing vehicular antennas, for measuring the radiation patterns of scale-model antennas mounted on model ships, and for locating the *phase centers* of feed antennas for reflector antennas. The first two uses require very large arches of the type illustrated in Fig. 6 [10, 11, 12], which is known as an image-plane range (see discussion under "Outdoor Ranges"). A range employing a horizontal rotating arm is more appropriate for locating phase centers. To accomplish this, the antenna under test is moved longitudinally in increments until the measured phase is nearly constant over a cut through the peak of its *main beam*.

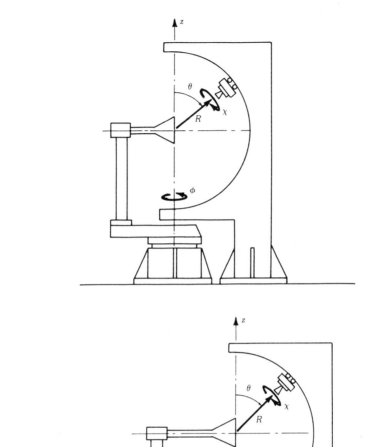

Fig. 5. Possible movable line-of-sight ranges. (*a*) AUT mounted on turntable. (*b*) With AUT fixed. (*c*) Source antenna on rotating arm.

The fixed line-of-sight ranges are by far the most commonly used ranges. They require a spherical-coordinate positioner for the antenna under test and a polarization positioner for the source antenna (unless its polarization can be controlled electronically). Each type of positioner has a coordinate system associated with it. The antenna under test should be mounted on the positioner in such a manner that its coordinate system properly coincides with that of the positioner; otherwise, errors will result.

There are three basic positioner configurations that provide the θ and ϕ

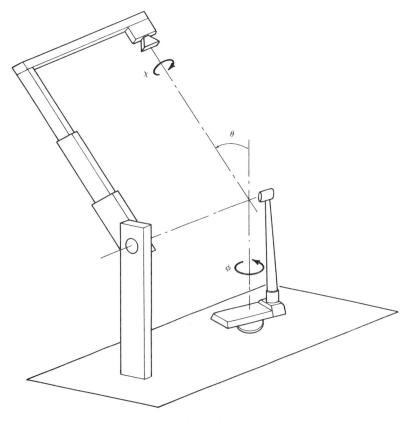

Fig. 5, *continued.*

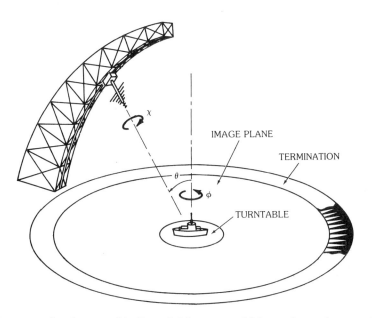

Fig. 6. An example of a movable line-of-sight range which employs a large arch. This is known as an image-plane range.

rotations. They are (*a*) azimuth-over-elevation, (*b*) elevation-over-azimuth, and (*c*) roll-over-azimuth. These positioners are illustrated in Figs. 7a, 7b, and 7c along with their respective coordinate systems. There is a practical restraint on the first two types, which employ elevation motions, namely, the antenna under test can not be mounted at the axes-crossing point which is the center of the positioner's coordinate system. In many instances the resulting error is negligible. The roll-over-azimuth positioner does not present this difficulty and hence is the most widely used of the three.

More elaborate systems are possible. For example, a three-axis positioner such as a roll-over-azimuth-over-elevation system can be operated in two modes. The roll axis can be auxiliary and the azimuthal positioner can provide the motion about the ϕ rotational axis, while the elevation positioner provides the motion about the θ rotational axis. Alternatively, the roll positioner can provide the ϕ motion, with the θ motion being provided by the azimuthal positioner. The elevation positioner can conveniently provide a tilt axis for alignment purposes.

Reference to Fig. 4 reveals that for fixed line-of-sight ranges, the θ rotational axis of the positioner's coordinate system must be the lower of the two axes; that is, it is the axis whose orientation is independent of the motion about the other axis. This is especially important when assigning the coordinate axes to a three-axis positioner.

A polarization positioner is one which provides motion about the roll axis of the source antenna. It is used to rotate the plane of polarization of a linearly

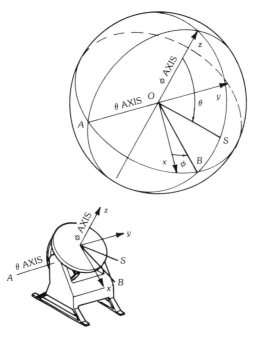

a

Fig. 7. The three basic spherical-coordinate positioners. (*a*) Azimuth-over-elevation positioner. (*b*) Elevation-over-azimuth positioner. (*c*) Roll-over-azimuth positioner.

Measurement of Antenna Radiation Characteristics on Far-Field Ranges

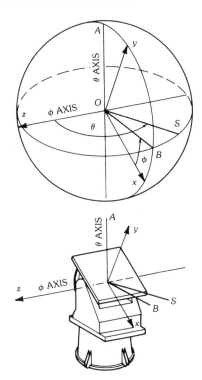

b

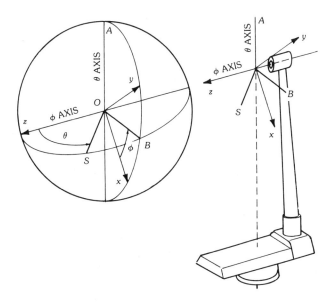

c

Fig. 7, *continued.*

polarized source antenna. A means of tilting the source antenna is usually provided for use in alignment.

Antenna Range Design Criteria

When the far field of an antenna is being measured in its receive mode, the ideal incident wave at the antenna is a uniform plane wave. Antenna ranges are designed to provide a close approximation of an incident plane wave at the test region. To accomplish this within acceptable error limits, the following must be considered ([5], pp. 14.1–14.41).

Reactive Field Coupling—The spacing between the source antenna and the antenna under test, R, must be large enough so that the reactive part of the field at the test region is negligible. This is rarely a problem except for electrically small antennas operating at low frequencies. From the field equations for a *Hertzian electric dipole*, it can be shown that the ratio of its inductive field to its radiation field is $\lambda/2\pi R$. This ratio is -36 dB at a separation of 10λ; this is the basis of the commonly accepted criterion

$$R \geqq 10\lambda \tag{1}$$

for antenna ranges to reduce reactive coupling to acceptable levels.

Reradiative Coupling—This type of coupling occurs because part of the energy received by the antenna under test is radiated back toward the source antenna. In turn, part of this energy is received by the source antenna and is reradiated back to the antenna under test. These multiple reflections can produce errors in the measurement of radiation patterns, especially in the measured level of *side lobes*. For a typical case it has been shown that the signal returning to the antenna under test after multiple reflection will be at least 45 dB below that of the original incident signal, provided the amplitude taper of the incident field across the aperture of the antenna under test is no greater than 0.25 dB ([5], pp. 14.3–14.5). An amplitude taper of no more than 0.25 dB is considered good practice. Actually, reducing the amplitude taper below 0.25 dB may produce a more severe problem of reflections from surrounding objects and the range surface since it usually is achieved by either increasing the beamwidth of the source antenna or increasing the separation of the source antenna and the antenna under test.

Transverse Amplitude Taper—Independent of the reradiative coupling effect, an amplitude taper of the incident field can produce an error in the measured radiation pattern of an antenna. The effect can be evaluated by recognizing that an amplitude taper over the aperture of the antenna under test on receive is analogous to a modification of the aperture excitation on transmit. For a typical microwave antenna with a circular aperture, Hollis, Lyon, and Clayton ([5], p. 14.13) have shown that a symmetrical amplitude taper of 0.5 dB produces an apparent reduction in gain of 0.15 dB and a taper of 0.25 dB produces a reduction in gain of 0.10 dB.

Longitudinal Amplitude Taper—A longitudinal amplitude taper has an effect similar to a transverse one; however, for an aperture-type antenna it is primarily the side lobes in the vicinity of 90° from the beam axis that are affected. The effect is most pronounced when high-gain end-fire antennas are being tested. In this case the main beam and the near-in side lobes are affected. If the range length is at least 10 times the length of the antenna (width in the case of an aperture antenna), the longitudinal taper will be less than 1 dB, which is considered to be satisfactory ([5], pp. 14.11–14.13, [13]).

Phase Taper—The measured radiation patterns of most antennas are considerably more sensitive to a deviation (from planar) of the phase of the incident wave over their apertures than they are to amplitude deviations. Thus the phase taper across the aperture of the antenna under test is the dominant factor in the selection of the separation distance R between the source and test antennas.

Hollis, Lyon, and Clayton ([5], pp. 14.5–14.7) have shown that for a typical source antenna whose aperture diameter is D meters, the phase front over its main beam, at least to the -0.5-dB level, is approximately spherical at range distances greater than about D^2/λ, where λ is the wavelength in meters. This means that the phase deviation over an aperture, in the absence of reflected signals, is determined solely by the range length R. With reference to Fig. 8, one finds that the phase deviation $\Delta\Psi$ across the test antenna's aperture is given by

$$\Delta\Psi = \frac{\pi D^2}{4\lambda R} \tag{2}$$

A criterion commonly employed is to limit this phase deviation to a maximum of $\pi/8$ rad. This leads to the restriction on range length,

$$R \geqq \frac{2D^2}{\lambda} \tag{3}$$

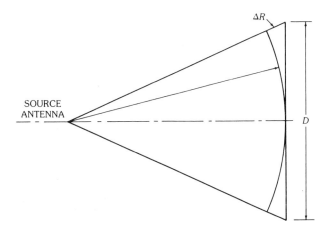

Fig. 8. A spherical phase front incident on a plane aperture.

While this criterion is adequate for many measurements, it is inadequate for others [14, 15, 16]. Equation 3 should therefore be interpreted simply as a reference criterion.

The effects of a phase taper on the measured radiation pattern of an antenna with a sum pattern include a change in levels of the first few side lobes and a filling of the nulls as shown in Fig. 9. As the phase taper is made more severe as a result of reducing the range length, the first side lobes on either side of the main beam begin to merge with the main beam, thereby forming a *shoulder lobe*. Further reduction in range length causes successive side lobes to merge with the main beam.

This effect is more pronounced for low and ultralow side lobe levels as demonstrated in Fig. 10. The distance at which the first side lobe begins to merge

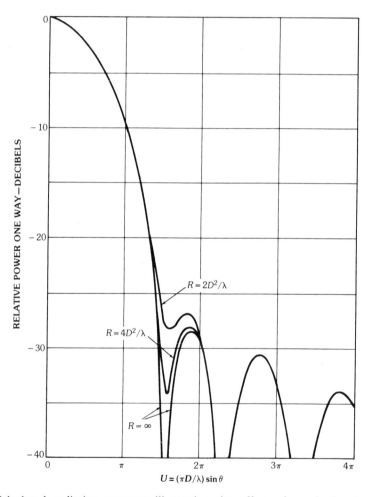

Fig. 9. Calculated radiation patterns illustrating the effect of quadratic phase errors encountered in measuring patterns at the ranges indicated. A 30-dB Taylor distribution is assumed. (*After ANSI/IEEE [2], © 1979 IEEE; after Hollis, Lyon, and Clayton [5], © 1970 Scientific-Atlanta*)

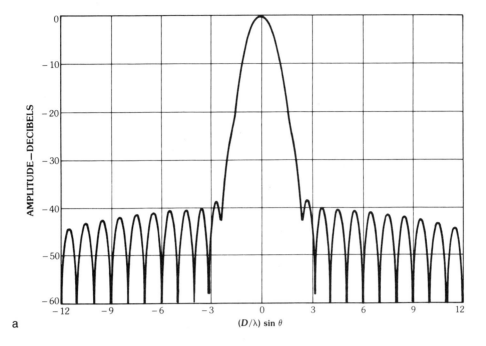

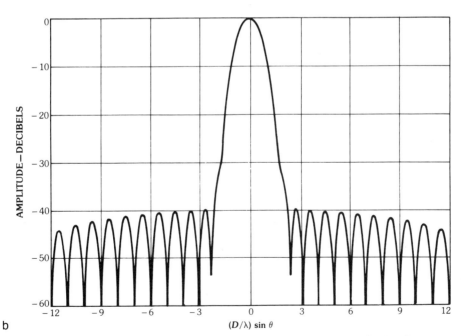

Fig. 10. Calculated radiation patterns illustrating the effect of quadratic phase errors encountered when measuring patterns at the ranges indicated. A 40-dB Taylor $\bar{n}$ distribution is assumed with $n = 11$. (a) $R = D^2/\lambda$. (b) $R = 2D^2/\lambda$. (c) $R = 4D^2/\lambda$. (d) $R = 6D^2/\lambda$. (*After Hansen [16], © 1984 IEEE*)

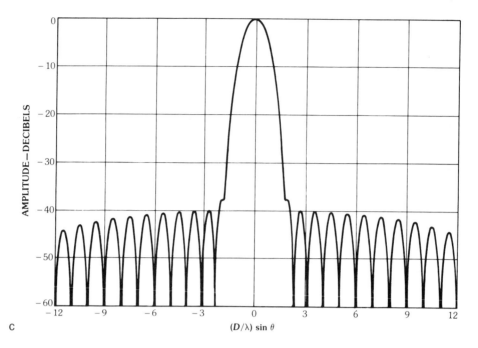

c

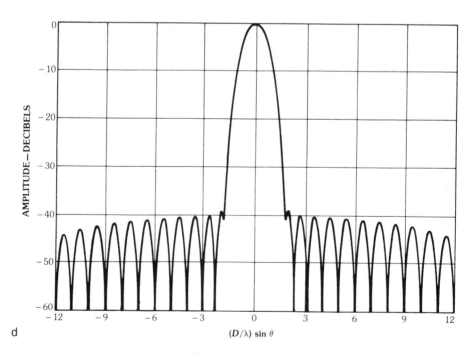

d

Fig. 10, *continued.*

with the main beam is dependent on the side lobe level of the antenna. In general, the lower the side lobe level of the antenna under test, the larger the range length must be made in order to prevent the merging of the first side lobes into the main beam [15, 16].

Shaped-beam antennas which consist of a reflector fed by an *array* also exhibit measured radiation pattern distortions as a result of finite range lengths [14]. These distortions decrease with increasing range lengths. As a rule of thumb the distance criterion for such antennas is usually taken to be about $8D^2/\lambda$ to $12D^2/\lambda$.

Whenever one designs a measurement program which involves the measurement of radiation patterns, the distance criterion should be based on the characteristics of the antenna to be tested.

Spatial Variation Caused by Reflections—Reflections of the radiated signal from the source antenna by range surfaces, nearby objects, or antenna positioners that reach the test region will cause a distortion of the incident field which illuminates the antenna under test. This can result in significant errors in radiation pattern measurements.

When an outdoor antenna range is designed, sufficient land should be cleared of reflecting objects so that it is principally the range surface (ground) and the positioner that must be considered. The latter problem can usually be reduced or eliminated with the use of a good-quality microwave-absorbing material covering the important reflecting surfaces.

The effect of reflections from the range surface can be reduced by proper range design, as will be discussed in the next section.

Interference from Sources Other Than the Range Source—Extraneous signals can produce errors in measurements if they are recorded along with the desired signal. Such signals can only be controlled by the proper selection of instrumentation. For example, a narrow-bandwidth receiver will discriminate against signals outside its beamwidth, whereas a simple detector and amplifier will not provide any discrimination.

Outdoor Ranges

There are three major types of outdoor ranges: elevated or free-space, ground-reflection, and slant ranges. In addition there are special-purpose ranges such as those used for measuring the radiation patterns of model ships and those used to measure the patterns of antennas on terrestrial vehicles [10, 11, 12]. These two particular examples of special-purpose ranges might be called image-plane or ground-plane ranges since the model ships and vehicles operate over an image plane which simulates seawater and earth, respectively.

Hollis, Lyon, and Clayton ([5], pp. 14.1–14.70) have provided detailed design and evaluation procedures for elevated and ground-reflection ranges. Much of their work appears in reference 2. The reader is referred to these references for design details not included here.

Elevated Antenna Ranges—An elevated range is one that simulates a free-space condition by judicious choices of the range length, height of the source and test antennas, and the radiation characteristics of the source antenna. In addition the reflections from the range surface are suppressed to be below an acceptable level.

The geometry of the elevated range is shown in Fig. 11. The range length R is usually determined by the maximum allowable phase taper of the incident field over the aperture of the test antenna as discussed under "Antenna Range Design Criteria." The value of R is usually chosen to be between $2D^2/\lambda$ and $12D^2/\lambda$, depending on the characteristics of the antenna being tested.

The height of the antenna under test should be chosen so that the main beam of the source antenna does not illuminate the range surface. If the 0.25-dB amplitude taper criterion is adhered to also, then the height of the antenna under test, h_r, should be at least four times its maximum dimension, D, i.e.,

$$h_r \gqq 4D \qquad (4)$$

This height requirement is usually met by mounting the test and source antennas on towers; however, it can be achieved by mounting them on the roofs of neighboring buildings or other permanent structures, or by mounting them on adjacent hilltops.

If the range surface is irregular or if there are objects nearby that cannot be removed, one should construct *Fresnel zones* over a topographic map of the range surface in order to locate the specular regions from which significant reflections can occur. A plaster scaled model of the terrain is often employed for this purpose. Once the specular regions are located, remedial measures can be taken to reduce the effect of reflections on the measurements.

If the range surface is flat, diffraction fences can be employed to redirect the reflected signals away from the test region. There are two cautions that should be emphasized. First, the diffraction fences should not intercept the main beam of the source antenna. Secondly, the top edge of the fences should not be straight but rather serrated. The reason for this is to reduce the effect of knife-edge diffraction, which could actually cause the indirect signal to increase as a result of placing the fences on the range. In some measurement situations it is impractical to elevate the test antenna to a height of four times its largest dimension. In these cases a

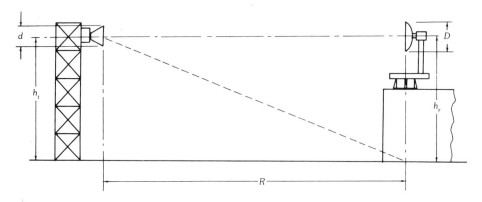

Fig. 11. An elevated range.

compromise must be made between (*a*) illuminating the range surface with energy contained in the main beam of the source antenna and (*b*) making the source antenna's beamwidth narrower. The former leads to increased reflections from the range surface, and the latter to an increased amplitude taper as well as increased radiation coupling. An error analysis is usually required in order to make a decision.

The height of the source antenna is usually chosen to be the same as that of the antenna being tested. If, however, one wishes to further reduce reflections from the range surface in order to measure near-in side lobes more accurately, then one could elevate the source antenna slightly higher than the test antenna. Some improvement in the measurement of a back lobe can be achieved by positioning the source antenna slightly lower than the antenna under test.

Finally, the source antenna should be chosen so that the amplitude taper over the antenna under test is no greater than 0.25 dB and the direction of its first null is toward the base of the test tower. This will ensure that the range surface intercepts only side lobe energy. Reducing the side lobe level of the source antenna will in turn reduce errors due to reflections.

Ground-Reflection Ranges—It should be apparent that for elevated ranges, if the range length is even moderately long, the source antenna must have a rather narrow beamwidth, which may not be easily achievable at lower frequencies (vhf-uhf). The restrictions on beamwidth can be relaxed for ground-reflection ranges which utilize the reflections from the range surface to approximate the desired plane wave at the antenna under test. Thus ground-reflection ranges find their greatest use in measurement of lower frequencies.

With reference to Fig. 12, the specularly reflected signal from the range surface combines with the direct signal from the source antenna to produce an interference pattern in the vertical direction at the test antenna. The heights of the source and test antennas are chosen so that the test antenna is centered on the first

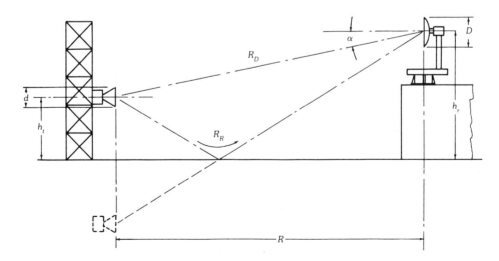

Fig. 12. A ground-reflection range.

interference lobe and the amplitude taper is no more than 0.25 dB. For a given frequency the approximate height of the source (transmitting) antenna, h_t, is given by

$$h_t \cong \frac{\lambda R}{4h_r} \qquad (5)$$

If the 0.25-dB amplitude taper criterion is also adhered to, the height of the test antenna should be $3.3D$, where D is the diameter of the antenna under test. Usually, it is recommended that the criterion be taken as

$$h_r \gtrsim 4D \qquad (6)$$

since there are approximations in the derivation. Note that the height of the source antenna is frequency sensitive; that is, if the frequency changes, so must the height of the source antenna.

It is extremely important that the range surface be smooth so that energy is specularly reflected. This means that "smoothness" must be specified for the range surface. The Rayleigh criterion can be employed. A useful form of it is given by

$$\Delta h = \leqq \frac{\lambda}{M \sin \psi} \qquad (7)$$

where Δh is the root-mean-squared deviation of the irregularities relative to a median surface, ψ is the angle of incidence measured from the horizontal, and M is a smoothness factor. Smoothness factors ranging from 8 to 32 or greater are commonly used. This corresponds to surfaces that range from tolerable to very smooth. A value of 20 is typical.

A possible ground-reflection range configuration is depicted in Fig. 13. It is the primary surface along the range axis that must have the most stringent requirement for smoothness. It should be about 20 Fresnel zones wide as shown. (The numerical subscript represents the number of Fresnel zones across that dimension.) The secondary surface does not need to be nearly as smooth as the primary one; however, there should not be an abrupt change in the surface roughness between the two surfaces, but rather a gradual change. Reference 17 contains an excellent example of a ground-reflection range.

Image-Plane Range—An image-plane range is by necessity a movable line-of-sight range. It consists of a large image plane with a turntable at its center which provides the ϕ rotation of the antenna under test. The source antenna is moved along an arc either by employing a rotating arm (see Fig. 5c) or an arch (see Fig. 6), thus providing the θ rotation.

The image plane is usually a good conductor such as aluminum; however, in some cases a material is chosen to simulate the operational environment of the antenna under test.

Two of the principal concerns in the design of an image-plane range are surface smoothness and edge diffraction around the image plane. The Rayleigh criterion,

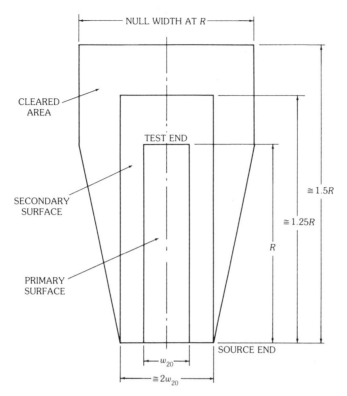

Fig. 13. A possible layout for a ground-reflection range. (*After ANSI/IEEE [2]*, © *1979 IEEE; after Hollis, Lyon, and Clayton [5]*, © *1970 Scientific-Atlanta*)

(7), is useful in establishing a limit on the root-mean-squared surface irregularities. Because the image plane is relatively small compared with a ground-reflection range, one can usually choose a larger value of smoothness factor for it. If the image plane is subjected to large extremes in temperature, there is the possibility that the surface will buckle, thus producing an unacceptable waviness in the surface. This should be taken into account when designing such a range.

Edge diffraction can also produce errors in image-plane measurements. It occurs in two ways. First, energy from the source antenna incident on the edges can be diffracted toward the antenna under test, and, secondly, energy reradiated by the antenna under test can be backscattered from the edges to the antenna under test.

The first effect can be minimized by designing the source antenna in such a way that, for a given-size image plane, the edge illumination is sufficiently low so that the diffracted signal at the antenna under test is at an acceptably low level. Since the beamwidth of the source antenna must also be broad enough to produce a wave with a sufficiently planar phase front over the antenna under test, a compromise must be made between edge illumination and the planarity of the incident phase front. It is good practice for the source antenna to extend over the image plane at grazing angles of incidence, thus eliminating the possibility of diffraction from the edge closest to the source antenna.

The effect of the backscattering of the reradiated energy by the edges can be reduced to an acceptable level by making the ground plane sufficiently large. Terminating the edges with serrations or by rolling them can give a further improvement. The range depicted in Fig. 6 employs a circular image plane. This shape has the disadvantage of having the edge equidistant in all directions from the antenna under test. This results in all of the backscattered energy returning in phase. Because of this, terminations are usually required for circular image planes. Most image-plane ranges, however, have rectangular shapes, thus eliminating the equal path lengths and usually the need for edge termination.

Many of the existing image-plane ranges are designed for use over the frequency range of 100 MHz to 2 GHz; hence they are physically quite large. The distance from the source antenna to the antenna under test is determined by the far-field criteria. These will also determine the dimension of the image plane between the antenna under test and the base of the arch since the source antenna must extend over the plane at grazing angles. The dimension beyond the antenna under test and to each side is a compromise. Values of 2.5 wavelengths to 8 wavelengths at the lowest frequency have been reported [12].*

Slant Range—A slant range [18] is one in which the source antenna is placed close to the ground and the antenna under test is mounted on a tower along with its positioner, as shown in Fig. 14. The beam axis of the source antenna points toward the center of the test antenna and ideally its first null points toward the base of the test antenna tower. It is desirable that the tower be constructed of nonconducting material.

If the antenna under test is highly directive and only the main beam and first few side lobes need to be measured, it is possible to reverse the positions of the source and test antennas. In this way the test antenna remains at ground level, which is a requirement for some types of antennas, such as satellite antennas, which must be protected by a radome housing from contamination.

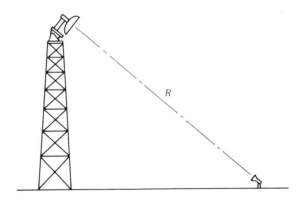

Fig. 14. A slant range.

---

*The US Naval Ocean Systems Center, San Diego, California, has a circular image-plane range that has a 160-ft (48.76-m) diameter.

Indoor Ranges

Environmental conditions such as adverse weather and electromagnetic interference can materially impede measurements made on outdoor ranges. This has led to the development of indoor ranges, all of which make use of microwave-absorbing material [18].

A room completely lined with absorbing material is called an anechoic chamber. Such chambers may also be electrically shielded to further prevent electromagnetic interference. There are two basic forms of chambers designed for antenna pattern measurements: the rectangular anechoic chamber and the tapered anechoic chamber.

These chambers must be long enough to satisfy the distance criteria presented under "Antenna Range Design Criteria." Therefore there is a restriction on their utility since it is impractical to construct a chamber of great length. One can, however, construct a partial anechoic chamber to house the antenna under test and its positioner by removing one wall of an otherwise enclosed chamber. In this way it can be combined with an outdoor range [19]. The distance criteria can be circumvented by the use of a compact range [3]. This type of range is one in which the antenna under test is illuminated by the collimated energy in the aperture of a larger point or line-focus reflector antenna.

Indoor ranges have other desirable features. For example, classified antennas can be tested without special shrouds covering them. Also, a relatively clean environment can be provided for satellite antennas.

Rectangular Chambers—A longitudinal-sectional view of a rectangular chamber is depicted in Fig. 15. The source antenna is located at the center of one of the end walls. The test antenna is located at the other end of the chamber at a point approximately equidistant from the side and back walls along the center line of the chamber.

Fig. 15. A longitudinal-sectional view of a rectangular anechoic chamber.

Although the chamber is completely lined with microwave-absorbing material, there will be reflections from the walls, floor, and ceiling. Of principal concern are the specular regions from which the reflected signals can reach the antenna under test. These regions are located midway between the source and test antennas on the side walls, floor, and ceiling. The center region of the back wall is also such a specular region.

Typically, absorbing materials produce the smallest reflection for normal incidence. As the angle of incidence θ_i increases from the normal, the reflection coefficient increases. It is good practice for the chamber width and height to be such that $\theta_i < 60°$. This requirement leads to the restriction that the length-to-width ratio of a chamber be about 2:1 ([8], pp. 669–671). In some instances it is necessary to increase this ratio to as much as 3:1 at the expense of higher levels of reflections from the walls, floor, and ceiling.

The region in which the test antenna is located is called the quiet zone. For a given chamber the size of the quiet zone depends on the specified or allowable deviation of the incident field from a uniform plane wave. The deviation is due to the finite distance from the source antenna and the specular reflections from the various surfaces of the chamber. There is no standardized figure of merit for anechoic chambers. What is often done is to establish the ratio of an "equivalent" reflected wave, which is the aggregate of all reflected waves incident on a probe antenna used to test the chamber, to the direct wave. The directivity of the probe antenna will affect the result [20].

Tapered Chambers—Rectangular anechoic chambers designed for frequencies below 1 GHz are very costly because of the increased size of the absorbing material. Furthermore, it is difficult to obtain accurate measurements in them. This is due in part to the fact that, at these frequencies, it is usually not possible to obtain a source antenna with a sufficiently narrow beam to avoid illumination of the walls, floor, and ceiling with the main beam.

The tapered anechoic chamber was introduced to overcome these difficulties. This type of chamber consists of a tapered section opening into a rectangular section. The taper has the shape of a pyramidal horn that tapers from a small source end to the large rectangular test region, as shown in Fig. 16. The tapered section is usually about twice the length of the rectangular section and the

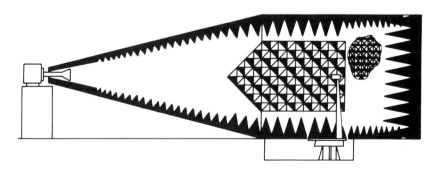

Fig. 16. A longitudinal-sectional view of a tapered anechoic chamber.

rectangular section is approximately cubical. The tapered chamber inherently requires less absorbing material; hence there is a substantial economic advantage compared with the rectangular chamber.

The low-frequency mode of operation requires the source antenna to be electrically very close to the apex of the tapered section. As a result, a slowly varying amplitude taper is produced in the test region. This can be understood by ray tracing. With reference to Fig. 17, note that the specular reflections that reach the test region occur close to the source antenna. The path lengths of the reflected signals are not very different, electrically, from that of the direct signal. The phasor sum of the reflected and direct signals thus produces the slowly varying amplitude variation shown. Another benefit is the fact that, as with the ground-reflection range, one is making use of the specularly reflected signal from the walls to create a constructive interference with the direct signal. This means that thinner absorbing material can be used over the walls in the tapered section.

The source antenna must be properly located to achieve the desired amplitude taper. If it is not close enough to the apex, deep nulls can appear in the test region. Furthermore, this location is frequency sensitive. Usually one must experimentally determine the proper location of the source antenna. This adds to the time required for a given measurement. When the source antenna is properly located the tapered chamber usually exhibits performance superior to that of the rectangular chamber at lower frequencies.

Because the source antenna must be moved closer to the apex as the frequency increases, the use of a log-periodic source antenna which has the reverse property is precluded. Also, as the frequency of operation increases, it becomes more difficult to place the source close enough to the apex to achieve proper operation. The chamber can then be operated in a free-space mode by moving the source antenna far enough forward so that it behaves as a rectangular chamber.

One can expect a certain amount of depolarization due to asymmetries in the chamber. This poses a problem when circularly polarized measurements are desired. One can adjust the polarization of the source antenna to compensate for the chamber at a single frequency of operation. For swept-frequency measurements one must sample the incident field and compensate for the axial ratio of the chamber [21].

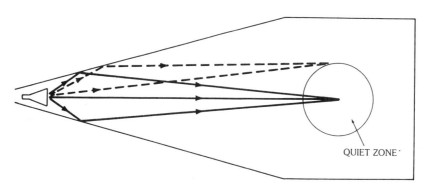

Fig. 17. Specular reflections that reach the quiet zone of a tapered anechoic chamber.

It should be noted that the field of the source antenna does not spread in the manner of a spherical wave; therefore one cannot make measurements based on the Friis transmission formula [22]. This means that absolute gain measurement cannot be made in a tapered chamber.

Compact Range—The distance criteria for the spacing between the source and test antennas are based on a maximum allowable deviation of the incident field's phase front over the test aperture from that of a plane wave. It is explicitly assumed that the incident phase front is spherical. Thus, for a given-size test aperture the spacing between the source and test antennas must be increased in order to decrease the phase deviation over the test aperture.

A large precision paraboloidal reflector [3, 6] or lens [23] provides a means of circumventing the distance criteria since either device can be used to convert a spherical phase front into a planar one as shown in Fig. 18. Because such a device is necessarily finite in size, its planar phase front exists only in a region near its aperture. This means that the test antenna will have to be located near the aperture of the reflector or lens if it is to be illuminated with a wave that exhibits a planar phase front. Thus, by using this technique a relatively short antenna test range can be constructed, hence the name compact range.

Practical considerations limit the usefulness of lenses for this purpose. One reason is that each lens exhibits an inherent mismatch at its surface, giving rise to unwanted reflections. The plano-convex dielectric lens shown in Fig. 18a is a case in point. To minimize this effect, a low-density dielectric must be used which results in a very thick, bulky structure that is difficult to manage. For these reasons, lenses are rarely employed for compact ranges, although they have found use, in a few cases, for anechoic chambers that are too short for a given measurement.

There are a number of factors that affect the performance of a compact range and which must be taken into account in the design and installation. They are:

aperture blockage
reflector surface accuracy
reflector edge diffraction
reflector focal length
direct coupling between the feed and test antenna
feed radiation pattern
room reflections

Aperture blockage is reduced by using an offset feed as shown in Fig. 18a. The reflector must have a precision surface. It should be emphasized that the surface accuracy requirement for a compact-range reflector is much more stringent than for a typical reflector antenna. Surface deviations from the ideal paraboloid must be less than a small fraction of a wavelength over any square wavelength of surface in order to ensure a sufficiently planar phase front at the test region. The condition of the surface is the controlling factor in establishing the upper limit on the frequency at which the range can be used.

The lower frequency limit is dictated by the reflector size and the edge diffraction. Serrations over the edges of the reflector [24] or rolled edges [25] provide means of controlling edge diffraction effects. Serrations are designed in

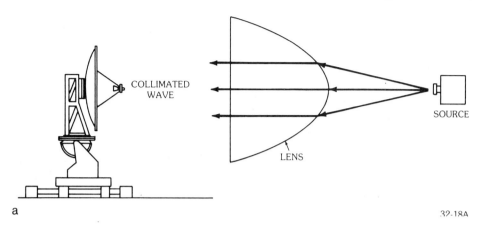

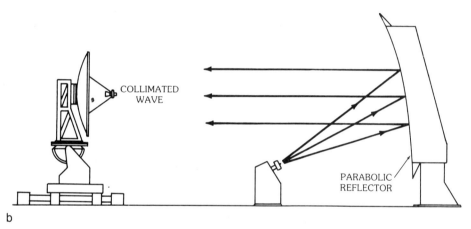

Fig. 18. Prime focus compact ranges. (*a*) Lens-type compact range. (*b*) Single-reflector or Johnson compact range.

such a manner that the cone of diffracted rays from the various edges of the serrations is in a direction away from the test antenna and feed. The appropriate shape of the rolled edge is elliptical; however, this shape should be blended with the parabolic shape of the reflector to minimize any discontinuities in the shape of the resulting surface [26]. In either case the effect of the two approaches is to direct energy from the feed incident at the edges away from the test antenna. This means that the absorbing material over the anechoic chamber in which the range is located must be designed to adequately absorb the redirected energy. Either approach when properly designed will yield excellent results for antenna pattern measurements. The rolled edge seems to yield a smoother amplitude variation over the test zone, which is required for low radar cross-section measurements [27].

The focal length of the reflector is chosen to be relatively long in order to reduce depolarization effects of the curved reflector surface. This also allows the feed to be located below the test antenna so that direct coupling between them can be controlled using absorbing material. The feed antenna must be designed to minimize side lobe radiation at wide angles. Finally, the range should be located in an anechoic chamber with absorbing material covering the positioner and the feed mount. The wall behind the reflector, however, need not be completely covered.

Typically, the test region is approximately one-third to one-half of the size of the reflector if an amplitude taper less than 0.5 dB is specified. Compact ranges have been used over a frequency range of 1 to 100 GHz. The compact range has proved to be a near-ideal antenna range. In addition to antenna pattern and radar cross-section measurements [28], it has been employed for boresight measurements [29] and can be used as an electromagnetic field simulator by moving the feed off focus [30].

In addition to the single-reflector or Johnson compact range described above, there are also dual-reflector ranges. The Vokurka compact range [31] consists of two cylindrical reflectors, one with a parabolic shape in the horizontal plane and the other in the vertical plane, as shown in Fig. 19. The arrangement is such that the rays from the feed to the first reflector cross those from the second reflector to the test antenna as shown. Another dual-reflector system is the shaped-reflector compact range [32], which yields high aperture efficiency.

Instrumentation

There are four basic subsystems of antenna range instrumentation. They are transmitting, positioning, receiving, and recording. The actual choice of equipment depends on the types of measurements that are required, the accuracy required, and the capital and operating budgets allotted.

The simplest system is a manually operated one with manual data reduction. The next level of sophistication might be the inclusion of a data storage and reduction system. One could add to that system a positioner programmer, thus resulting in a semiautomatic system. The highest level of sophistication is a completely automated system, Automated antenna ranges have become a necessity for measurements that require very large amounts of data to be taken and reduced efficiently and with a relatively high degree of accuracy. Simplified block diagrams of the manual and automated systems are given in Fig. 20.

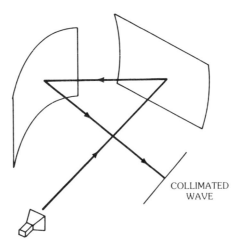

Fig. 19. The Vokurka compact range.

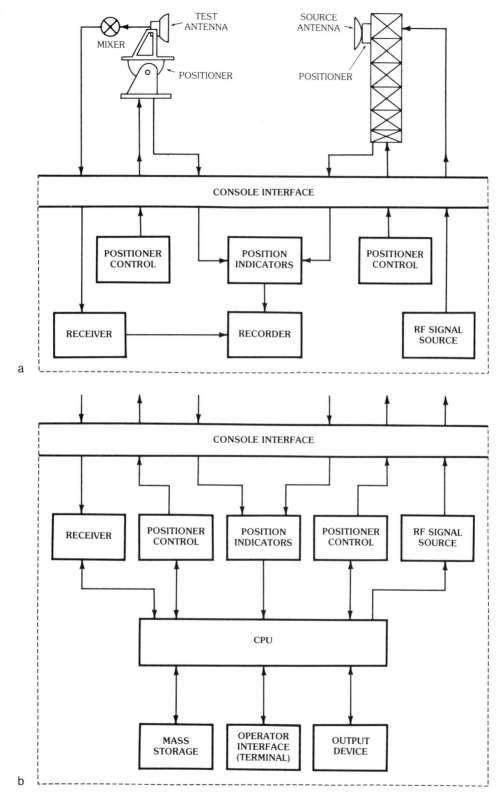

Fig. 20. Simplified block diagrams of manual and automated antenna instrumentation. (*a*) Manual system. (*b*) Automated system.

The specific measurement to be performed usually dictates the frequency of operation, polarizations, and the pattern cuts required. There are a variety of decisions that the engineer must make in the design of a range and the choice of components for the four basic subsystems of pattern range instrumentation. A tabulation of the characteristics that should be specified for each subsystem is given in Chart 2. The actual specifications will be based on the accuracy required, the budget, and the state of the art of instrumentation at the time the measurement is being planned.

Since it is a system being designed, there are interrelationships between the subsystems. For example, the specifications for the signal source are dictated by the

Chart 2. A Checklist of Antenna Range Instrumentation Characteristics That Must Be Specified

1. Signal Sources
    Frequency range
    Frequency accuracy
    Frequency stability
    Power output
    Modulation
    Temperature stability
    Operating temperature range
    Operating environment (indoor or outdoor)
    Control (local or remote)
    Interfaces

2. Positioners
    Axes required
    Load
    Bending moment
    Torque
    Speed
    Accuracy
    Limit switches
    Slip rings
    Rotary joints
    Control (automatic or manual; open loop or closed loop)

3. Receiving Subsystem
    Frequency range
    Sensitivity
    Selectivity
    Dynamic range
    Linearity
    Bandwidth
    Programmability
    Speed

4. Recording Subsystem
    Recording formats (rectangular, polar, computer generated)
    Chart drive
    Pen drive
    Amplitude response (linear, square root, or logarithmic)
    Phase response

choice of the receiver. If a simple detector-amplifier is employed for the receiver, then modulation would be required. Compared with the use of a superheterodyne receiver, higher output power would be required, but there could be a less stringent requirement for frequency stability.

Where high accuracy and precision are required, it is usually necessary to develop an error budget. It is at this point that trade-offs can be made in order to design a realistic experiment.

In addition to the basic elements of an antenna range instrumentation system, there is auxiliary equipment that must be available. This includes standard-gain antennas and range evaluation and alignment instrumentation.

Range Evaluation

An important part of antenna range measurements is the alignment and evaluation of the range. The first step in the procedure is that of alignment of the axes of the positioner. The vertical axis (θ axis for the elevation-over-azimuth and the roll-over-azimuth positioners) can be aligned using a clinometer or a precision level. Also, a plumb bob can be used for the roll-over-azimuth positioner. The horizontal axis is usually aligned optically [33].

The source antenna must also be aligned. A convenient method of accomplishing this is by the use of a field probe ([5], pp. 14.53–14.54). A typical field probe is shown in Fig. 21. The probe is first mounted on the test antenna positioner in such a manner that it can be oriented for a horizontal pattern cut in a plane perpendicular to the range axis. The amplitude pattern obtained should have a clearly defined maximum. Equilevel points on either side of the maximum are located

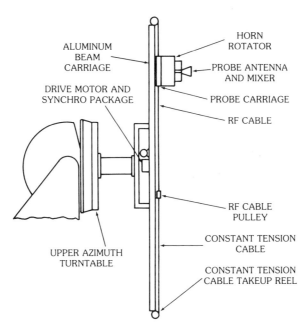

Fig. 21. A field-probe mechanism. (*After ANSI/IEEE [2], © 1979 IEEE; after Hollis, Lyon, and Clayton [5], © 1970 Scientific-Atlanta*)

for reference. The orientation of the source antenna is adjusted until the geometrical mean of these points is located on the range axis. The source antenna is then rotated 180° and another horizontal cut is made. If the pattern obtained is not symmetrical about the range axis, then the source antenna must be adjusted with respect to the polarization positioner until the pattern is symmetrical for both cases. In the case of a reflector antenna being used as a source antenna, the position of the feed might need adjusting.

For elevated ranges the procedure is repeated for the vertical direction. Ground-reflection ranges utilize the reflected signal from the range surface to obtain the vertical amplitude distribution. For this type of range the height of the source antenna is adjusted for symmetry in the vertical direction.

If there is a severe problem with range reflections, it may be necessary to take remedial measures prior to the completion of the source antenna alignment.

Once alignment has been accomplished, the state of the incident field over the test region must be determined. The principal reasons for the incident field to deviate from that predicted from an idealized range are reflections from the range surface, surrounding objects, the positioner, and the cables used for the test antenna. Sometimes signals from external sources also pose a problem.

To illustrate how a reflected signal can affect a radiation pattern, consider the situation depicted in Fig. 22. There is a direct signal from the source antenna and a reflected one from an angle measured from the range axis. The phasor sum of the two signals depends not only on their amplitudes but also their relative phase. The in-phase and out-of-phase conditions give the maximum and minimum levels. A plot of the errors due to the in-phase and out-of-phase conditions as a function of the decibel ratio E_R/E_D is given in Fig. 23. The plus and minus errors are essentially equal for ratios of -25 dB or less and are given in the logarithmic plot for ratios down to -75 dB.

The graph can be used to assess the possible errors in antenna pattern measurements due to the presence of reflected signals. Suppose that the main beam of the test antenna is pointing toward the source antenna, as depicted in Fig. 22a. At the same time a reflected signal, whose level is 35 dB below the direct signal received by the main beam, is received by a -25-dB side lobe. For this case the decibel ratio $E_R/E_D = -60$ dB. From the graph the error bounds in the measurement of the peak of the main beam are just under ± 0.01 dB. On the other hand, if the -25-dB side lobe receives the direct signal and the main beam the reflected signal, as shown in Fig. 22b, the ratio is only -10 dB. The error bounds on the measurement of the peak level of the side lobe are -3.3 dB to $+2.4$ dB.

The purpose of range evaluation is to determine the levels of the reflected signals and to take remedial measures if they are too high for the accuracy required. For an outdoor range it is convenient to consider reflected signals which arrive at relatively small angles with respect to the range axis separate from these which arrive at wide angles.

The near-axis reflected signals can be assessed by the use of a field probe mounted in place of the test antennas and oriented for transverse cuts in the plane perpendicular to the range axis. The beamwidth of the probe antenna should be broad enough to receive signals reflected from the specular region of the range surface.

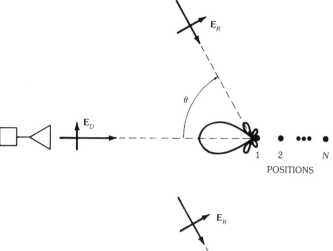
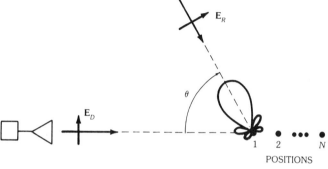

Fig. 22. Illustration of how the side lobe level of the test antenna is affected during antenna-pattern comparison measurement. (*a*) Test antenna pointing toward source. (*b*) Side lobe pointing toward source. (*After ANSI/IEEE [2], © 1979 IEEE*)

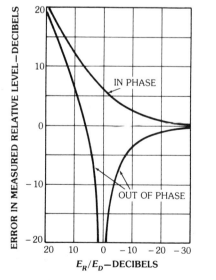
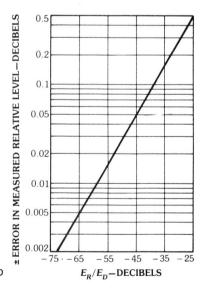

Fig. 23. Possible error in measured relative pattern level due to coherent extraneous signals. (*a*) Linear scales are employed for signal ratios of +20 to −30 dB. (*b*) Plus-or-minus errors are essentially equal for ratios of −25 dB or less, as indicated in this logarithmic plot for ratios down to −75 dB. (*After ANSI/IEEE [2], © 1979 IEEE; after Hollis, Lyon, and Clayton [5], © 1970 Scientific-Atlanta*)

If a reflected signal is present, a standing-wave pattern will be measured, as illustrated in Fig. 24. The spatial period P of the resultant waveform is given by

$$P = \frac{\lambda}{\sin \theta} \tag{8}$$

This assumes that the pattern cut was made along the intersection of the plane containing the propagation vectors of both the direct and reflected signals and the transverse plane. If the probe were rotated α radians about its axis, the period becomes

$$P = \frac{\lambda}{\sin \theta \cos \alpha} \tag{9}$$

From such measurements the direction from which the reflected signal arrives can be determined.

The wide-angle reflected signals can be measured by either a longitudinal field probe or by a pattern comparison technique. The longitudinal probe is one that moves along the range axis with the probe antenna pointing toward the source antenna. The spatial period P_l is given by

$$P_l = \frac{\lambda}{2 \sin^2(\theta/2)} \tag{10}$$

The direction from which the reflected signal arrives can be deduced from this equation.

The pattern comparison method makes use of the fact that the side lobes of the measured antenna patterns are drastically affected by reflected signals, as illustrated above. By making 360° azimuthal cuts of the test antenna's radiation pattern at several locations along the range axis and superimposing them, one can deduce the level of the reflected signal from various directions. (See Fig. 22.) For a given direction the difference between the maximum and minimum values in

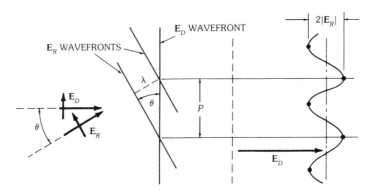

Fig. 24. Spatial interference pattern due to a reflected wave. (*After ANSI/IEEE [2]*, © *1979 IEEE; after Hollis, Lyon, and Clayton [5]*, © *1970 Scientific-Atlanta*)

decibels can be used to determine the level of the reflected signal from the graph of Fig. 25.* To illustrate this, consider the data shown in Fig. 26. Note that there is a variation of approximately 12 dB in the patterns at an azimuth angle of 120°. If the direct and major reflected waves were both received on the side lobe, this would mean that the reflected wave is 4.5 dB below the level of the direct wave. However, the reflected wave was received on the main beam and the direct wave on the side lobe. Since the variation occurred approximately 30 dB below the peak of the main beam, the reflected wave is at least −34.5 dB relative to the level of the direct wave.

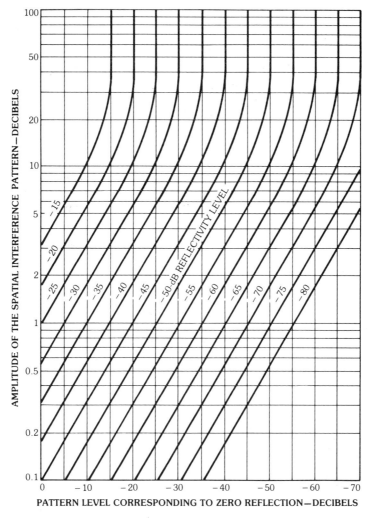

Fig. 25. Amplitude of spatial interference pattern for a given reflectivity level and antenna-pattern level. (*After ANSI/IEEE [2], © 1979 IEEE*)

---

*The curves plotted as straight lines below the 5.7-dB ordinate value are actually very slightly concave upward. All curves are correct as plotted above the 5.7-dB ordinate value.

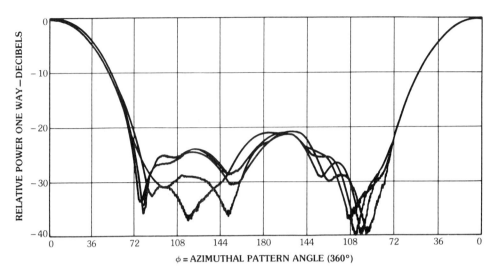

Fig. 26. Azimuthal pattern comparisons (360° cuts) for incremental longitudinal displacements of the center of rotation. (*After ANSI/IEEE [2], © 1979 IEEE; after Hollis, Lyon, and Clayton [5], © 1970 Scientific-Atlanta*)

Anechoic chambers are usually evaluated by means of the free-space voltage-standing-wave-ratio (vswr) method [34, 35]. This method is similar in principle to the field-probe method used for outdoor ranges. For transverse cuts the probe antenna is adjusted for several aspect angles ϕ, as shown in Fig. 27. Typically the aspect angle is varied in 10 steps from 0° to 360°. This approach allows one to choose a probe antenna with a higher gain than one could with the outdoor field-probe method.

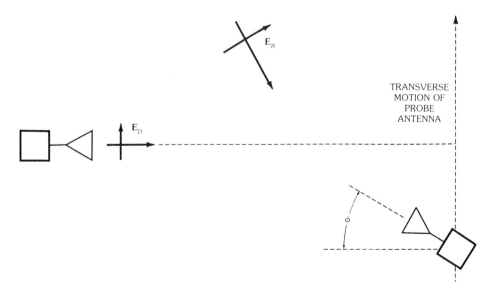

Fig. 27. Geometry for the free-space vswr method. (*After ANSI/IEEE [2], © 1979 IEEE*)

For a complete evaluation a horizontal vertical and longitudinal cut is required for each orientation of the probe antenna. Sometimes cuts in other directions are required.

Because of the six surfaces of the chamber it is desirable that the motion of the probe antenna be controlled at the base of the positioner. For example, the positioner can be mounted on rails at its base.

The antenna-pattern comparison method can also be applied; however, there is the possibility that the maximum reflectivity level will not be detected.

4. Measurement of Radiation Characteristics

The requirements of the system in which an antenna is to be incorporated usually dictate the extent of the measurements. In some instances only the principal-plane amplitude patterns and, perhaps, the peak gain of the antenna are all that are needed. Often, however, it is necessary to measure both the copolar and cross-polar amplitude patterns over the entire radiation sphere of the antenna. Sometimes a separate measurement of the antenna's polarization properties is also required. Both amplitude and phase patterns are needed for certain types of antennas, such as feeds for reflector and lens antennas and for antennas used as probes in near-field measurements where probe correction is required. Most systems require a knowledge of the gain of an antenna. Generally, for those cases where modeling techniques are employed, the gain can only be obtained from the directivity, which can be computed from the measured amplitude patterns and an estimation of the antenna's radiation efficiency. Conversely, there are cases where both the gain and directivity can be measured. The radiation efficiency of the antenna can then be determined from these results since it is the ratio of the gain to directivity.

Special measurements are usually required for antennas employed in radar systems. These include boresight and, for those cases where the antenna is enclosed within a radome, radome measurements. These measurements are beyond the scope of this chapter.

Amplitude Patterns and Directivity Measurements

The decisions that must be made before one begins a measurement program include which pattern cuts are required. As discussed in Section 2, this consideration influences the manner in which the antenna under test is mounted on the positioner. Other decisions include the frequencies of operation and the increments of the coordinate chosen as a parameter (ϕ increments for θ-cuts and θ increments for ϕ-cuts) and the mode of display.

Once the antenna is mounted on the positioner, it must be carefully aligned as described in Section 3, under "Range Evaluation." A few exploratory patterns should be measured before the production run begins. For example, if the general characteristics of the antenna's pattern are known, then a few cuts through the main beam will reveal whether or not there are any obvious problems. One might also perform a pattern comparison measurement using the antenna under test (see Section 3, under "Range Evaluation") to determine if the range reflections are at the same level as they were when the range was last evaluated.

The antenna's measured radiation patterns give the spatial distribution of relative power density contained in a specified polarization that would be obtained if the antenna operated in the transmit mode. The directivity of the antenna can be computed from the measured patterns provided that they are measured over the entire radiation sphere.

The directivity is usually determined for the direction of the peak value of the amplitude pattern (θ', ϕ'). From the definition of directivity, the maximum directivity, $D_m(\theta', \phi')$ can be written as

$$D_m(\theta', \phi') = \frac{\Phi_m(\theta', \phi')}{1/4\pi \int_0^\pi \left[\int_0^{2\pi} \Phi(\theta, \phi) d\phi \right] \sin\theta \, d\theta} \tag{11}$$

where $\Phi(\theta, \phi)$ is the radiation intensity as a function of direction, (θ, ϕ), and $\Phi_m(\theta', \phi')$ is its maximum value. Since only relative values of the radiation intensity are measured, it is necessary to normalize the radiation intensity by dividing by the value of the radiation intensity which corresponds to that used for the measured radiation patterns.

The radiation intensity can be decomposed into two components corresponding to the two orthogonal polarizations employed in the measurement of the radiation patterns. These are usually the copolarization and the cross polarization, in which case the normalization will be with respect to the peak radiation intensity of the copolarization.

This decomposition of the radiation intensity allows the directivity to be written as the sum of two partial directivities corresponding to the two orthogonal polarizations, that is,

$$\begin{aligned} D_m(\theta', \phi') &= D_1(\theta', \phi') + D_2(\theta', \phi') \\ &= \frac{\overline{\Phi}_1(\theta', \phi')}{1/4\pi(\overline{P}_{1t} + \overline{P}_{2t})} + \frac{\overline{\Phi}_2(\theta', \phi')}{1/4\pi(\overline{P}_{1t} + \overline{P}_{2t})} \end{aligned} \tag{12}$$

The subscripts 1 and 2 denote the two orthogonal polarizations; the bars indicate normalization, and $\overline{P}_{1t}$ and $\overline{P}_{2t}$ are given by

$$\overline{P}_{1t} = \int_0^\pi \left[\int_0^{2\pi} \overline{\Phi}_1(\theta, \phi) d\phi \right] \sin\theta \, d\theta \tag{13}$$

and

$$\overline{P}_{2t} = \int_0^\pi \left[\int_0^{2\pi} \overline{\Phi}_2(\theta, \phi) d\phi \right] \sin\theta \, d\theta \tag{14}$$

Note that both $\overline{P}_{1t}$ and $\overline{P}_{2t}$ are required in order to compute either $D_1(\theta', \phi')$ or $D_2(\theta', \phi')$.

Since the polarization of an antenna changes with direction, it is necessary to determine the partial directivities with respect to two specified orthogonal polarizations and then add them to obtain the (total) directivity of the antenna. It is desirable that the copolarization and cross polarization be the specified polarizations. If polarizations 1 and 2 are the copolarization and cross polarization, respectively, then in (12), $\overline{\Phi}(\theta', \phi') = 1$. Also, the normalized radiation intensities $\overline{\Phi}_1(\theta, \phi)$ and $\overline{\Phi}_2(\theta, \phi)$ are equal to the relative power patterns that are measured.

If the radiation patterns are measured using ϕ-cuts with the θ increments chosen to be π/M rad, the partial directivities are given by

$$D_1(\theta', \phi') = \frac{4M}{\sum_{i=1}^{M}\left\{\int_0^{2\pi}[\overline{\Phi}_1(\theta_i, \phi) + \overline{\Phi}_2(\theta_i, \phi)]d\phi\right\}\sin\theta_i} \quad (15)$$

$$D_2(\theta', \phi') = \frac{4M\overline{\Phi}_2(\theta', \phi')}{\sum_{i=1}^{M}\left\{\int_0^{2\pi}[\overline{\Phi}_1(\theta_i, \phi) + \overline{\Phi}_2(\theta_i, \phi)]d\phi\right\}\sin\theta_i} \quad (16)$$

where $\theta_i = i\pi/M$ rad.

If θ-cuts are employed with the ϕ increments chosen to be $2\pi/N$ rad, the partial directivities are given by

$$D_1(\theta', \phi') = \frac{2N}{\sum_{j=1}^{N}\int_0^{\pi}[\overline{\Phi}_1(\theta, \phi_j) + \overline{\Phi}_2(\theta, \phi_j)]\sin\theta\, d\theta} \quad (17)$$

$$D_2(\theta', \phi') = \frac{2N\overline{\Phi}_2(\theta', \phi')}{\sum_{j=1}^{N}\int_0^{\pi}[\overline{\Phi}_1(\theta, \phi_j) + \overline{\Phi}_2(\theta, \phi_j)]\sin\theta\, d\theta} \quad (18)$$

In this case each $\overline{\Phi}(\theta, \phi_j)$ must be multiplied by $\sin\theta$ during the process of integration.

These equations can be evaluated numerically using a computer [36]. On-line computers incorporated into the measurement system can be used to yield essentially real-time computations. Alternatively, the data can be stored and entered into a computer at a later time.

For pencil-beam antennas an estimate of the directivity can be obtained from the principal-plane patterns since it is inversely proportional to the *beam area* or *areal beamwidth*. The beam area is the product of the two half-power beamwidths, θ_{HP} and ϕ_{HP}. Therefore one can write

$$D_m(\theta', \phi') = \frac{K}{\theta_{HP}\phi_{HP}} \quad (19)$$

The value of K has been given as being between 25 000 and 41 253. For most cases a value of 32 400 yields an acceptable estimate of directivity [37, 38].

Antenna Gain Measurements

Antenna gain measurements are generally classified as absolute gain measurements or gain-transfer measurements ([2], pp. 94–112). Absolute gain measurements require no *a priori* knowledge of the gains of the antennas being tested, whereas the gain-transfer method requires the use of a gain standard whose gain is accurately known.

Absolute gain measurements have found their greatest use in the calibration of gain standards. The gain-transfer method is the more commonly employed method of measuring antenna gain.

Absolute Gain Measurements—Absolute gain measurements are based on the Friis transmission formula for coupling between two antennas. To illustrate this, consider two antennas, one of which is transmitting and the other receiving. They are separated by R meters as depicted in Fig. 28.

The power received by antenna B, P_R, is given by the Friis transmission formula

$$P_R = P_0 G_A G_B \left(\frac{\lambda}{4\pi R}\right)^2 q_B p \qquad (20)$$

where G_A and G_B are the gains of antennas A and B, respectively, P_0 is the power accepted by antenna A from its generator, q_B is the impedance mismatch factor evaluated at a reference plane between antenna B and its load, and p is the *polarization efficiency*.*

This formulation implicitly contains the assumptions that free-space conditions prevail and that the separation R is such that the incident field produced by antenna A over the aperture of antenna B is planar.

The impedance mismatch factor is given by

$$q_B = \frac{(1 - |\Gamma_B|^2)(1 - |\Gamma_L|^2)}{|1 - \Gamma_B \Gamma_L|^2} \qquad (21)$$

where Γ_B and Γ_L are the reflection coefficients of antenna B and the load evaluated at a common reference plane. Note that both the amplitude and phase of Γ_B and Γ_L are required in order to evaluate q_B.

Fig. 28. Two-antenna system illustrating the Friis transmission formula.

---

*The polarization efficiency p is also called the *polarization mismatch factor*.

For measurements performed on a well-designed free-space range, one can usually assume that there are no depolarization effects. In that case the polarization efficiency p can be computed from a knowledge of the polarizations of the two antennas ([2], pp. 76–85). Various formulas for p are given in the following subsection on polarization measurements.

The accepted power P_0 cannot be measured directly. If one measures the available power, P_A, of the generator by conjugately matching a power meter to it, then

$$P_0 = P_A q_A \tag{22}$$

where

$$q_A = \frac{(1 - |\Gamma_G|^2)(1 - |\Gamma_A|^2)}{|1 - \Gamma_G \Gamma_A|^2} \tag{23}$$

Here Γ_G and Γ_A are the complex reflection coefficients of the generator and antenna A, respectively, measured at the same reference plane as shown in Fig. 28. Often a power meter which is matched to the transmission line is used to measure the power without a tuner to conjugately match it to the generator. This measurement will yield the line-matched power P_M and

$$P_0 = P_M q'_A \tag{24}$$

in which

$$q'_A = \frac{1 - |\Gamma_A|^2}{|1 - \Gamma_A \Gamma_G|^2} \tag{25}$$

In either case, if the generator were matched to the transmission line, the impedance mismatch factor becomes

$$q''_A = 1 - |\Gamma_A|^2 \tag{26}$$

and

$$P_0 = P_A q''_A \tag{27}$$

Here the power P_A is also known as the incident power at the antenna.

Since (22) contains all of the special cases, the Friis transmission formula will be written in terms of the available power P_A, that is,

$$P_R = P_A G_A G_B \left(\frac{\lambda}{4\pi R}\right)^2 q_A q_B p \tag{28}$$

Two-Antenna Method—For what follows, it is convenient to express (28) in logarithmic form which, when rearranged, becomes

$$(G_A)_{\text{dB}} + (G_B)_{\text{dB}} = 20\log\left(\frac{\lambda}{4\pi R}\right) - 10\log\left(\frac{P_A}{P_R}\right) - 10\log C \quad (29)$$

where

$$C = q_A q_B p \quad (30)$$

The expression C can be considered a correction factor which requires the measurement of Γ_G, Γ_A, Γ_B, Γ_R, and p. Of course, there will remain a residual measurement error in the measurement of these quantities.

If the gain of one of the antennas is known, then the gain of the other antenna can be determined from a measurement of λ, R, C, P_A, and P_R. If the gain of neither antenna is known but $G_A = G_B$, then the gains are given by

$$(G_A)_{\text{dB}} = (G_B)_{\text{dB}}$$
$$= 1/2\left[20\log\left(\frac{4\pi R}{\lambda}\right) - 10\log\left(\frac{P_A}{P_R}\right) - 10\log C\right] \quad (31)$$

This method is called the two-antenna method since only two antennas are required.

Three-Antenna Method—If the two antennas are not identical, or nearly so, and neither gain is known, the two-antenna method fails. In this case a third antenna is required in order to determine the gain of any one of the antennas. Since three antennas are required, the procedure is called the three-antenna method. Three sets of measurements are performed using the three possible combinations of antennas A, B, and C. The resulting set of equations is

$$(G_A)_{\text{dB}} + (G_B)_{\text{dB}} = 20\log\left(\frac{4\pi R}{\lambda}\right) - \left[10\log\left(\frac{P_A}{P_R}\right) + 10\log C\right]_{AB}$$
$$(G_A)_{\text{dB}} + (G_C)_{\text{dB}} = 20\log\left(\frac{4\pi R}{\lambda}\right) - \left[10\log\left(\frac{P_A}{P_R}\right) + 10\log C\right]_{AC} \quad (32)$$
$$(G_B)_{\text{dB}} + (G_C)_{\text{dB}} = 20\log\left(\frac{4\pi R}{\lambda}\right) - \left[10\log\left(\frac{P_A}{P_R}\right) + 10\log C\right]_{BC}$$

where the subscripts indicate the specific combinations of two antennas. The simultaneous solution of these equations gives the gains of all three antennas.

Insertion-Loss Method—There is an alternative method in which the insertion loss between the reference planes shown in Fig. 28 is measured [39]. This can be accomplished by first removing the antennas and connecting the load directly to the generator as shown in Fig. 29. The power transferred to the load, P_R^i, is measured. Secondly, the antennas are inserted and separated to a distance of R as before and the power transferred to the load for this case, P_R^f, is measured.

The sum of the two antennas' gains in decibels can be shown to be given by

$$(G_A)_{\text{dB}} + (G_B)_{\text{dB}} = 20\log\left(\frac{4\pi R}{\lambda}\right) + 10\log\left(\frac{P_R^f}{P_R^i}\right) + 10\log C'' \quad (33)$$

where

$$C'' = \frac{|1 - \Gamma_G\Gamma_A|^2|1 - \Gamma_B\Gamma_R|^2}{|1 - \Gamma_G\Gamma_R|^2(1 - |\Gamma_A|^2)(1 - |\Gamma_B|^2)} + p \qquad (34)$$

Again, if one of the gains is known, the other can be computed using (33). If neither gain is known but G_A is known to equal G_B, then the gain of each antenna is equal to one-half the right side of (33). Three antennas are required as before if the gains are not equal and neither G_A or G_B is known.

One of the advantages of the insertion-loss method is that the measurement of the ratio of power (P_R^f/P_R^i) can be accomplished with the use of a precision attenuator. The attenuator can be placed in the line at either the generator or load end. It is adjusted so that the signal at the load is the same for both measurements. The difference in the settings for the direct measurement and measurement with antennas inserted is precisely the ratio of power in decibels.

The basic instrumentation required for the two- and three-antenna methods is shown in Fig. 30. The small squares at each device are included as a reminder of the need for connectors and adapters, which are potential sources of impedance mismatch. Furthermore, the devices themselves, as well as the antennas under test, are not perfectly matched at any given frequency within the band of operation; therefore tuners are often used. They must be used with care, however, since the resulting network becomes very narrow band. Thus a small amount of detuning will cause a great change in input impedance.

The transmitting and receiving systems should be bench tested and calibrated prior to assembly on the antenna range. Also, great care must be exercised in handling the interconnecting transmission lines. These must also be calibrated in order to account for losses and impedance mismatches. Wherever possible, transmission lines should be eliminated because they introduce calibration errors and they deteriorate or may be damaged through use. For example, the directional coupler might be connected directly to the transmitting antenna. Also, if power meters are used to measure the powers at the transmitting and receiving test ports, it may be possible to connect their power sensors directly to the directional coupler/attenuator and to the transmitting antenna, respectively.

The measurement system may be automated provided that the instruments employed are computer controllable [40]. Fig. 31 illustrates the basic instrumentation required. The system must be calibrated as in the manual case. With the automated system the calibration data are stored in the computer so that corrections can be made on the data. This type of system can easily be programmed to measure the gain over a band of frequencies.

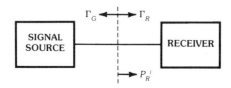

Fig. 29. An illustration of the direct measurement of power before the antennas are inserted in the system.

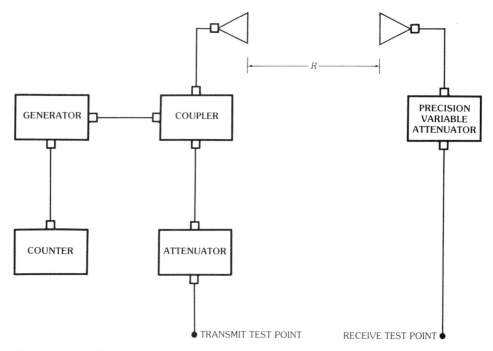

Fig. 30. Typical instrumentation for two-antenna and three-antenna methods of power-gain measurement.

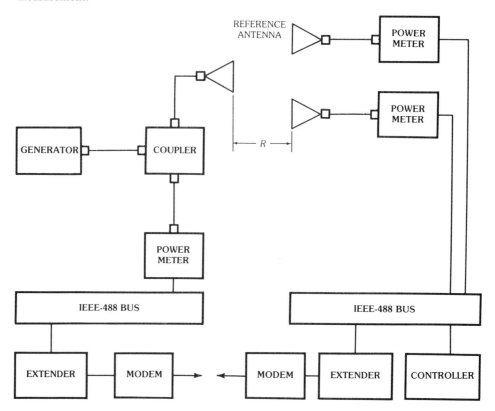

Fig. 31. Block diagram of an automated gain-measurement system.

Extrapolation Method—There are two significant sources of error in absolute gain measurements which can be avoided by the use of the extrapolation technique [41]. These are the errors caused by the finite spacing between the antennas and multipath interferences.

Because of finite spacing, the uncorrected measured value of an antenna's gain will be lower than its actual value because of the presence of near-field components which cause a broadening of the beam of the antenna.

Multipath interference causes an undulating received signal when the spacing between the transmitting and receiving antennas is varied. This results in the uncorrected measured gain exhibiting a similar characteristic.

The details of the extrapolation technique are given in references 8 and 41 and will not be given here. However, an examination of the principal result of the extrapolation theory yields insights into conventional far-field gain measurements. The extrapolation theory states that, for any two antennas used in the gain measurement, the received signal as a function of spacing can be represented by the double series

$$V(R) = C \sum_{p=0}^{\infty} \frac{\exp[-j(2p + 1)kR]}{R^{2p+1}} \sum_{q=0}^{\infty} \frac{A_{pq}}{R^q} \qquad (35)$$

in which R is the spacing, k the free-space wave number, and C a constant which is determined by the incident signal at the transmitting antenna and reflection coefficients of the receiving antenna and its load. To interpret this equation it is convenient to rewrite it in the following manner:

$$V(R) = [\exp(-jkR)/R](A_{00} + A_{01}/R + A_{02}/R^2 + \cdots)$$
$$+ [\exp(-j3kR)/R^3](A_{10} + A_{11}/R + A_{12}/R^2 + \cdots) + \cdots \qquad (36)$$

The coefficients of $\exp(-jkR)/R$ are identified as the direct transmission terms with A_{00} being the one used to determine the far-field properties of the antenna, including its gain. The coefficients of $\exp(-j3kR)/R^3$ give the first-order reflection or multipath terms. All other terms represent higher-order reflections.

In a conventional far-field measurement of gain one must choose the spacing R to be sufficiently large to ensure that A_{00} is the only significant term. This results in very long ranges for some types of antennas. As one might expect, ground reflections will increase with increased spacing unless a corresponding increase in antenna height above the ground is provided.

The extrapolation procedure consists of the measurement of the received signal as a continuous function of the spacing between the transmitting and receiving antennas. The resulting cyclic variation in the received signal due to multipath is averaged out either mathematically or electronically. Next, a curve based on (36) is fit to the averaged measured data. From this result, A_{00} can be determined and hence the gain.

This method allows one to perform the actual measurement at spacings less than $2D^2/\lambda$, where D is the largest dimension of the antenna. The procedure is in effect an extrapolation of the fitted curve from the near field to the far field of the antenna.

Just as in the conventional far-field technique, the method can be employed as a two-antenna method or three-antenna method.

By measuring both the amplitude and phase of the received signal the method can be generalized to give both the gains and polarizations of the antennas under test. There is the restriction, however, that no two antennas be circularly polarized. If one antenna is circularly polarized, only that antenna's characteristics can be determined.

The extrapolation technique requires a special range consisting of two towers, at least one of which is movable. The range must be such that boresight between the transmitting and receiving antennas is maintained as their separation varies over the length of the range. Because of this requirement the extrapolation method seems to be only practical as a specialized technique for calibrating gain standards by a standards agency. Indeed, the technique is capable of yielding gain measurements with uncertainties as low as ±0.05 dB and more routine measurements to ±0.08 dB [7].

Absolute-Gain Measurement on a Ground-Reflection Range—At frequencies below about 1 GHz, ground-reflection ranges are commonly used. With some restrictions and modifications the two- or three-antenna method can be implemented on a ground-reflection range [42].

The method is limited to linearly polarized antennas which couple to the electric field only. It is recommended that the antennas be oriented for horizontal polarization in order to avoid the rapid variation of the earth's reflection coefficient as a function of the angle of incidence that is associated with vertical polarization.

The criteria for ground-reflection ranges, outlined in Section 3, under "Outdoor Ranges," should be satisfied. The geometry of the ground-reflection range for gain measurements is shown in Fig. 12. It is desirable that the range length R_0 be such that $R_0 \gg 2h_r$, where h_r is the height of the receiving antenna. When the height of the transmitting antenna is adjusted so that the field at the receiving antenna is at the first maximum closest to the ground, then the gain sum equation for the two-antenna or the three-antenna method can be modified to read

$$(G_A)_{dB} + (G_B)_{dB} = 20 \log\left(\frac{4\pi R_D}{\lambda}\right) - 10 \log\left(\frac{P_0}{P_r}\right)$$
$$- 20 \log\left[(D_A D_B)^{1/2} + \frac{rR_d}{R_R}\right] \quad (37)$$

where D_A and D_B are the directivities along R_D relative to the peak directivities of antennas A and B, respectively. The factor r is to be determined. It is a function of the electrical and geometrical properties of the antenna range, the radiation patterns of the antennas, and the frequency of operation. Directivities D_A and D_B are obtained from amplitude patterns of the two antennas which should be measured prior to performing the gain measurement. The quantities R_D, R_R, and λ and the ratio P_0/P_r are measured directly. Once r is determined, the gain sum can be evaluated.

To obtain r, the preceding measurement is repeated, but this time with the height of the transmitting antenna adjusted to a position such that the field at the receiving antenna is at a minimum. To distinguish the quantities measured with this geometry from those of the previous one, let all the quantities associated with the latter geometry be represented by the same letters, except with primes. With this notation the equation for r can be written as

$$r = \left(\frac{R_R R'_R}{R_D R'_D}\right) \frac{[(P_r/P'_r)(D'_A D'_B)]^{1/2} R_D - (D_A D_B)^{1/2} R'_D}{(P_r/P'_r)^{1/2} R_R + R'_R} \tag{38}$$

where R_R, R_D, and P_r are obtained with the transmitting antenna adjusted for a maximum signal at the receiving antenna, and R'_R, R'_D, and P'_r are obtained with the antenna adjusted for a minimum signal at the receiving antenna. The relative directivities D_A, D'_A, D_B, and D'_B are obtained from the amplitude patterns of the two antennas.

The instrumentation for this measurement is essentially the same as that for the free-space range measurement (see Fig. 30). Accuracies of ± 0.3 dB are attainable with this method.

Gain-Transfer Method—The gain-transfer method, also called the gain-comparison method, requires the use of a gain standard with which the gain of the test antenna is compared. Once the comparison has been performed, the gain of the standard is said to have been transferred to the test antenna. The measurement can be performed on a free-space range or ground-reflection range in an anechoic chamber, or *in situ*; however, in any case the far-field distance criteria must be satisfied.

Typically the test antenna is operated in the receive mode. With it mounted and properly oriented, the received power P_T is measured. The test antenna is then replaced with a gain-standard antenna, leaving all other conditions the same. The received power P_S is then measured. The gain of the test antenna expressed in decibels is given by

$$(G_T)_{dB} = (G_S)_{dB} + 10 \log\left(\frac{P_T}{P_S}\right) - 10 \log\left(\frac{q_T}{q_S}\right) - 10 \log\left(\frac{p_T}{p_S}\right) \tag{39}$$

in which G_S is the known gain of the gain-standard antenna, q_T and q_S are the impedance mismatch factors of the test antenna and the gain-standard antenna, respectively, and p_T and p_S are their polarization efficiencies.

The instrumentation for this method is essentially the same as that for the absolute gain measurement. A convenient method of achieving the exchange of the test antenna and the gain-standard antenna is to mount them back to back on the positioner so that rotating 180° in azimuth effects the positional exchange. Usually absorbing is required between them to reduce coupling.

Measurement of Circularly and Elliptically Polarized Antennas—In the preceding it was tacitly assumed that the source antenna, the test antenna, and the gain-standard antenna are nominally polarization matched; that is, the ratio p_T/p_S is very nearly equal to unity. Often, however, one must measure the gains of circularly or elliptically polarized antennas using a nominally linearly polarized gain-standard antenna.

This measurement can be accomplished by measuring the partial gains of the test antenna with respect to two orthogonal orientations of the gain-standard antenna. For example, the gain-standard and the source antennas are first oriented for vertical polarization and then for horizontal polarization. This results in the measurement of G_{TV} and G_{TH}, the partial gains with respect to vertical and horizontal polarization, respectively. The gain of the test antenna is then given by

$$(G_T)_{\text{dB}} = 10\log(G_{TV} + G_{TH}) \tag{40}$$

Note that the partial gains must be in ratio form when added; that is,

$$G_{TV} = G_{SV}\frac{P_{TV}\,q_{SV}\,p'_{SV}}{P_{SV}\,q_{TV}\,p'_{TV}}$$
$$G_{TH} = G_{SH}\frac{P_{TH}\,q_{SH}\,p'_{SH}}{P_{SH}\,q_{TH}\,p'_{TH}} \tag{41}$$

where the subscripts V and H indicate vertical and horizontal polarizations. The previously defined impedance mismatch factors apply here; however, the polarization efficiencies must be modified; hence, the primes.

Polarization mismatches enter in a more complicated fashion for this procedure. The factors p'_{TV}, p'_{TH}, p'_{SV}, and p'_{SH} are *not* the usual polarization efficiencies but rather factors that account for the deviations of the source and gain-standard antennas from linear polarization. This is best illustrated by an example.

Consider the case where the gain of a circularly polarized test antenna is being measured by this method. Suppose that the test and gain-standard antennas are purely circularly and linearly polarized, respectively, but the source antenna has a finite axial ratio. For this example p'_{SV} and p'_{SH} become the usual mismatch factors so that the primes can be dropped. The factors p'_{TV} and p'_{TH}, on the other hand, are the factors by which the usual mismatch factors deviate from 1/2. The reason for this is that a circularly polarized receiving antenna has a polarization efficiency of 1/2 when the incident wave is linearly polarized and the measurement requires a partial gain with respect to linear polarization. Table 1 gives the errors in measured

Table 1. Errors in the Measured Gain of a Purely Circularly Polarized Antenna Due to a Finite Axial Ratio of the Linearly Polarized Sampling Antenna

Sampling Antenna Axial Ratio (dB)	Measurement Error (dB)	
	Same Sense	Opposite Sense
20	+0.828	−0.915
25	+0.475	−0.503
30	+0.270	−0.279
35	+0.153	−0.156
40	+0.086	−0.109
45	+0.049	−0.049
50	+0.027	−0.028

gain for this example as a function of the source antenna axial ratio. Note that the error can be quite significant.

Polarization Measurements

The polarization of an antenna is defined as the polarization of the wave it radiates [1]. As discussed in Chapter 1, what is implied is that in a localized region in the far field of the antenna the radiated wave may be considered to be a uniform plane wave, and that the polarization of the antenna is the same as that of the plane wave. Furthermore, polarization is defined in such a way that it is dependent on the ratio and relative phase of any two orthogonal field components into which the plane wave is resolved. Therefore, polarization, like gain and directivity, is only a function of direction, (θ, ϕ). To completely specify an antenna's polarization, one would have to determine the distribution of its polarization over its radiation sphere.

Polarization can be specified in a number of different ways, the most common being by the *axial ratio r* and *tilt angle τ* of the polarization ellipse and the *sense of polarization*. The tilt angle requires the establishment of a reference direction in the *plane of polarization*, and because polarization is a function of direction, reference directions are required over the radiation sphere of the antenna. A common choice is in the direction of $\hat{\theta}$ as depicted in Fig. 32.

It is convenient to define a local Cartesian coordinate system $(1, 2, 3)$ at the point on the radiation sphere where the polarization is to be determined. One coordinate axis, 3, is the direction of propagation, thus placing the 1 and 2 axes in the plane of polarization. The orientation of the $(1, 2)$ axes is arbitrary as long as their relationship to the (θ, ϕ) axes of the antenna's radiation sphere is known. Since the 1 axis will be the reference direction for measuring the tilt angle, it is desirable that it be in the $\hat{\theta}$ direction. For the antenna range measurements discussed here, the 1 axis will be chosen to be in the horizontal direction and the 2 axis in the vertical direction, which results in the horizontal direction being the reference direction for the tilt angle as shown in Fig. 33. In this figure $\mathbf{E}(t)$ is the instantaneous electric-field vector.

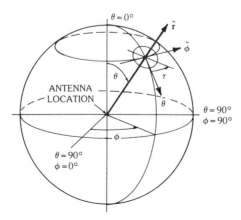

Fig. 32. The polarization ellipse shown in relation to the antenna's coordinate system.

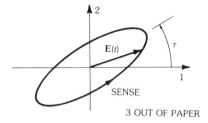

Fig. 33. A local coordinate system $(1, 2, 3)$ at a point (θ, ϕ) on the radiation sphere.

To express a plane wave in a form useful for antenna measurements, it is decomposed into the linear combination of two orthogonally polarized wave components. The orthogonal polarizations useful for antenna measurements are horizontal and vertical linear (E_H, E_V), diagonal linear $(E_{45°}, E_{135°})$, and right-hand and left-hand circular (E_R, E_L), where the symbol **E** denotes the electric-field vector and the subscript indicates the particular component. The polarization of a plane wave can be expressed in terms of *complex polarization ratios* as follows:

$$\hat{\varrho}_\ell = \varrho_\ell \exp(j\delta_\ell)$$
$$\hat{\varrho}_d = \varrho_d \exp(j\delta_d) \tag{42}$$
$$\hat{\varrho}_c = \varrho_c \exp(j\delta_c)$$

where

$$\varrho_\ell = E_V/E_H$$
$$\varrho_d = E_{135°}/E_{45°} \tag{43}$$
$$\varrho_c = E_R/E_L$$

and δ_ℓ, δ_d, and δ_c are the relative phases of the corresponding pairs of field components. Circular polarization components require a reference direction for the definition of phase. For our purposes the horizontal direction is chosen; hence δ_c is defined by the angle of the instantaneous electric-field vector of the right-hand circularly polarized component with respect to the horizontal direction at the instant that the electric-field vector of the left-hand circularly polarized component is in the horizontal direction. By this definition the right-hand and left-hand components are in phase if their instantaneous electric-field vectors are in the same horizontal direction at the same time.

Of particular interest is the complex circular polarization ratio $\hat{\varrho}_c$, because of its relationship to the axial ratio and the tilt angle of the wave's polarization, i.e.,

$$r = \frac{\varrho_c + 1}{\varrho_c - 1} \tag{44}$$

and

$$\tau = \delta_c/2 \qquad (45)$$

The sign of the denominator of (44) gives the sense of the polarization, with a positive value indicating right-hand sense and a negative value indicating a left-hand sense. Also, since the reference direction for defining phase is in the horizontal direction, the tilt angle is also referenced to the horizontal direction as required.

If linear polarization ratios are to be measured rather than circular ones, they can be converted to circular ones by use of

$$\hat{\varrho}_c = \varrho_c \exp(j\delta_c) = \frac{1 + j\hat{\varrho}_\ell}{1 - j\hat{\varrho}_\ell} \qquad (46)$$

or, if diagonal components are employed,

$$\hat{\varrho}_c = \varrho_c \exp(j\delta_c) = -\frac{1 + j\hat{\varrho}_d}{j + \hat{\varrho}_d} \qquad (47)$$

When making polarization measurements on an antenna range it is useful to establish not only the coordinate system for the antenna's polarization as defined above, but also the antenna's *receiving polarization*. Fig. 34 depicts these two coordinate systems. Note that the tilt angle of the antenna's polarization and that of the receiving polarization are supplements of one another; that is,

$$\tau_r = 180° - \tau_a$$

where τ_r and τ_a are the tilt angles of the receiving polarization and the (transmitting) polarization of the antenna, respectively. Note that the receiving polarization of the antenna is the polarization of an incident plane wave which is *polarization matched* to the antenna.

The receiving polarization of an antenna is introduced because, unlike the antenna's (transmitting) polarization, it is in the same coordinate system as an

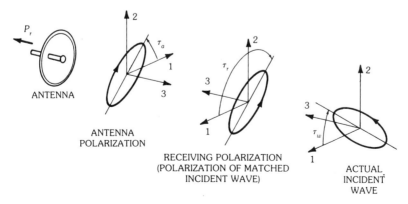

Fig. 34. The relationship between the antenna's (transmitting) polarization and receiving polarization.

incident plane wave of arbitrary polarization as shown in Fig. 34. It follows that the various mathematical representations of these polarizations will be in the same coordinate system and hence can be combined to give the coupling between the wave and the antenna. This is particularly important in polarization measurements since the antenna under test can be operated in either the transmit or receive mode.

The *Poincaré sphere* provides a means of visualizing polarization measurements. It is defined so that there is a one-to-one correspondence between all of the possible polarizations of a single-frequency plane wave and the points over its surface. This is illustrated in Fig. 35. Deschamps [43] introduced its use for antenna work. Hollis ([5], Chapters 3 and 10) greatly elaborated on its development and use, providing numerous relationships useful for polarization calculations and measurement. Much useful material appears in references 2 and 7. A thorough treatment is presented in reference 44. The reader is referred to these sources for detailed expositions on the Poincaré sphere and its use, as well as polarization theory and measurements.

Fig. 35 depicts the representation of all possible polarizations of a single-frequency plane wave in free space on the Poincaré sphere. The 1 axis is labeled H, which indicates horizontal polarization. The polarization of a specific plane wave, **W**, can be located on the sphere as shown in Fig. 36. Note the role the relative phases of the complex polarization ratios play in locating **W**. The polarization ratios are related to the angles α, β, and γ shown in Fig. 36 as follows:

$$\varrho_\ell = \tan \alpha$$
$$\varrho_d = \tan \beta \qquad (48)$$
$$\varrho_c = \tan \gamma$$

The polarization box ([5], pp. 3.23 to 3.24) is a useful tool in establishing further relationships between the various defined quantities. It is in fact an extension of

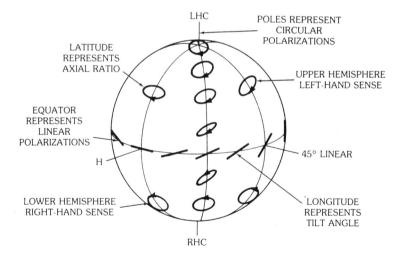

Fig. 35. Representation of polarization on Poincaré sphere.

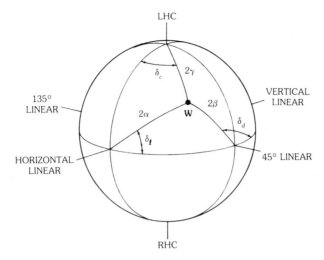

Fig. 36. The Poincaré sphere representation of the polarization of plane wave **W**.

the usual Stokes's parameters for a single-frequency wave of unit intensity. The polarization box is shown in Fig. 37 in relation to the Poincaré sphere.

The Poincaré sphere can also be used to determine the polarization efficiency when a plane wave is incident on an antenna. If the antenna's receiving polarization A_r is located on the sphere along with that of the incident wave, **W**, then the polarization efficiency is given by

$$p = \cos^2 \zeta \qquad (49)$$

where ζ is one-half the angular separation of **W** and A_r on the Poincaré sphere as shown in Fig. 38. If the law of cosines for spherical triangles is applied to the one

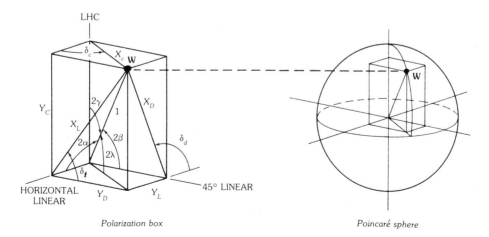

Polarization box   *Poincaré sphere*

Fig. 37. The polarization box and its relation to the Poincaré sphere. (*After ANSI/IEEE [2], © 1979 IEEE; after Hollis, Lyon, and Clayton [5], © 1970 Scientific-Atlanta*)

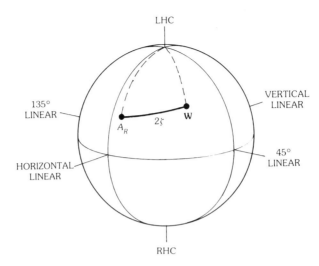

Fig. 38. The incident wave's polarization **W** and the antenna's receiving polarization A_r located on the Poincaré sphere.

formed by A_r, **W**, and the north pole of the sphere, a form for the polarization efficiency in terms of the polarization ratios results, i.e.,

$$p = \frac{1 + \varrho_w^2 \varrho_r^2 + 2\varrho_w \varrho_r \cos\Delta}{(1 + \varrho_w^2)(1 + \varrho_r^2)} \tag{50}$$

where, although the construction implies circular polarization ratios, ϱ_w and ϱ_r can be any of the three polarization ratios ϱ_ℓ, ϱ_d, or ϱ_c for the wave and antenna, respectively, and the angle Δ is the corresponding phase angle $[(\delta_\ell)_w - (\delta_\ell)_r]$, $[(\delta_d)_w - (\delta_d)_r]$, or $[(\delta_c)_w - (\delta_c)_r]$.

The polarization efficiency can also be written in terms of the axial ratios r_w and r_r of the polarization ellipses of the wave and receiving antenna polarizations as follows:

$$p = \frac{(1 + r_w^2)(1 + r_r^2) + 4r_w r_r + (1 - r_w^2)(1 - r_r^2)\cos\Delta_c}{2(1 + r_w^2)(1 + r_r^2)} \tag{51}$$

where $\Delta_c = 2(\tau_w - \tau_r)$.

Remember that the axial ratio carries a sign which must be used. The sign is positive for right-hand sense and negative for left-hand sense. If either axial ratio approaches infinity, the limit must be taken in order to evaluate p.

The various techniques for polarization measurement can be classified as absolute measurements or polarization-transfer measurements just as was done for gain measurements. In addition there are techniques which only yield partial polarization information.

Absolute measurements require no detailed *a priori* information about the polarizations of the antennas being tested. This method requires three antennas, no two of which can have circular or near-circular polarizations. This method can be combined with the three-antenna method of gain measurements and the extrapolation method to yield both gain and polarization of antennas [41].

The polarization-transfer method requires a polarization standard if highly accurate measurements are to be performed. The standard can be used as a sampling antenna or to calibrate the sampling antennas employed. Unfortunately, there is no universally accepted polarization standard. A carefully designed and fabricated linearly polarized antenna often can be used without proof of its polarization purity. When greater accuracy is required, the three-antenna absolute method can be used to calibrate the standard.

Generally, the sampling antennas used for polarization measurements are dual polarized with either orthogonal linear or orthogonal circular polarizations. With modern automated systems the antenna's polarization does not have to be purely linear or circular, provided the polarization is known with sufficient accuracy, since a computer can be programmed to make the necessary corrections. An alternate procedure is to adjust the polarization by use of a polarization-adjustment network [44]. The system can be designed to operate at radio, intermediate, or audio frequencies, and it can also be realized digitally.

The Phase-Amplitude Method of Polarization Measurement—Fig. 39 is a block diagram of the basic instrumentation required for the phase-amplitude method of polarization measurement. The dual-polarized sampling antenna and the two channels of the phase-amplitude receiver must be calibrated. As previously discussed this usually can be accomplished by employing a well-designed linearly polarized antenna as the standard.

First, assume that the sampling antenna has right- and left-hand circular polarizations. Then, with the antenna under test replaced by the linearly polarized standard, the amplitude responses at the outputs of the receiver should be equal and independent of the orientation of the standard in the transverse plane. The phase should be zero when the linearly polarized antenna is oriented for horizontal polarization. As the standard is rotated from the horizontal to vertical position in a positive angular direction with respect to the local coordinate system $(1, 2, 3)$, the output phase δ_c should increase from 0 to π rad. A helical antenna can be used to determine the senses of the polarization if required.

Once the system has been calibrated the linearly polarized standard antenna

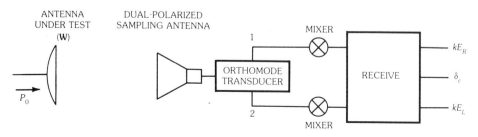

Fig. 39. Instrumentation for the phase-amplitude method of polarization measurement.

is replaced by the antenna under test. The ratio of the responses will give the circular polarization ratio, that is,

$$\varrho_c = \frac{|\hat{V}_R|}{|\hat{V}_L|} = \frac{E_R}{E_L} = \tan \gamma \qquad (52)$$

and the phase δ_c is given directly. The axial ratio r and the tilt angle τ are given by (44) and (45), respectively. The measurement is illustrated by means of the Poincaré sphere as shown in Fig. 40.

If extreme accuracy is required, the standard antenna should be measured using an absolute measurement technique so that its polarization is precisely known. However, usually polarization measurements are more limited by range errors than by the standard antenna, provided that care was exercised in the design of the standard.

Next, let the dual-polarized sampling antenna have horizontal and vertical linear polarizations (see Fig. 41). The accuracy of the calibration in this case is more sensitive to a deviation of the standard's polarization from pure linear than it was in the previous case, where circularly polarized sampling antennas were being used. Therefore, great care should be taken in the design and construction of the standard.

With the antenna under test replaced by the linearly polarized standard, the two polarizations of the sampling antenna are adjusted for optimum nulls when the standard is oriented for horizontal polarization and for vertical polarization, respectively.

Next, the standard is rotated to 45° with respect to the horizontal. The two responses $\hat{V}_V$ and $\hat{V}_H$ should be equal. Also the phase angle δ_ℓ should be zero for

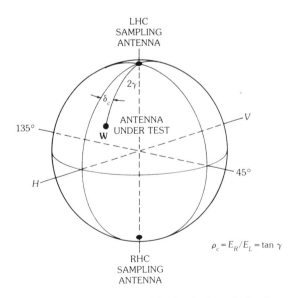

Fig. 40. Phase-amplitude measurements with circularly polarized sampling antennas.

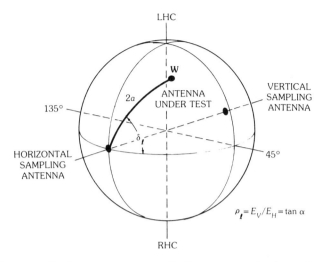

Fig. 41. Phase-amplitude measurements with linearly polarized sampling antennas.

all orientations of the standard. Finally, a helical antenna can be employed to check the sense of the phase angle.

The standard is replaced by the antenna under test. The ratio of the responses gives the linear polarization ratio ϱ_ℓ, i.e.,

$$\varrho_\ell = \frac{|\hat{V}_V|}{|\hat{V}_H|} = \frac{E_V}{E_H} = \tan\alpha \tag{53}$$

and the phase angle δ_ℓ is given directly. The complex circular polarization ratio and, hence, the axial ratio and tilt angle of the antenna under test can be computed with use of (46).

An alternate approach is to use a linearly polarized sampling antenna and a single-channel receiver. The two orthogonal polarizations are obtained by rotating the sampling antenna 90°, from horizontal to vertical, as illustrated in Fig. 42. In this figure $\hat{V}_{0°}$ is the relative response with the sampling antenna oriented to 0° and $\hat{V}_{90°}$ is the relative response with it oriented to 90°. For this measurement the complex circular polarization ratio of the sampling antenna must be accurately known.

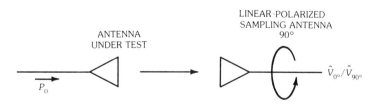

Fig. 42. The phase-amplitude method using a single linearly polarized sampling antenna; $\hat{V}_{0°}$ and $\hat{V}_{90°}$ are the relative responses when the sampling antenna is oriented at 0° and 90°, respectively.

It can be shown [44] that the ratio of the response for 0° orientation to that of 90° is given by

$$\hat{K} = \frac{\hat{V}_{0°}}{\hat{V}_{90°}} = j\frac{1 - \hat{\varrho}_{ca}\hat{\varrho}_{cs}}{1 + \hat{\varrho}_{ca}\hat{\varrho}_{cs}} \tag{54}$$

where the subscripts a and s refer to the antenna under test and the sampling antenna, respectively. The complex circular polarization ratio of the antenna under test is given by

$$\hat{\varrho}_{ca} = \frac{j - \hat{K}}{\hat{\varrho}_{cs}(j + \hat{K})} \tag{55}$$

If the sampling antenna is purely linearly polarized, $\varrho_{cs} = 1$.

Three-Antenna, Absolute Polarization Measurement—The three-antenna method does not require the use of a polarization standard; however, there is the restriction that no two of the antennas be circularly or near-circularly polarized; the third antenna's polarization can be arbitrary. A total of six measurements are required in order to determine the polarization of the three antennas. The method is illustrated in Fig. 43, where $\hat{V}_{12,0°}$ is the relative response at antenna 2 with its orientation at 0°. The response is measured twice for each combination of antennas, once with the receiving antenna oriented for 0° and once with it rotated 90°. The ratio of the two responses for each combination of antennas is determined. This gives the following three complex equations:

$$\hat{K}_{12} = j\frac{1 - \hat{\varrho}_{c1}\hat{\varrho}_{c2}}{1 + \hat{\varrho}_{c1}\hat{\varrho}_{c2}}$$

$$\hat{K}_{23} = j\frac{1 - \hat{\varrho}_{c2}\hat{\varrho}_{c3}}{1 + \hat{\varrho}_{c2}\hat{\varrho}_{c3}} \tag{56}$$

$$\hat{K}_{31} = j\frac{1 - \hat{\varrho}_{c3}\hat{\varrho}_{c1}}{1 + \hat{\varrho}_{c3}\hat{\varrho}_{c1}}$$

These equations can be combined to give

$$\hat{\varrho}_{c1} = (\hat{X}_{12}\hat{X}_{31}/\hat{X}_{23})^{1/2}$$
$$\hat{\varrho}_{c2} = (\hat{X}_{12}\hat{X}_{23}/\hat{X}_{31})^{1/2} \tag{57}$$
$$\hat{\varrho}_{c3} = (\hat{X}_{23}\hat{X}_{31}/\hat{X}_{12})^{1/2}$$

in which

$$\hat{X}_{mn} = \frac{1 + j\hat{K}_{mn}}{1 - j\hat{K}_{mn}}, \qquad mn = 12, 23, 31$$

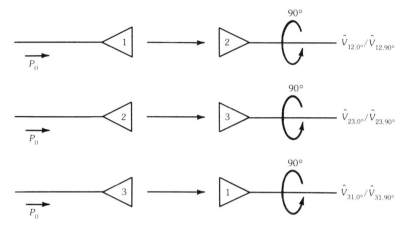

Fig. 43. The three-antenna absolute method of polarization measurement. (a) $\hat{V}_{12,0°}$ is the relative response at antenna 2 with its orientation at 0°. $\hat{V}_{12,90°}$ is the relative response with its orientation at 90°, etc.

There are two sets of solutions because of the square-root operation. One set is the negative of the other. This corresponds to an ambiguity in the resulting tilt angles, with those of one solution set rotated 90° from the other. This obliquity can be removed by a knowledge of at least one of the tilt angles. If none of the tilt angles is known, then an auxiliary measurement of one of the tilt angles may be required [44, 45].

Newell [46] has devised an improved procedure for this measurement. It entails rotating the receiving antenna from 0° to the angle that results in a minimum response, which is then recorded. The antenna is rotated 90° from that angle to where a maximum response is obtained. The ratio of the minimum to maximum responses is formed. This time the phase angle of the ratio is known to be 90°. This results in all of the $\hat{X}_{mn}$s being pure real; but then the phases of the circular polarization ratios are shifted by twice the total angular rotation from 0° to the angle where the maximum response is recorded. The ambiguity in the tilt angles remains, however.

The Multiple-Amplitude Component Method—Clayton and Hollis [47] have shown that the polarization of an antenna can be determined without measuring phase. It requires the measurement of the three polarization ratios, ϱ_ℓ, ϱ_d, and ϱ_c. From these data, 2α, 2β, and 2γ are determined [see (48)] since they define the loci of all possible polarizations on the Poincaré sphere which correspond to the polarization ratios ϱ_ℓ, ϱ_d, and ϱ_c, respectively. This is illustrated in Fig. 44.

The axial ratio r is determined by the use of (44). The phase angle of the complex circular polarization from which the tilt angle is determined is given by

$$\delta_c = \tan^{-1}(Y_D/Y_L) \tag{58}$$

where, from the polarization box,

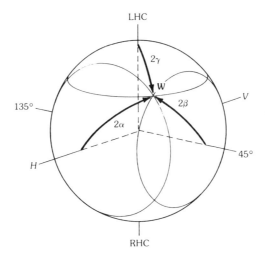

Fig. 44. Multiple-amplitude component method of polarization measurement.

$$Y_D = \frac{1 - \varrho_d^2}{1 + \varrho_d^2} \tag{59}$$

and

$$Y_L = \frac{1 - \varrho_\ell^2}{1 + \varrho_\ell^2} \tag{60}$$

A modified form of the multiple-amplitude component method attributed to Hollis [2] only requires the use of a linearly polarized sampling antenna which is used to measure ϱ_ℓ and ϱ_d. The method will yield the magnitude of the axial ratio and the tilt angle but not the sense of polarization. The tilt angle is determined from δ_c, which is computed using (58) as before, since only ϱ_ℓ and ϱ_d are required. The axial ratio is given by

$$r = -\cot \lambda \tag{61}$$

in which

$$\lambda = \tfrac{1}{2} \cos^{-1}(Y_L^2 + Y_D^2)^{1/2} \tag{62}$$

The terms Y_L and Y_D are obtained with use of (59) and (60). The square-root operation in the computation of λ results in the sense of polarization being indeterminate; however, the sense can be determined using a helical antenna.

Polarization Pattern Method—This method yields the tilt angle and the magnitude of the axial ratio, but it does not determine the sense of polarization. The antenna under test may be operated in either the receive or transmit mode. Assume that it is in the transmit mode. The method consists of plotting the magnitude of the voltage response $\hat{V}$ of a linearly polarized sampling antenna as its tilt angle is rotated in a plane perpendicular at the range axis. The graphical result is called a *polarization pattern*. A typical pattern is shown in Fig. 45. Note that the polarization ellipse of the antenna is inscribed in the pattern, it being tangent to the polarization pattern at the end of its major and minor axes.

Rotating-Source Method—In some cases a knowledge of the axial ratio over an entire pattern cut is required. Usually, it is a circularly polarized antenna for which this is a requirement. This can be accomplished by rotating the tilt angle of a linearly polarized sampling antenna operating as a source antenna while a pattern cut of the antenna under test is being recorded. The rate of rotation of the source antenna must be much greater than the angular rate of rotation of the positioner orientating the antenna under test. An *axial ratio pattern* results. A typical pattern is shown in Fig. 46. Note that an inner and an outer envelope can be discerned. At any given angle the axial ratio can be determined, it being the ratio of the outer to inner response levels. If the pattern is plotted in decibels, the axial ratios in decibels are simply the excursions of the responses recorded on the pattern. This method does not yield either tilt angle or sense of polarization.

5. Modeling and Model Measurements

Modeling techniques are used extensively in antenna measurements [2, 36, 48]. One reason for this is that it is often impractical or impossible to test an antenna in its operational environment. Antennas mounted on aircraft, ships, missiles, and satellites are examples of such situations. The design of the antennas for these vehicles usually begins in the preliminary design phase at which time only the external shape of the vehicle is available to the antenna designer. One of the first tasks is to determine the location of the various antennas subject to the structural and operational constraints imposed. While the experienced antenna engineer usually can predict the gross effect of the vehicular shape on the antenna's radia-

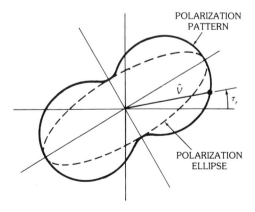

Fig. 45. The polarization pattern and its relation to the polarization ellipse of a wave.

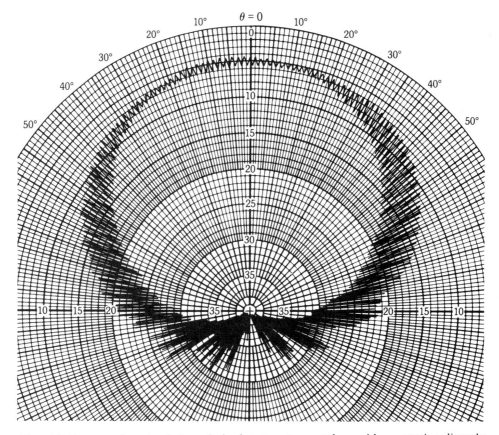

Fig. 46. Pattern of a circularly polarized test antenna taken with a rotating, linearly polarized, source antenna. (Courtesy of Tecom Industries, Inc.)

tion pattern, more precise information is ultimately required. The use of scale models provides a convenient means of obtaining such data.

Scale models are also used in radar cross-section (RCS) measurements [49]. Often such measurements are performed to verify theoretical computations of RCS; however, if the system is physically complex, an analytic solution may be impossible to obtain and numerical solutions prohibitively expensive. One must then rely solely on measurements to obtain the RCS. In many instances the targets to be tested are too large to be conveniently tested on a range or they are simply not available. Model measurements provide an appropriate solution.

Modeling techniques have been employed in various other situations. Both the radiation patterns and impedance properties of large hf antennas have been investigated using modeling techniques [50, 51]. The properties of these antennas are greatly affected by the earth; hence for these cases the earth is usually modeled or simulated also. Another application is the investigation of the effects that large metallic structures, such as bridges, towers, buildings, or other large antennas, have on communication or navigational systems [52, 53]. Scale-model systems are also used in the study of electromagnetic pulse effects [54], geological explorations

[55, 56], and radio-wave propagation [57, 58, 59]; these studies are beyond the scope of this chapter. References 60 and 61 contain thorough treatments of the use of modeling techniques for radio-wave propagation studies and experimental investigations of antennas in matter, respectively, and hence will not be considered here.

Modeling techniques find their greatest use in the study of physically large systems operating in free space by scaling the system to a smaller physical size. This allows controlled experiments to be performed on antenna ranges or in laboratories. Furthermore, large amounts of data can be conveniently obtained using scale models, whereas this is usually not possible if full-scale systems are tested. In almost all cases, however, it is desirable, if not required, that limited full-scale measurements be performed to substantiate the model measurements.

In addition to physical models there is widespread use of mathematical models. In the broadest sense of the word any mathematical equation that relates electromagnetic fields to sources may be considered a mathematical model. For our purpose a more restrictive usage is necessary. Here a mathematical model will refer to the case where a physical structure is described mathematically in order to compute electromagnetic fields in its presence. Perhaps "computer model" is a more apt term since the digital computer has made such modeling techniques possible. The method of moments [62, 63] and the geometrical theory of diffraction (GTD) [64] can be used to compute the fields and input impedances of antennas as well as RCS.

These methods find their greatest use in those cases where relatively simple shapes can be employed to represent the structures to be modeled. For example, an antenna mounted on the fuselage of an aircraft away from any of its complex structures, such as the cockpit, nose radar, or landing gear, lends itself to computer modeling [65–67]. The use of simple computer models to explore the sensitivity of the radiation pattern of the structure in various angular regions is very helpful in determining the requirements for a physical model or for a more sophisticated computer model.

Both physical and computer modeling are important and widely used tools in the study of electromagnetic systems.

Theory of Electromagnetic Models

Whenever the medium in which a field exists is linear, then Maxwell's equations for that medium will also be linear. It is the linearity of Maxwell's equations that suggests that scale models of electromagnetic systems are possible. Nonlinear modeling has been proposed [68]; however, its implementation has not been developed [54]. Only linear modeling will be considered here, which means that all systems having nonlinear or time-varying constituents are excluded.

The theory of electromagnetic models has long been known and used for antenna research and development [69–80]. Sinclair [81] laid the foundation for most modern work.* What follows was largely extracted from his work except that whereas he used time-domain quantities we shall use frequency-domain ones.

---

*References 70 and 81 contain rather complete bibliographies of early papers on modeling theory and techniques up to 1947.

To arrive at the condition for a scale model of an electromagnetic system: consider two electromagnetic systems: a full-scale, or actual, system and a scale model of that system. The complex electric- and magnetic-field vectors in the full-scale system shall be denoted as

$$\mathbf{E}(\omega, \mathbf{r}) \quad \text{and} \quad \mathbf{H}(\omega, \mathbf{r})$$

where ω is the angular frequency and $\mathbf{r}$ is the position vector. All variables in the model shall be denoted by primed quantities and they are related to the full-scale quantities in the following manner:

$$\begin{aligned} \mathbf{E}(\omega, \mathbf{r}) &= \alpha \mathbf{E}'(\omega', \mathbf{r}') \\ \mathbf{H}(\omega, \mathbf{r}) &= \beta \mathbf{H}'(\omega', \mathbf{r}') \end{aligned} \tag{63}$$

The position vectors are related by

$$\mathbf{r} = p\mathbf{r}' \tag{64}$$

and the angular frequencies by

$$\omega = \omega'/\gamma \tag{65}$$

The factors $\alpha, \beta, p,$ and γ are the electric, magnetic, spatial, and temporal (inverse frequency) scale factors, respectively.* Four factors are required since there are four fundamental quantities required to express the electromagnetic fields, mass, length, time, and electric current.

Both sets of fields must satisfy Maxwell's equations and the constitutive relations. This leads to the relationship between the permittivity, permeability, and conductivity of the media that constitute the two systems.

In the full-scale system, Maxwell's equations can be written in the frequency domain as

$$\begin{aligned} \nabla \times \mathbf{H}(\omega, \mathbf{r}) &= \sigma \mathbf{E}(\omega, \mathbf{r}) + j\omega\hat{\epsilon}\mathbf{E}(\omega, \mathbf{r}) \\ \nabla \times \mathbf{E}(\omega, \mathbf{r}) &= -j\omega\hat{\mu}\mathbf{H}(\omega, \mathbf{r}) \end{aligned} \tag{66}$$

in which $\hat{\mu}$ and $\hat{\epsilon}$ are the complex permeability and permittivity, respectively. In the scale-model system they can be written as

$$\begin{aligned} \nabla' \times \mathbf{H}'(\omega', \mathbf{r}') &= \sigma_m \mathbf{E}'(\omega', \mathbf{r}') + j\omega'\hat{\epsilon}_m \mathbf{E}'(\omega', \mathbf{r}') \\ \nabla' \times \mathbf{E}'(\omega', \mathbf{r}') &= -j\omega'\hat{\mu}_m \mathbf{H}'(\omega', \mathbf{r}') \end{aligned} \tag{67}$$

where $\hat{\mu}_m$ and $\hat{\epsilon}_m$ are the complex permeability and permittivity for the scaled system.

---

*King and Smith [61] and King [60] use scale factors that are reciprocal to the ones used here.

If the model is to be an exact simulation of the full-scale system, then the substitution for the scaled quantities given in (63)–(65) should transform (67) into (66). Since

$$\nabla' \times \mathbf{H}' = \frac{p}{\beta} \nabla \times \mathbf{H}$$
$$\nabla' \times \mathbf{E}' = \frac{p}{\alpha} \nabla \times \mathbf{E} \tag{68}$$

one obtains from the substitution

$$\nabla \times \mathbf{H} = \frac{\beta}{p\alpha} \sigma_m \mathbf{E} + j\omega \frac{\gamma\beta}{\alpha p} \hat{\epsilon}_m \mathbf{E}$$
$$\nabla \times \mathbf{E} = -j \frac{\gamma\alpha}{p\beta} \hat{\mu}_m \mathbf{H} \tag{69}$$

A comparison of (66) and (69) reveals that

$$\sigma_m = \frac{p\alpha}{\beta} \sigma$$
$$\hat{\epsilon}_m = \frac{p\alpha}{\beta\gamma} \hat{\epsilon} \tag{70}$$
$$\hat{\mu}_m = \frac{p\beta}{\alpha\gamma} \hat{\mu}$$

A tabulation of the equations relating the various quantities of a model system to the corresponding ones of the full-scale system is given in Table 2.

There are practical considerations that restrict one's choices for p, α, β, and

Table 2. The General Scale-Model System

Quantity	Transformation	Quantity	Transformation
Length	$\ell' = \ell/p$	Voltage	$V' = V/\alpha p$
Wavelength	$\lambda' = \lambda/p$	Current	$I' = I/\beta p$
Time	$t' = t/\gamma$	Power	$P' = P/\alpha\beta p^2$
Frequency	$f' = \gamma f$	Charge	$Q' = Q/\beta\gamma p$
Electric-field strength	$E' = E/\alpha$	Propagation constant	$k' = pk$
Magnetic field strength	$H' = H/\beta$	Resistance	$R' = \beta R/\alpha$
Conductivity	$\sigma' = p\alpha\sigma/\beta$	Reactance	$X' = \beta X/\alpha$
Permittivity	$\hat{\epsilon}' = p\alpha\hat{\epsilon}/\beta\gamma$	Impedance	$Z' = \beta Z/\alpha$
Permeability	$\hat{\mu}' = p\beta\hat{\mu}/\alpha\gamma$	Electric-flux density	$D' = pD/\beta\gamma$
Current density	$J' = pJ/\beta$	Capacitance	$C' = \alpha C/\beta\gamma$
Power density	$S' = S/\alpha\beta$	Inductance	$L' = \beta L/\alpha\gamma$
Charge density	$\varrho' = p^2\varrho/\beta\gamma$	Magnetic-flux density	$B' = pB/\alpha\gamma$
Angular frequency	$\omega' = \gamma\omega$	Scattering cross section	$A' = A/p^2$
Phase velocity	$v' = \gamma v/p$	Antenna gain	$G' = G$

γ. Perhaps the most common restriction is imposed because it is, in general, impractical to scale the permeability of a good conductor or a dielectric. These materials generally have a permeability equal to that of free space. This implies that for such cases one must choose

$$p\beta/\gamma\alpha = 1 \tag{71}$$

There are also practical considerations that dictate the choices of p and γ. These include restrictions on the physical size of the model system and the availability of suitable equipment. It should be noted that p and γ can be chosen independently. Thus the ratio α/β usually cannot be independently chosen because of these restrictions.

Note that the ratio α/β is the scale factor for impedance. Of particular importance in modeling are surface impedances at the interfaces between two different media. With each vector component of the fields tangential to an interface denoted by a subscript t, one can write

$$Z_s = \frac{E_t}{H_t} = (\alpha/\beta)\frac{E'_t}{H'_t} = (\alpha/\beta)Z_{sm} \tag{72}$$

in which Z_s is the surface impedance in the full-scale system and Z_{sm} the corresponding one in the model system.

Wave impedances are important in modeling because there are instances where it is desirable to simply model the surface impedances over the surface bounding the region of interest rather than those media adjacent to it [60]. The uniqueness theorem [82] can be imposed to show that this procedure is valid.

A further restriction on the choices of p and γ occurs when the region of interest of the full-scale system is air and it is desired that the measurement on the model system also be in air. This results in

$$p\alpha/\gamma\beta = 1 \tag{73}$$

in addition to (71). A comparison of (71) and (73) reveals that

$$\alpha = \beta \tag{74}$$

$$p = \gamma \tag{75}$$

Therefore,

$$\hat{\epsilon}_m = \hat{\epsilon} \tag{76}$$

$$\hat{\mu}_m = \hat{\mu} \tag{77}$$

and hence

$$Z_s = Z_{sm} \tag{78}$$

However, the conductivity must be scaled, since from (70)

$$\sigma_m = p\sigma \tag{79}$$

Note that $\hat{\epsilon}$, $\hat{\mu}$, $\hat{\epsilon}_m$, and $\hat{\mu}_m$ are complex quantities.

Equation 74 states that the electric and magnetic scale factors are equal. This can be understood by recognizing that the equality results in the wave impedances being equal at corresponding points in the two systems. This is a consequence of making the media the same in both systems. The restriction imposed by (75), which states that the spatial and temporal scale factors are equal, results in the frequency of operation in the scaled system being inversely scaled from the full-scale system compared to the mechanical dimensions. That is, if the physical sizes of structures are reduced by a factor of p, then the frequency of operation is increased by a factor p.

Notice that both (74) and (75) merely require that the ratios α/β and p/γ be equal to unity. Thus the value of α can be left arbitrary and only p, hence γ, be specified. If this is done the model is said to be a *geometrical model*. The flux lines will be similar in the two systems but not the power levels. If both α and p, hence β and γ, are specified, the model is called an *absolute* one in which the power levels in the two systems are related by a scale factor given by the product of α and β. This means that for an absolute model, quantitative data can be obtained directly.

Table 3 gives the scaling relations for the simpler geometrical model system in which only one scale factor, p, needs to be specified. This system will yield all of the important parameters that are usually required for antenna and RCS investigations and is therefore the system most often employed.

Materials for Electromagnetic Models

Geometrical modeling is conceptionally straightforward. One first chooses an appropriate scale factor which can be such that the model is made smaller or larger than, or, in some cases, the same size as, the full-scale system. A common case is to reduce the physical dimensions by the scale factor. The frequency of operation will then be increased by the scale factor. All structures are constructed of materials which have the same complex permittivities and permeabilities as those of the full-scale system. The conductivities, however, are increased by the scale factor.

A difficulty arises because of the availability of materials. For example, if good conductors such as aluminum, copper, silver, or gold are to be modeled, it will not

Table 3. The Geometrical Scale-Model System

Quantity	Transformation	Quantity	Transformation
Length	$\ell' = \ell/p$	Permittivity	$\hat{\epsilon}' = \hat{\epsilon}$
Time	$t' = t/p$	Permeability	$\hat{\mu}' = \hat{\mu}$
Wavelength	$\lambda' = \lambda/p$	Phase velocity	$v' = v$
Capacitance	$C' = C/p$	Resistance	$R' = R$
Inductance	$L' = L/p$	Reactance	$X' = X$
Impedance	$Z' = Z$	Conductivity	$\sigma' = p\sigma$
Frequency	$f' = pf$	Propagation constant	$k' = pk$
Antenna gain	$G' = G$	Scattering cross section	$A' = A/p^2$

be possible to find conductors with higher conductivity. Also, it is difficult, and in some cases impossible, to find dielectric and magnetic materials that have complex permittivities and permeabilities which have both their real and imaginary parts at the scaled frequency equal to those in the full-scale system. This is especially true of magnetic materials.

Good Conductors and Dielectrics—In most cases the conductivities of good conductors do not have to be scaled in order to produce an acceptable model. To show this, consider a plane wave incident on the interface between a good nonmagnetic dielectric and a good nonmagnetic conductor as shown in Fig. 47. The angle of incidence is denoted by θ_i, the angle of reflection by θ_r, and that of transmission by θ_t.

Snell's law states that

$$\sin \theta_t = \frac{k_d}{k_m} \sin \theta_i \tag{80}$$

in which k_d and k_m are the propagation constants for the dielectric and conductor regions, respectively. They are given by

$$k_d = \omega \sqrt{\mu_0 \epsilon} \tag{81}$$

and

$$k_m = \sqrt{-j\omega\mu_0\sigma} \tag{82}$$

Substitution of (81) and (82) into Snell's law, (80), gives

$$\sin \theta_t \cong \sqrt{\frac{j\omega\epsilon}{\sigma}} \sin \theta_i \tag{83}$$

This is an extremely small quantity for good conductors even at microwave frequencies; hence, for all practical purposes the wave can be considered to prop-

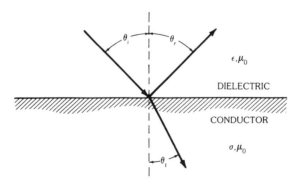

Fig. 47. The interface between a good nonmagnetic dielectric and a good nonmagnetic conductor.

agate normally ($\theta_t \cong 0$) into the conductor, regardless of the angle of incidence θ_i. This means that the surface impedance at the interface looking into the conductor is just the intrinsic impedance of the conductor, η_m; namely,

$$\eta_m \cong \left(\frac{\omega\mu_0}{2\sigma}\right)^{1/2}(1+j) \tag{84}$$

which is independent of θ_i. In most cases η_m can be assumed to be equal to zero. One can conclude that scaling the conductivity σ is not necessary as long as a good nonmagnetic conductor is used for the model. For our purposes a good conductor is one which has a conductivity such that

$$\sigma \gg \omega\epsilon_0 \tag{85}$$

Copper, gold, silver, aluminum, and brass are examples of good conductors at microwave frequencies.

Similarly, it is often the case that only the real part of a good dielectric needs to be invariant to scaling in order to produce an acceptable model. A good dielectric here is one for which the real part of the complex permittivity is much larger than its imaginary part, i.e.,

$$\epsilon' \gg \epsilon'' \tag{86}$$

where ϵ' is the real part or dielectric constant and ϵ'' is the imaginary part.

The principal effect of not scaling the conductivities of good conductors and ignoring the loss factor of good dielectrics is that the system losses will not be properly scaled. This means that while the radiation patterns of the modeled antennas may be reproduced to an acceptable accuracy, the gain will not. One can, however, obtain the directivity by pattern integration. Therefore the data from such model measurements are qualitative; that is, at least one measurement on the full-scale system is required to obtain the gain. Alternatively, if the losses in the full-scale antenna can be determined, the gain can be obtained since it is the product of the radiation efficiency and the directivity.

Lossy Dielectrics—A great deal of ingenuity is required to model lossy dielectrics. It is indeed fortuitous for one to find a commercially available dielectric that satisfies a modeling need. Usually composite materials must be employed. These include mixtures of powered dielectric and conductive materials which behave as lossy dielectrics. To establish a nomenclature for describing such materials consider the complex permittivity of a homogeneous, lossy dielectric,

$$\hat{\epsilon} = \epsilon' - j\epsilon'' \tag{87}$$

The relative values of ϵ' and ϵ'' with respect to the free-space permittivity ϵ_0 are usually given in the literature. The quantity ϵ'/ϵ_0 is usually called the relative dielectric constant and ϵ''/ϵ_0 is usually called the loss factor. Often one uses the loss tangent in computing losses in dielectrics; it is defined as

$$\tan \delta = \frac{\epsilon''}{\epsilon'} \qquad (88)$$

where δ is called the loss angle.

For composite materials there will be a conductivity as well as a complex permittivity. One can consider an equivalent conductivity that includes the dielectric's loss factor. To show this, consider the admittivity $\hat{y}(\omega)$ of the material,

$$\begin{aligned}\hat{y}(\omega) &= \sigma + j\omega\hat{\epsilon} \\ &= (\sigma + \omega\epsilon'') + j\omega\epsilon'\end{aligned} \qquad (89)$$

The equivalent conductivity is defined as

$$\sigma_{eq} = \sigma + \omega\epsilon'' \qquad (90)$$

Note that because of the factor ω multiplying ϵ'', ϵ_{eq} scales properly.

When a composite material is measured, it is more convenient to treat the equivalent conductivity as though it were a part of the loss factor; for example, one can write (89) as

$$\begin{aligned}\hat{y}(\omega) &= j\omega\left(\epsilon' - j\frac{\sigma_{eq}}{\omega}\right) \\ &= j\omega(\epsilon' - j\epsilon''_{eq})\end{aligned} \qquad (91)$$

where

$$\epsilon''_{eq} = \frac{\sigma + \omega\epsilon''}{\omega} \qquad (92)$$

The equivalent loss factor in $\epsilon''_{eq}/\epsilon_0$ and the equivalent loss tangent is given by

$$\tan \delta_{eq} = \frac{\epsilon''_{eq}}{\epsilon'} \qquad (93)$$

Generally these materials are measured using techniques that yield the complex permittivity, the imaginary part of which is the equivalent loss factor. Because of this, the adjective equivalent is usually dropped. A chart displaying the relative permittivity and loss tangent of the principal classes of solid materials at a frequency of 100 MHz and at room temperature is given in Fig. 48. Since the permittivities of most of these materials are reasonably independent of frequency up to the low microwave frequencies, these data can be considered typical for the classes of materials represented. For an exhaustive list of permittivities of specific materials, the reader is referred to references 83 and 84. Table 4 gives the conductivities of several conductors useful in modeling.

To obtain a specified complex permittivity it is usually necessary to use a mixture of dielectric and conductive materials. For example, a carbon-loaded

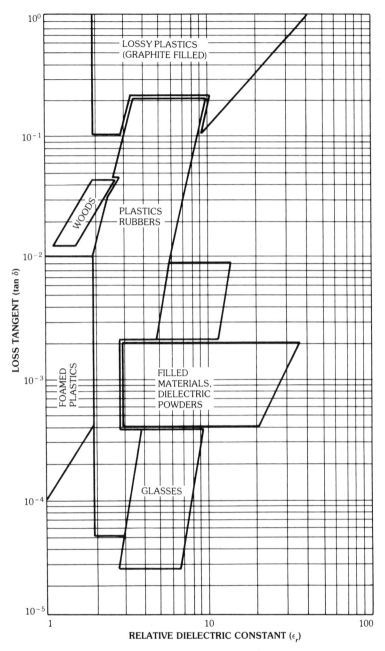

Fig. 48. The relative dielectric constant and loss tangent of solid materials at a frequency of 100 MHz and room temperature. (*After King and Smith [61], © 1981 MIT Press*)

polyurethane elastomer has been used to model land [85], as has a mixture of sand and carbon powder [58, 59].

There are a number of formulas for estimating the dielectric properties of mixtures [86, 87]; however, in all cases the complex permittivity must be measured.

Table 4. Conductivities of Metals and Alloys

Material	Conductivity (S/m at 20°C)
Aluminum	3.82×10^7
Aluminum, commercial alloys	$1.22\text{--}3.60 \times 10^7$
Brass	$1.2\text{--}1.6 \times 10^7$
Bronze, phosphor	$0.97\text{--}1.06 \times 10^7$
Carbon	$2.0\text{--}2.8 \times 10^4$
Copper, annealed	5.80×10^7
Copper, beryllium	1.72×10^7
Gold	4.10×10^7
Graphite	$0.7\text{--}1.2 \times 10^5$
Iron	1.03×10^7
Iron, gray cast	$0.05\text{--}0.20 \times 10^7$
Lead	0.457×10^7
Nickel	1.45×10^7
Silver	6.14×10^7
Steel	$0.5\text{--}1.0 \times 10^7$
Zinc	1.67×10^7

It is most useful to construct a graph to show how the dielectric constant and loss tangent vary with the constituents of the mixture. An excellent example has been given by Hall, Chambers, and McInnes [51]. They were modeling land using sand, salt, and water. Their results are given in Fig. 49 and Table 5.

Freshwater can be modeled by adding salt to water to obtain the scaled conductivity [61]. It should be noted that mixtures which include liquids are subject to change because of evaporation. This effect can be minimized by covering the mixture with a thin film of a plastic such as Mylar.

An attractive alternative to actually modeling a lossy dielectric is to simulate its presence by creating the required surface impedance over the surface corresponding to the boundary of the dielectric [88]. Of course, this approach can only be applied when the fields within the dielectric are not to be measured.

Scale-Model Construction, Instrumentation, and Measurement

A measurement program involving scale modeling requires careful planning. The first step in the process is to select a scale factor and to determine the extent of the modeling. The scale factor selection is based on a number of practical considerations, including the full-scale frequency of operation, the physical size and complexity of the structure to be modeled, the availability of instrumentation and any limitations imposed by the instrumentation, and cost.

The most convenient range of frequencies for model measurements is 100 MHz to 18 GHz. Instrumentation is usually readily available for these frequencies. Since radio frequencies for the system that require modeling are, in general, allocated internationally, it is not surprising that there is a striking uniformity throughout industry in the choice of scale factors for similar projects. For radiation pattern measurements, aircraft models range from 1/4 to 1/60 scale, ship models range from 1/48 to 1/200 scale with 1/48 or 1/50 being quite standard, and ground-based hf antennas are usually of the order of 1/100 scale.

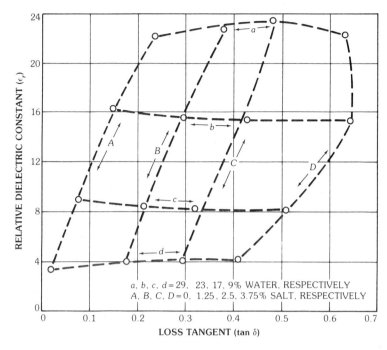

Fig. 49. Relative dielectric constant and loss tangent of various mixtures of water, salt, and sand. (*After Hall, Chambers, and McInnes [51], © 1978 IEE; published by Peter Peregrinus Ltd., London*)

Table 5. Comparison of the Dielectric Constant and Conductivity of a Sand, Salt, and Water, 1/80 Scale Simulation of Wet and Dry Ground With Those of the Full-Scale Ground (*After Hall, Chambers, and McInnes [51], © 1978 IEE; published by Peter Peregrinus Ltd., London*)

Ground Type	Wet	Dry
Full-scale (25 MHz)	$\epsilon_r = 23.0$ $\sigma = 5$ mS/m	$\epsilon_r = 5.0$ $\sigma = 1.0$ mS/m
Model (2 GHz)	$\epsilon_{rm} = 23.0$ $\sigma_m = 0.4$ S/m	$\epsilon_{rm} = 5.0$ $\sigma_m = 0.08$ S/m
Mixture (%) Sand Water Salt	 68.5 30.0 1.5	 87 12 1

The extent of the modeling refers to the amount of detail required for the model and how much of the antenna's environment must be included. This decision is based on the general radiation characteristics of the antenna to be tested and its location relative to other antennas, sharp discontinuities, and any protuberances on the surface of the structure on which it is mounted. For example, the radiation pattern of a directive antenna located on the nose bulkhead of an aircraft will be relatively unaffected by the fuselage and wings; therefore only the nose of the aircraft would need to be included in the model. On the other hand, the radiation pattern of an omnidirectional vhf antenna located on the fuselage of an aircraft will be greatly affected by the entire structure of the aircraft and other antennas in its vicinity. For this case the entire aircraft must be precisely modeled. It is very important that models be well designed mechanically. Any deformation of the model surface or movement during a measurement will result in unreliable data.

Ground-Based HF Antennas—The hf band is designated by the ITU and the IEEE to be 3 to 30 MHz; however, the usable frequency band extends from 2 MHz to above 30 MHz, with corresponding wavelengths of 150 m to less than 10 m. Directive antennas designed to operate at the low end of this band are, therefore, necessarily physically quite large. They are usually constructed of stranded, steel-cored aluminum wire which is supported by masts or towers [89, 90]. In addition there must be a feed system to excite the antenna and guy wires to support the towers, both of which must be included in the model.

The radiation patterns of these antennas are greatly affected by the ground, the properties of which can vary from day to day. The water content and salinity, are the dominant factors which affect the conductivity of soil [91]. Conductivities ranging from 0.001 to 0.05 S/m, depending on the amount of water present and its salinity, are normal. The dielectric constant is also affected by the water content; it varies from about 10 to 20. Because of the difficulties of modeling earth, scaled hf antennas are often tested on an image-plane range, the surface of which is constructed of a good conductor. The measured data are then adjusted to account for imperfect earth [89].

There are inherent difficulties in obtaining an exact scale model of an hf antenna. The principal reason is that when the wire diameters are exactly scaled, they become too fragile to be used. Also the ohmic loss becomes excessive. It is usual practice to use a copper wire of sufficient diameter to maintain mechanical integrity during the measurement process. A low–dielectric-constant foam can be used to support the modeled wires. The same situation occurs for guy wires and feed lines. Guy wires are usually designed to have little effect on the radiation pattern of the antenna; therefore one would not expect a problem because of the guy wire size.

The radiation pattern of some types of antennas is affected by the choice of wire diameter. The Yagi-Uda antenna as well as the log-periodic antenna are notable examples [51, 92]. If one is primarily interested in the effect of the antenna's environment on its radiation pattern, then it is appropriate to redesign the scaled model to account for the change in wire diameter so that its radiation pattern in free space duplicates that computed for the full-scale antenna alone. When the redesigned scaled antenna is placed in the modeled environment, the measured radiation patterns should replicate those of the full-scale system. Fig. 50 depicts an experimental apparatus for measuring the radiation pattern of an hf antenna where the earth is also simulated.

It is impractical to make input impedance measurements on a 1/100 scale-modeled hf antenna. Not only is the antenna's input impedance affected by the wire size and losses in the image plane, but also by the design of the feed of the

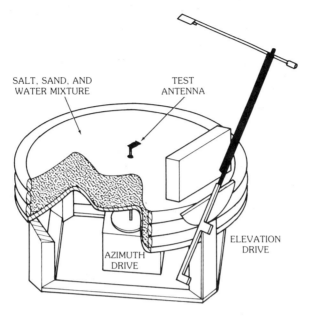

Fig. 50. An apparatus for the simulation of ground and measurement of radiation patterns of antennas over ground. (*After Hall, Chambers, and McInnes [51]*, © *1978 IEE; published by Peter Peregrinus Ltd., London*)

antenna. All of these are virtually impossible to precisely model with such a large scale factor; therefore impedance models are usually 1/2 to 1/7 of the full-scale antenna. It is often the case that designers of such structures make an engineering model with a scale factor within this range of values for mechanical and scheduling purposes. Such a model can also be used for impedance measurements.

Shipborne Antennas—The electronics systems aboard a large modern warship may require over 100 antennas, all of which must be located in the very restricted space of the ship's topside (79). This results in perhaps the most complex environment in which antenna engineers must place antennas. Modeling techniques are extensively used in the design of the hf antennas required and have been employed for omnidirectional vhf and uhf antennas as well [80]. The superstructure of a ship being physically complex, it must be modeled accurately since it provides many electrical paths commensurate with resonant lengths at hf. Model ships are usually scaled to 1/48 or 1/50 size. To obtain the precision required, the models must be constructed from the ship's actual plans and include all detail above the waterline of the full-scale ship. The construction is usually of wood covered with brass sheet. Turrets, masts, and other such objects are usually machined from brass. The model must include provisions for duplicating the motions of movable structures such as radar antennas, missile launchers, and landing-pad guardrails so that their effect on the characteristic of the antenna being tested can be measured. A strip of brass screen around the base of the model (the waterline of the full-scale ship) provides a means of making an electrical connection to the image-plane turntable. Figs. 51 and 52 picture models of ships illustrating the detail that is required.

Fig. 51. A 1/48 scale model ship for hf antenna radiation pattern measurements. (*Courtesy Naval Ocean Systems Center, San Diego, Calif.*)

Fig. 52. A scale model of a ship soldered to the turntable of an image-plane range. Note the gantry in the background used to rotate the source antenna. (*Courtesy Naval Sea Systems Command, Washington, D.C.*)

Like the model ground-based hf antenna, wire diameters cannot be modeled exactly; however, the radiation patterns for the types of wire antennas used aboard ships are not greatly affected by wire size.

Model shipborne antennas are tested on image-plane ranges such as that depicted in Fig. 6. In that figure a large arch is used to provide the elevation motion of the source antenna. An alternative approach is to use a large nonmetallic boom to provide the semicircular motion required for the source antenna [80, 94, 95]. A corner reflector can be used as a source antenna when propagation along the surface ($\theta = 90°$) is required [96].

The ground plane simulates the saltwater of the ocean. Because saltwater is, in fact, a good conductor, aluminum or wire mesh image planes are usually employed. While it is possible to model the ocean using composite materials, the cost can become prohibitive. This is particularly true in harsh climates where the surface is exposed to extremes in temperature which can cause damage to the surface.

By using a good conductor, such as aluminum, for the surface, the Brewster's angle phenomena will not be duplicated. This, however, is of little consequence at high frequencies since, for seawater, the Brewster's angle occurs at a grazing angle of less than 1°, as shown in Fig. 53.

The instrumentation for radiation patterns is the same as that for other ranges. Impedance measurements can also be performed on the modeled hf antennas. Usually a network analyzer is employed for this purpose. The model antenna is fed from below the image plane. Often all of the required instrumentation is also located under the image plane. Modeled shipborne antennas can also be used to perform electromagnetic compatibility studies.

Aircraft Antennas—Aircraft, like ships, carry many antennas. Table 6 lists the standard communication, navigational, and identification systems found on most aircraft. All of these systems require antennas. In addition, there are many more special-purpose antennas, especially on military aircraft.

For radiation pattern measurements, models 1/7 to 1/10 of the full-scale aircraft size are typical for those antennas which operate at frequencies above about 50 MHz. Models 1/50 or 1/60 full scale are typical for hf antennas. Model aircraft are usually fabricated from patternmaker's white pine or basswood [97]. The model contours are obtained from the aircraft designer's loft drawings photographically reduced to the proper scale and applied to metal sheets from which templates are made. Models are made hollow to accommodate the required instrumentation. The hollow interior also helps to reduce the weight of the model. Removable sections are required to allow access to the interior of the model. Where weight is an important factor the model can be fabricated from a glass cloth–reinforced plastic laminate with a honeycomb core. This type of construction is usually more expensive than wood since a plastic mold is required for the lay-up of the laminate. In some instances it may be advantageous to construct part of the model from wood and part from glass cloth–reinforced plastic laminate [48].

The model must be covered with a good conductor to simulate the metallic skin of the aircraft. Sheets of soft copper can be formed over the model, or the model can be flame-sprayed with copper or aluminum. A copper surface has the

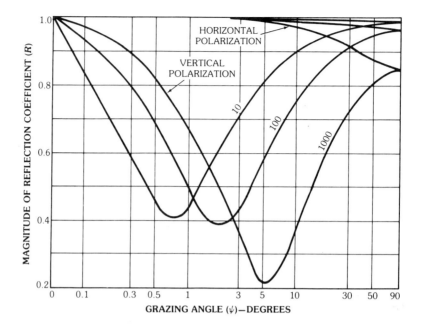

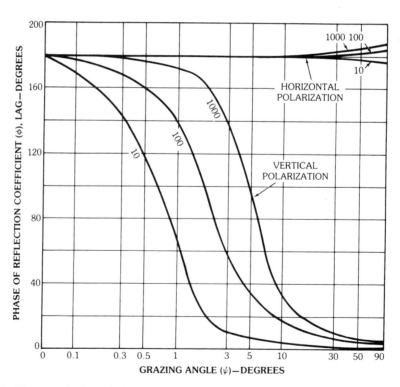

Fig. 53. The magnitude and phase of the reflection coefficient as a function of grazing angle for a plane wave incident on smooth seawater ($\epsilon_r = 80$ and $\sigma = 4.64$ S/m) at 10, 100, and 1000 MHz. (*a*) Magnitude of reflection coefficient. (*b*) Phase of reflection coefficient.

Table 6. Communication/Navigational/Identification Systems Requiring Antennas on Aircraft

System	Frequency Range (MHz)
Hf communication	2–30
Marker beacon	75
Localizer	108–112
VOR	108–122
Vhf communications	116–156
Uhf communications	225–400
Glide slope	329–335
Tacan	960–1020
DME	962–1024
Control radar beacon system	1030–1090
Global positioning system	1227.60–1575.42
Radio altimeter	4300

advantage of being readily solderable, which is important when mounting antennas onto the model. A metal thickness of about 0.05 cm is usually satisfactory. A model aircraft for radiation pattern measurements is shown in Fig. 54.

Models that are physically large are more conveniently fabricated with wooden formers covered by thin plywood veneer and metal foil. Hot-dipped hard-ware cloth with 1/8-in (3.175-mm) mesh can also be used to simulate the metallic surface of the aircraft. Copper screen is not recommended because the screen wires are not bonded together, and therefore, after oxidation takes place, it will not adequately simulate a continuous conducting surface. If a section of the model has a constant cross section, then an aluminum sheet can be used.

As a rule of thumb, the dimensional tolerances required for the external surface of the model is a sixteenth of a wavelength at the highest model frequency used. This is usually about 0.064 cm. All protuberances on the full-scale vehicle which are a tenth of a wavelength or larger should be included in the model if they are in the vicinity of the antenna to be tested. These include other antennas, which should be appropriately terminated to simulate the actual full-scale environment of the test antenna. It is very important that all thin wires, sharp edges, and surface discontinuities be modeled if they are within two or three wavelengths of the antenna under test. Flaps, retracting landing gear, and other movable structures must be modeled.

Generally, if the dielectric in the full-scale system is electrically very thin, it can be neglected in the model; or, more properly stated, it can be modeled as free space (air). For example, an aircraft model for the study of scaled hf, vhf, or uhf antennas would not have to include the plastic canopy or nose radome. However, all metallic parts such as frames, pitot tubes, and lightning diverters, which are built into the canopy or radome, must be modeled along with the metallic structures enclosed by canopy and radome. One has to be especially careful if the enclosed parts are of resonant dimensions at the frequency of operation.

There are situations in which even electrically thin dielectrics affect the radiation pattern of antennas. These are cases where surface waves are excited or lens effects are prominent [98, 99]. One such example is a radome of circular cross

Fig. 54. A copper-clad model airplane mounted on a model tower in a large microwave anechoic chamber. (*Courtesy MBB, Munich, West Germany*)

section enclosing an end-fire surface-wave antenna [100]. If the average diameter of the radome is on the order of a wavelength or less, then it is, in fact, an integral part of the antenna even though it may be electrically thin and hence should be included in the model. The lenticularly shaped rotadome enclosing a uhf antenna is another such example [101].

Radomes are sometimes of a sandwich-type construction with a honeycomb core and glass fiber–reinforced epoxy or polyester laminates on either side. This type of radome can be satisfactorily modeled using a single sheet of dielectric which has the same electrical thickness [99], although, if the scale factor is not too great or the physical dimensions too small, one may model the sandwich construction using low-loss foam in place of the honeycomb for the model [101].

Aircraft antennas usually cannot be modeled exactly. This can be due to the complexity of their structures or a lack of detailed information from their manufacturers. To ensure that model antennas will reproduce the radiation characteristics of the full-scale antennas, a preliminary measurement, if possible, is advisable. For example, the radiation pattern of a stub-type antenna, which is to be modeled, can be measured with the antenna mounted on a flat, circular, ground plane. The measurement is repeated with the corresponding modeled antenna mounted on a scaled, circular, ground plane. A comparison of the patterns will reveal the adequacy of the scale-model antenna. Convenient sizes of the full-scale ground planes range from 0.85 to 3.5 m in diameter, depending on the frequency used. The modeled ground plane usually requires a quarter-wave choke on its periphery or a ring of absorbing material on its back side to reduce the currents that flow around its edge.

Once the model antenna's radiation patterns have been shown to replicate those of the full-scale antenna mounted on the flat ground plane, it is then ready for installation on the aircraft model. The use of a standard mounting system for model antennas is highly desirable. One approach is to use a standard brass receptacle which can be soldered in place on the model aircraft. The model antenna is mounted on a mating plug terminated with a miniature coaxial connector. The antenna-plug assembly can be inserted into the receptacle from the outside of the model. This method is most useful when one is making design changes in an antenna structure or when standard antennas are used on successive models. Fig. 55 depicts one type of receptacle and plug arrangement.

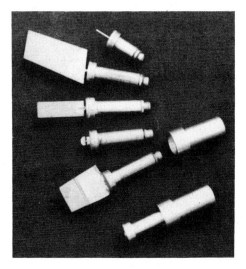

Fig. 55. A receptacle and plug assembly for mounting antennas onto model aircraft. (*Courtesy Lockheed-California Corp.*)

To perform radiation pattern measurements, the model aircraft is mounted on a special roll-over-azimuth positioner with a nonmetallic mast. These positioners are called model towers. The remote mixer or any other instrumentation that must be located at the test antenna is placed inside the model. Care has to be taken to mount the devices directly to the model structure so that there will be no unwanted movement. The coaxial cable between the remote mixer and the receiver exits the model at the point where the model is mounted to the model tower. The antenna being measured must be on the opposite side of the model from the mount. Care must be exercised to ensure that the cable does not affect the measurements. Two blade-type antennas mounted on a model aircraft are shown in Fig. 56.

A severe problem arises when the model aircraft is small compared with the wavelength at the test frequency. This situation arises when hf antennas are being modeled, in which case the model may be physically small also. Thus the head of the model tower and the cable from the model to the receiver can greatly affect the radiation patterns. To reduce the effect of the model tower head, the model can be mounted on a plastic rod and located above the model tower. The cable effect can be eliminated by use of a battery-operated transmitter located within the model. For this approach the receiver is connected to the source antenna. Another method is to use a fiber-optic waveguide between the model and the base of the model tower [97]. The optical transmitter must be mounted inside the model. Finally, one can use an amplitude-modulated rf signal of sufficient power to illuminate the model. The modulation frequency is usually 1 kHz. The detector is then located in the model aircraft. Instead of using coaxial cable to couple the audio signal to the amplifier, a carbon-loaded, high-impedance cable can be employed. This type of cable has little effect on the radiation pattern of the model. A system of this type requires higher power than does the conventional mixer-receiver system and does not have the selectivity of a receiver to discriminate against unwanted signals.

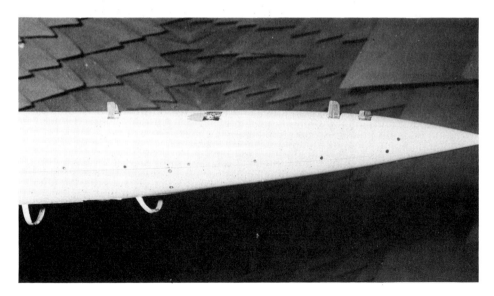

Fig. 56. Model blade antennas mounted on a model aircraft. (*Courtesy Northrop Corp.*)

It is standard practice to determine the directivity of the test antenna by pattern integration. The isotropic level can then be located on the graphical display of the radiation pattern. The radiation patterns measured on the model aircraft are ultimately compared to measurements made on the aircraft flying over a prescribed path.

With the exception of the hf antennas, impedance measurements are made

Fig. 57. A 1/24 scale-model helicopter mounted on a model tower. (*Courtesy National Research Council, Canada*)

at full scale on partial mock-ups of the aircraft. To measure the impedances of hf antennas, 1/4 to 1/7 scale models are usually employed. The model should be located at least a wavelength above the ground. All instrumentation must be located within the model and remotely controlled.

In addition to airplanes, helicopters, satellites, spacecraft, rockets, and projectiles of various types have systems that require antennas. Modeling techniques are often used in the design and study of these antennas [102]. Figs. 57 and 58 depict two such models.

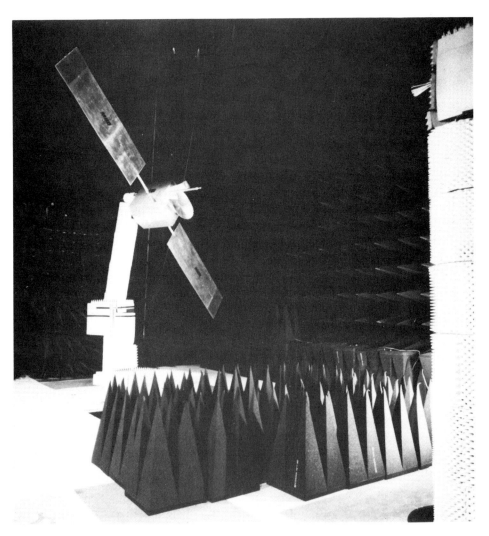

Fig. 58. A 1/2.5 scale model of the German Communication Satellite DFS for testing the telemetry and telecommand antenna. The measurement was performed on the spherical near-field range of the Technical University of Denmark. (*Courtesy MBB, Munich, West Germany*)

6. References

[1] *IEEE Standard Definitions of Terms for Antennas*, IEEE Std. 145-1982, IEEE, New York, NY.

[2] *ANSI/IEEE Standard Test Procedures for Antennas*, ANSI/IEEE Std. 149-1979, IEEE, New York: John Wiley distributors.

[3] R. C. Johnson, H. A. Ecker, and R. A. Moore, "Compact range techniques and measurements," *IEEE Trans. Antennas Propag.*, vol. AP-17, pp. 563–576, September 1969.

[4] C. C. Cutler, A. P. King, and W. E. Kock, "Microwave antenna measurements," *Proc. IRE*, vol. 35, pp. 1462–1471, December 1947.

[5] J. S. Hollis, T. J. Lyon, and L. Clayton, Jr., eds., *Microwave Antenna Measurements*, Atlanta: Scientific-Atlanta, 1970.

[6] R. C. Johnson, H. A. Ecker, and J. S. Hollis, "Determination of far-field antenna patterns from near-field measurements," *Proc. IEEE*, vol. 61, pp. 1668–1694, December 1973.

[7] W. H. Kummer and E. S. Gillespie, "Antenna measurements—1978," *Proc. IEEE*, vol. 66, pp. 483–507, April 1978.

[8] J. Appel-Hansen, J. D. Dyson, E. S. Gillespie, and T. G. Hickman, "Antenna measurements," chapter 8 in *The Handbook of Antenna Design*, vol. 1, ed. by A. W. Rudge et al., Stevenage, Herts., UK: Peter Peregrinus, 1982, pp. 584–694.

[9] A. C. Ludwig, "The definition of cross-polarization," *IEEE Trans. Antennas Propag.*, vol. AP-21, pp. 116–119, January 1973.

[10] A. J. Lombardi and M. S. Polgar, Jr., "Wide band antenna facility," *Elec. Commun.*, vol. 49, pp. 94–98, January 1974.

[11] L. G. Sturgill and S. E. Thomas, "Scale model shipboard antenna measurements with a computer automated antenna analyzer system," *Proc. Antenna Meas. Techniques Assoc. Mtg.*, Atlanta, October 17–18, 1979.

[12] D. E. Fessenden and D. C. Portofee, "An antenna pattern arch for measurements over sea water," *Proc. Antenna Meas. Techniques Assoc. Mtg.*, Las Cruces, October 5–7, 1982.

[13] C. G. Bachman, H. E. King, and R. C. Hansen, "Techniques for measurements of reduced radar cross-sections, pt. 1," *Microwave J.*, vol. 6, pp. 61–67, January 1963.

[14] D. F. DiFonzo and W. J. English, "Far-field criteria for reflectors with phased array feeds," *1974 Intl. IEEE/AP-S Symp. Dig.*, IEEE, New York.

[15] P. S. Hacker and H. E. Schrank, "Range distance requirements for measuring low and ultra-low sidelobe antenna patterns," *IEEE Trans. Antennas Propag.*, vol. AP-30, pp. 956–966, September 1982.

[16] R. C. Hansen, "Measurement distance effects on low sidelobe patterns," *IEEE Trans. Antennas Propag.*, vol. AP-32, pp. 591–594, June 1984.

[17] D. E. Baker, "Development and evaluation of the 500m ground-reflection antenna test range of CSIR, Pretoria, South Africa," *Proc. Antenna Meas. Techniques Assoc. Mtg.*, San Diego, October 2–4, 1984.

[18] P. W. Arnold, "The slant antenna range," *IEEE Trans. Antennas Propag.*, vol. AP-14, pp. 658–659, September 1966.

[19] W. H. Emerson, "Electromagnetic wave absorbers and anechoic chambers through the years," *IEEE Trans. Antennas Propag.*, vol. 21, pp. 484–490, July 1973.

[20] J. A. Strom and W. G. Mavroides, "RADC electromagnetics test facility at Ipswich, Ma.," *Proc. 1982 Antenna Meas. Techniques Assoc. Mtg.*, Las Cruces, October 5–7, 1982.

[21] J. Appel-Hansen, "Reflectivity level of radio anechoic chamber," *IEEE Trans. Antennas Propag.*, vol. AP-21, pp. 490–498, July 1973.

[22] W. J. English and R. W. Gruner, "Precision polarization measurement techniques for circularly polarized antennas," *Dig. IEEE AP-S Intl. Symp. Antennas Propag.*, University of Illinois (IEEE catalog no. 75CH0965-9AP).

[23] J. R. Mentzer, *Scattering and Diffraction of Radar Waves*, London: Pergamon Press,

1955, pp. 124–126.

[24] D. W. Hess, F. G. Willwerth, and R. C. Johnson, "Compact range improvements and performance at 30 GHz," *Dig. Intl. Symp. Antennas Propag.*, Stanford Univ., June 20–22, 1977, pp. 264–267.

[25] W. D. Burnside, M. C. Gilreath, and B. Kent, "A rolled edge modification of compact range reflector," *Proc. Antenna Meas. Techniques Assoc. Mtg.*, San Diego, October 2–4, 1984.

[26] W. D. Burnside, A. K. Dominek, and R. Barger, "Blended surface concepts for a compact range reflector," *Proc. Antenna Meas. Techniques Assoc. Mtg.*, Melbourne, Fla., October 29–31, 1985.

[27] W. D. Burnside and L. Peters, Jr., "Target illumination requirements for low RCS target measurements," *Proc. Antenna Meas. Techniques Assoc. Mtg.*, Melbourne, Fla., October 29–31, 1985.

[28] D. W. Hess and R. C. Johnson, "Compact ranges provide accurate measurement of radar cross section," *Microwave Syst. News*, vol. 12, no. 9, September 1982.

[29] T. G. Hickmann and R. C. Johnson, "Boresight measurements utilizing a compact range," *Abstracts, Spring USNC/URSI Mtg.*, Washington, D.C., April 13–15, 1972.

[30] R. C. Rudduck, W. D. Burnside, M. C. Liang, and T. H. Lee, "The compact range as an electromagnetic field simulator," *Proc. Antenna Meas. Techniques Assoc. Mtg.*, Melbourne, Fla., October 29–31, 1985.

[31] V. J. Vokurka, "9.3 compact range," in *Satellite Communication Antenna Technology*, ed. by R. Mittra et al., Amsterdam: Elsevier Science Publishers B.V. (North-Holland), 1983.

[32] J. K. Conn, C. L. Armstrong, and L. S. Gans, "A dual shaped compact range for ehf antenna measurements," *Proc. Antenna Meas. Techniques Assoc. Mtg.*, San Diego, October 2–4, 1984.

[33] S. W. Zieg, "A precision optical range alignment technique," *Proc. Antenna Meas. Techniques Assoc. Mtg.*, Las Cruces, October 5–7, 1982.

[34] J. Appel-Hansen, "Reflectivity level of radio anechoic chambers," *IEEE Trans. Antennas Propag.*, vol. AP-21, pp. 490–498, July 1973.

[35] E. F. Buckley, "Outline of evaluation procedures for microwave anechoic chambers," *Microwave J.*, vol. 6, pp. 69–75, August 1963.

[36] C. A. Balanis, *Antenna Theory: Analysis and Design*, New York: Harper & Row, 1982, pp. 37–42.

[37] R. J. Stegen, "The gain-beamwidth product of an antenna," *IEEE Trans. Antennas Propag.*, vol. AP-12, pp. 505–506, July 1966.

[38] R. S. Elliott, *Antenna Theory and Design*, Englewood Cliffs: Prentice-Hall, 1981, pp. 205–207.

[39] R. R. Bowman, "Field strength above 1 GHz: measurement procedures for standard antennas," *Proc. IEEE*, vol. 55, pp. 981–990, June 1967.

[40] J. Bellamy, J. Hill, and S. Wilson, "Automatic gain measurement system," *Proc. Antenna Meas. Techniques Assoc. Mtg.*, September 27–29, 1983.

[41] A. C. Newell, R. C. Baird, and P. F. Wacker, "Accurate measurement of antenna gain and polarization at reduced distances by an extrapolation technique," *IEEE Trans. Antennas Propag.*, vol. AP-21, pp. 418–431, July 1973.

[42] L. H. Hemming and R. A. Heaton, "Antenna gain calibration on a ground reflection range," *IEEE Trans. Antennas Propag.*, vol. AP-21, pp. 532–538, July 1973.

[43] H. G. Booker, V. H. Rumsey, G. A. Deschamps, M. I. Kales, and J. I. Bohnert, "Techniques for handling elliptically polarized waves with special reference to antennas," *Proc. IRE*, vol. 39, pp. 533–552, May 1951.

[44] J. S. Hollis, et al. *Microwave Antenna Polarization Measurements*, to be published, Scientific-Atlanta, Inc.

[45] J. R. Jones and D. W. Hess, "Automated three-antenna polarization measurements using digital signal processing," *Proc. Antenna Meas. Techniques Assoc. Mtg.*, Melbourne, Fla., October 29–31, 1985.

[46] A. C. Newell, "Improved polarization measurements using a modified three-antenna

technique," *Dig. IEEE Intl. Antennas Propag. Symp.*, Urbana-Champaign, Ill., pp. 337–340, June 2–4, 1975.

[47] L. Clayton, Jr., and J. S. Hollis, "Antenna polarization analysis by amplitude measurement of multiple components," *Microwave J.*, vol. 8, pp. 35–41, January 1965.

[48] E. L. Kane, "Scale model aircraft antenna measurements," *Paper 7002*, Douglas Aircraft Company, Long Beach, 1980.

[49] C. K. Krichbaum, "Radar cross-section measurements," in *Radar Cross-section Handbook, Vol. 2*, ed. by George T. Ruck, New York: Plenum Press, 1970, pp. 893–935.

[50] E. L. Kilpatrick and H. V. Cottony, "Scaled model measurements of performance of an elevated sloping vee antenna," *ESSA Tech. Rep. ERL 72-ITS62*, US Department of Commerce, ESSA Research Laboratories, Boulder, 1968.

[51] P. S. Hall, B. Chambers, and P. A. McInnes, "Microwave modelling of h.f. Yagi antennas over imperfectly-conducting ground," *Proc. IEE*, vol. 125, no. 4, pp. 261–266, 1978.

[52] S. Toyada and H. Hashimoto, "Scattering characteristics of vhf television broadcasting waves by steel towers of overhead power transmission lines," *IEEE Trans. Electromagnetic Compat.*, vol. EMC-21, no. 1, pp. 62–65, 1979.

[53] V. V. Liepa and R. L. Frank, "Experimental scale model study of loran-C signals near bridges," *1983 Intl. Symp. Dig. Antennas Propag.*, vol. 2, pp. 545–547 (IEEE catalog no. 83CH1860-6).

[54] T. B. A. Senior and V. V. Liepa, "Scale modeling," in *EMP Interaction: Principles, Techniques and Reference Data*, ed. by K. S. H. Lee, Air Force Weapons Laboratory, EMP Interaction 2-1, AFWL-TR-80-402, Albuquerque, 1980.

[55] G. V. Keller and F. C. Frischknecht, *Electrical Methods in Geophysical Prospecting*, New York: Pergamon Press, 1966, pp. 295–299.

[56] F. C. Frischknecht, "Electromagnetic scale modelling," in *Electromagnetic Probing in Geophysics*, ed. by J. R. Wait, Boulder: Golem Press, 1971, pp. 265–320.

[57] E. Bahar, "Propagation of vlf radio waves in a model earth-ionosphere waveguide of arbitrary height and finite surface impedance boundary: theory and experiment," *Radio Sci.*, vol. 1, no. 8, pp. 925–938, 1966.

[58] D. E. Winder, I. C. Peden, and H. M. Swarm, "A 3 GHz scale model of submerged vlf antenna using lossy ceramic powder," *IEEE Trans. Antennas Propag.*, vol. AP-14, no. 4, pp. 507–509, 1966.

[59] G. E. Webber and I. C. Peden, "Vlf mode analysis using a ceramic dielectric model of the earth-ionosphere waveguide," *IEEE Trans. Antennas Propag.*, vol. AP-17, no. 5, pp. 613–620, 1969.

[60] R. J. King, "Physical modeling of EM wave propagation over the earth," *Radio Sci.*, vol. 19, no. 5, pp. 1103–1116, 1982.

[61] R. W. P. King and G. S. Smith, *Antennas in Matter*, Cambridge: MIT Press, 1981, pp. 727–818.

[62] R. F. Harrington, *Field Computations by Moment Methods*, New York: Macmillan, 1968.

[63] J. Perini and D. J. Buchanan, "Assessment of MOM techniques for shipboard applications," *IEEE Trans. Electromagnetic Compat.*, vol. EMC-L4, no. 1, pp. 32–39, 1982.

[64] J. B. Keller, "Geometrical theory of diffraction," *J. Opt. Soc. Am.*, vol. 52, no. 2, pp. 116–130, 1962.

[65] W. D. Burnside, M. C. Gilreath, R. J. Marhefka, and C. L. Yu, "A study of KC-135 aircraft antenna patterns," *IEEE Trans. Antennas Propag.*, vol. AP-23, no. 3, pp. 309–316, 1975.

[66] C. L. Yu, W. D. Burnside, and M. C. Gilreath, "Volumetric pattern analysis of airborne antennas," *IEEE Trans. Antennas Propag.*, vol. AP-26, no. 5, pp. 636–641, 1978.

[67] W. D. Burnside, N. Wang, and E. L. Pelton, "Near-field pattern analysis of airborne

antennas," *IEEE Trans. Antennas Propag.*, vol. AP-28, no. 3, pp. 318–327, 1980.

[68] R. K. Ritt, "The modeling of physical systems," *IEEE Trans. Antennas Propag.*, vol. AP-4, no. 3, pp. 216–218, 1956.

[69] J. Tykocinski-Tykociner, "Investigation of antennas by means of models," *Bulletin 22, no. 39*, University of Illinois, Urbana, 1925.

[70] G. Sinclair, E. C. Jordan, and E. W. Vaughn, "Measurement of aircraft-antenna patterns using models," *Proc. IRE*, vol. 35, no. 12, pp. 1451–1462, 1947.

[71] J. V. N. Granger and T. Morita, "Radio-frequency current distributions on aircraft structures," *Proc. IRE*, vol. 39, no. 8, pp. 932–938, 1951.

[72] E. F. Harris, "Investigating antennas for uhf mobiles," *Electronics*, vol. 25, no. 11, pp. 127–129, 1952.

[73] J. T. Bolljahn and R. F. Reese, "Electrically small antennas and the low-frequency aircraft problem," *IRE Trans. Antennas Propag.*, vol. AP-1, no. 2, pp. 46–54, 1953.

[74] W. Sichak and J. J. Nail, "Uhf omnidirectional antenna systems for large aircraft," *IRE Trans. Antennas Propag.*, vol. AP-2, no. 1, pp. 6–15, 1954.

[75] I. Carswell, "Current distribution on wing-cap and tail-cap antennas," *IRE Trans. Antennas Propag.*, vol. AP-3, no. 4, pp. 207–212, 1955.

[76] R. E. Webster, "20–70 mc monopole antennas on ground-based vehicles," *IRE Trans. Antennas Propag.*, vol. AP-5, no. 4, pp. 363–368, 1957.

[77] R. L. Tanner, "Shunt and notch-fed hf aircraft antennas," *IRE Trans. Antennas Propag.*, vol. AP-6, no. 1, pp. 35–43, 1958.

[78] R. A. Burberry, "Progress in aircraft aerials," *Proc. IEE*, vol. 109, pt. B, no. 48, pp. 431–444, 1962.

[79] P. E. Law, "Accommodating antenna systems in the ship design process," *Naval Engineer's J.*, vol. 91, no. 2, pp. 65–75, 1979.

[80] K. F. Woodman and B. E. Stemp, "Shipborne antenna modeling," *Microwave J.*, vol. 22, no. 4, pp. 73–81, 1979.

[81] G. Sinclair, "Theory of models of electromagnetic systems," *Proc. IRE*, vol. 36, no. 11, pp. 1364–1370, 1948.

[82] R. F. Harrington, *Time-Harmonic Electromagnetic Fields*, New York: McGraw-Hill Book Co., 1961, pp. 100–103.

[83] A. R. von Hippel, *Dielectric Materials and Applications*, Cambridge: MIT Press, 1954.

[84] E. C. Jordan, ed., *Reference Data for Engineers: Radio, Electronics, Computer and Communications*, 7th ed., Indianapolis: Howard W. Sams, Inc., 1985.

[85] J. F. Ramsey, B. V. Popovich, and J. F. Gobler, "Research on compact and efficient antennas," *Report ECOM-02111-F*, Airborne Instruments Laboratory, 1967.

[86] L. S. Taylor, "Dielectric properties of mixtures," *IEEE Trans. Antennas Propag.*, vol. AP-13, pp. 943–947, 1965.

[87] A. M. Shutko and E. M. Reutov, "Mixture formulas applied in estimation of dielectric and radioactive characteristics of soils and grounds at microwave frequencies," *IEEE Trans. Geosci. Remote Sensing*, vol. GE-20, pp. 29–32, 1982.

[88] R. J. King, D. V. Theil, and K. S. Park, "The synthesis of a surface reactance using artificial dielectric," *IEEE Trans. Antennas Propag.*, vol. AP-31, pp. 471–476, 1983.

[89] M. F. Radford, "High-frequency antennas," chapter 16 in *Handbook of Antenna Design*, vol. 2, ed. by A. W. Rudge et al., London: Peter Peregrinus, 1982, pp. 663–724.

[90] R. Wilensky and W. Wharton, "High-frequency Antennas," chapter 26 in *Antenna Engineering Handbook*, 2nd ed., ed. by R. C. Johnson and H. Jasik, New York: McGraw-Hill Book Co., 1984, pp. 26-1 to 26-41.

[91] D. Wobschall, "A theory of the complex dielectric permittivity of soil containing water: the semidisperse model," *IEEE Trans. Geosci. Electron.*, vol. GE-15, pp. 49–58, January 1977.

[92] E. D. Sharpe and R. L. Tanner, "Scale modeling of high-frequency antennas," *IEEE Trans. Antennas Propag.*, vol. AP-17, pp. 810–811, November 1969.

[93] P. E. Law, Jr., *Shipboard Antennas*, 2nd ed., Norwood, Massachusetts: Artech House, 1986.
[94] J. Y. Wong and J. C. Barnes, "Design and construction of a pattern range for testing high-frequency shipborne antennas," *Trans. Eng. Inst. Canada*, vol. 2, pp. 2–8, January 1958.
[95] J. Y. Wong and J. C. Barnes, "A new facility for measuring radiation patterns of model shipborne antennas," *The Engineering J.*, the Engineering Institute of Canada, pp. 3–4, June 1961.
[96] K. M. Keen, R. R. Grime, and B. E. Stemp, "Improvements to a surface-wave antenna measurement range with troublesome site effects," *Electron. Lett.*, vol. 18, pp. 439–440, May 27, 1982.
[97] C. D. Widell, "Modeling techniques," *Lecture Notes Microwave Antenna Measurements, Short Course*, California State University, Northridge, 1977.
[98] E. S. Gillespie, "The effect of radome geometry on the radiation pattern of a retarded wave antenna," *Proc. Ninth Ann. Symp. USAF Antenna Res. Dev.*, Univ. of Illinois, October 1959.
[99] E. S. Gillespie and R. S. Elliott, "The effects of a radome on a surface wave antenna," *LR 15,297*, Lockheed Aircraft Corp., Burbank, September 1961.
[100] D. D. Batson, "Retarded wave antenna for AN/APG-30 installed in model PLV-7," *LR 12,258*, Lockheed Aircraft Corp., Burbank, April 1957.
[101] I. Alne, R. Taron, E. S. Gillespie, and F. R. Zboril, "Evaluation of the WV-2E airframe interference on the rotadome uhf and IFF patterns, pt. 2," *LR 12,618*, Lockheed Aircraft Corp., Burbank.
[102] J. H. Lemanczyk, D. Fasold, and H.-J. Steiner, "Development and measurements of the TM-TC antenna for the DFS satellite," *Proc. Fourth IEEE Intl. Conf. Antennas Propag.*, Univ. of Warwich, Coventry, UK, April 1985.

Chapter 33

Near-Field Far-Field Antenna Measurements

Jørgen Appel-Hansen
Electromagnetics Institute, Technical University of Denmark

CONTENTS

1. Introduction	33-3
2. General Concepts	33-3
3. Current Distribution Measurements	33-5
4. Planar Scanning	33-7
5. Cylindrical Scanning	33-10
6. Spherical Scanning	33-12
7. Error Budgets for Scanning Ranges	33-15
8. Scanning Ranges	33-18
9. Plane-Wave Synthesis Techniques	33-21
10. Compact Range	33-22
11. Defocusing Techniques	33-24
12. Extrapolation Techniques	33-26
13. References	33-28

Jørgen Appel-Hansen was born in Farendløse, Denmark, on March 13, 1937. He received an MSc degree in electrical engineering in 1962 from the Technical University of Denmark, Lyngby. From 1962 to 1964 he was in the Royal Danish Army, where part of his time was spent as a teaching associate at the Physics Laboratory II, Technical University of Denmark. In 1964 he joined the Electromagnetics Institute, Technical University of Denmark, where he received the PhD degree in 1966. In 1967 he was a visiting research associate at the University of Michigan, Ann Arbor. From 1968 to 1978 he worked as head of the Radio Anechoic Chamber, Electromagnetics Institute.

Also, in 1972 he gave a postgraduate course in antenna measurements at the National Physical Laboratory, New Delhi, India. In May 1978, he was a visiting professor at Universitá di Napoli, Italy. In 1979 and 1981 he was a lecturer at the Microwave Antenna Measurements Short Course at the California State University, Northridge. He gave a series of lectures at the Beijing Municipal Institute of Labour Protection, the Shanghai Electrical Apparatus Research Institute, and Jiao Tong University, China, in 1984.

Dr. Appel-Larsen is currently an associate professor at the Electromagnetics Institute. He is also a consultant in design of anechoic chambers, test ranges and their instrumentation, and the implementation of measurement procedures.

1. Introduction

Over the past 15 years major efforts in antenna measurement studies have been invested in improving measurement accuracy and shortening the required test distance. It is interesting to note that some of the techniques seem to have managed to meet both objectives. As a result these techniques, in particular the scanning techniques and the compact range, have been implemented in various forms on several test ranges.

The purpose of this chapter is to give a concise introduction to established near-field far-field measurement techniques. These are here broadly defined as techniques which use test distances shorter than those obtained by using the conventional far-field criterion as derived in the next section. The presentation is based on a previous work [1] published in Rudge et al. [2]. The present work differs from the previous one in two respects. First, several mathematical derivations and discussions of literature earlier than 1980 in the previous work are shortened or omitted. This, it is hoped, will facilitate the reader's understanding of the basic ideas of the techniques treated. Second, the present chapter includes references to works carried out after 1980. This means that the following sections include the major progress in the near-field techniques over the past three years.

In Section 2 some general concepts are presented and the treated techniques are related to each other with respect to their test distance. Basically, far-field properties may be determined from knowledge of the current distribution on the antenna under test. Therefore, although current measurement techniques are not often used, these are touched upon briefly in Section 3. In Sections 4 to 8, the scanning techniques are dealt with in some detail. In comparison with the previous work, a more condensed mathematical notation is used and some error budgets are presented.

In Section 9 the plane-wave synthesis technique is described. The commonly used compact range, defocusing, and extrapolation techniques are outlined in Sections 10, 11, and 12, respectively.

2. General Concepts

The radiating properties of an antenna depend on the distance from the antenna structure to the considered field point. From an examination of the field from a Hertzian dipole, it turns out that the field expressions involve terms which vary as r^{-1} (radiated field), r^{-2} (inductive field), and r^{-3} (static field), where r is the distance between the dipole and the field point [3]. It is easily shown that for

$$r > \lambda/2\pi \tag{1}$$

the terms involving the inverse distance dependence r^{-1} are the largest ones. Since any antenna structure may be considered as composed of small dipoles, the region around an antenna structure may be divided into two major regions, viz., a reactive near-field region close to the antenna and a radiating-field region. The boundary between the two regions is determined by (1). An analysis of the instantaneous power flow density in space shows that this can be expressed as a sum of terms. Only the radiated field contributes a constant term, which gives the average power flow density away from the antenna. All types of fields contribute to the other terms which indicate sinusoidal varying power flow in space. Terms which are due to inductive and static fields can usually be neglected 10 wavelengths away from the antenna.

As the distance tends to infinity, the angular dependence of antenna properties becomes independent of distance. In the limit at infinite distance the properties are characterized as true far-field parameters. In practice there are several advantages to measuring the antenna properties at short distances. However, usually antennas are used for large distances. Therefore, measurements have to be carried out at such distances that the measured parameters agree with the far-field parameters within a certain specified accuracy. In each case the minimum distance depends on the type of antenna, its dimension, the wavelength λ, the measurement technique, the manner in which the data are dealt with, the measured property, and the specified accuracy. Various criteria for far-field measurements have obtained general acceptance [4]. As an example, taking the most popular criterion, the required distance for broadside major lobe pattern measurements on uniformly illuminated aperture antennas, which have a largest cross-sectional diameter D of several wavelengths, the minimum range R is given by

$$R = 2D^2/\lambda \qquad (2)$$

Then, when the antenna is illuminated by a point source, the phase variation over the antenna aperture is less than $\pi/8$ rad. The criterion is easily derived from Fig. 1a and it is often referred to as the conventional far-field criterion. It subdivides the radiating-field region into two regions, the radiating near-field region and the radiating far-field region. The three major field regions are illustrated in Fig. 1b.

When the criterion (2) is satisfied, the measured data usually represent the far-field properties. However, at short wavelengths, demanded in particular for satellite communication, the criterion may require a test range which is long. This creates several disadvantages. Measurements may have to be carried out in open air with possible adverse weather conditions. They may also be disturbed by other signals, and obstacle reflections may give inaccurate results. Reduction of ground reflections requires high towers and special arrangements in the form of fences and absorbers around the specular reflection point. The long test distance requires high power or reduced dynamic range. The open-air range may also have to be placed inconveniently with respect to other activities. Thus, there is a need for indoor near-field measurements under fully controlled conditions. Mainly, two approaches have been followed to meet this objective. In one approach, test setups giving far-field data directly from measurements at short distances are invented. In

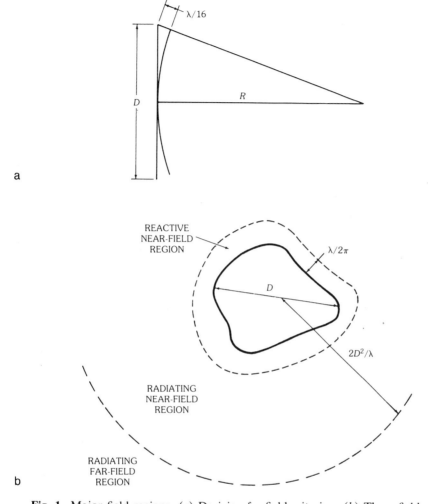

Fig. 1. Major field regions. (*a*) Deriving far-field criterion. (*b*) Three field regions.

the other approach the measured data are transformed mathematically to predict the true far-field data.

The measurement techniques to be described in this chapter are outlined in Fig. 2. The techniques are shown in a framework characterizing their typical test distances.

3. Current Distribution Measurements

Measurement of the surface current on an antenna structure is usually carried out with a probe placed in the reactive-field region as close to the surface as possible. For example, the magnetic field **H** may be measured in phase and amplitude with a circular loop as shown in Fig. 3. Then the surface current density **J** may be found from

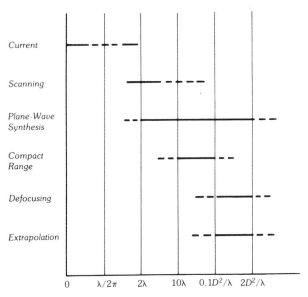

Fig. 2. Outline of treated measurement techniques as a function of test distance. (*After Rudge et al. [2]*)

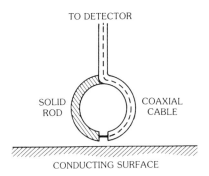

Fig. 3. Current measurement.

$$\mathbf{J} = \hat{\mathbf{n}} \times \mathbf{H} \qquad (3)$$

where $\hat{\mathbf{n}}$ is a unit vector pointing perpendicularly out from the surface toward the loop. The loop should be sufficiently small so it does not disturb the current seriously. From the measured current distribution the radiating properties of the antenna structure can be calculated. A detailed discussion of various probes and supporting mechanisms may be found in Dyson's paper [5]. Current distribution measurements are not generally used for antennas. However, the current distribution measurement is mentioned here because it is the near-field technique which uses the shortest distance.

4. Planar Scanning

The simplest scanning technique to treat mathematically is the planar scanning. It was first described by Kerns and Dayhoff [6] using a scattering matrix formulation. A detailed discussion has been given recently [7]. By using the Lorentz reciprocity theorem Brown and Jull [8] treated the two-dimensional cylindrical case. Following the steps of this derivation Jensen [9] and Leach and Paris [10] derived the coupling equation for scanning on a sphere and cylinder, respectively. In Sections 4 to 8 the three types of scanning shall be introduced by following the scattering matrix formulation. This has also been adopted by Yaghjian [11] for the cylindrical case and by Wacker [12] and Larsen [13] in the spherical case. The equivalence between the Lorentz reciprocity theorem formulation and the scattering matrix formulation was demonstrated by Appel-Hansen [14]. The scattering matrix formulation is preferred here because all derivations are based on considerations of modes as for waveguide techniques. In the treatment of planar, cylindrical, and spherical techniques in Sections 4, 5, and 6, respectively, the same steps are followed in the exposition, only the scanning surface is changed in accordance with the technique. A condensed notation like the one introduced by Jensen [9] is made use of.

In planar scanning a probe antenna is moved in a plane situated in front of the antenna under test as shown in Fig. 4. The position of the probe is characterized by the position vector $\mathbf{r}_0$ with coordinates (x_0, y_0, z_0) in a rectangular xyz-coordinate system associated with the test antenna. During scanning, z_0 is kept constant and y_0 is varied for fixed values of x_0. This is illustrated in the figure. However, other types of scanning, such as plane-polar, are also possible [15, 16].

The field radiated by the test antenna and incident on the probe may be expressed as

$$\mathbf{E}(\mathbf{r}_0) = \int_{-\infty}^{\infty} \int_{-\infty}^{\infty} \sum_{s=1}^{2} b^s(k_x, k_y) \mathbf{f}_{sk_x k_y}(\mathbf{r}_0) dk_x dk_y \qquad (4)$$

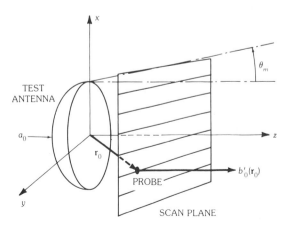

Fig. 4. Planar scanning.

where, for an arbitrary field point with position vector $\mathbf{r}$,

$$\mathbf{f}_{1k_xk_y}(\mathbf{r}) = j\frac{\hat{\mathbf{z}} \times \mathbf{k}}{|\hat{\mathbf{z}} \times \mathbf{k}|}e^{-j\mathbf{k}\cdot\mathbf{r}} \qquad (5)$$

and

$$\mathbf{f}_{2k_xk_y}(\mathbf{r}) = -\frac{1}{k}\frac{\hat{\mathbf{z}} \times \mathbf{k}}{|\hat{\mathbf{z}} \times \mathbf{k}|} \times \mathbf{k}\, e^{-j\mathbf{k}\cdot\mathbf{r}} \qquad (6)$$

are elementary planar vector wave functions. The time factor $e^{j\omega t}$ is omitted. The rectangular components (k_x, k_y, γ) of the propagation vector $\mathbf{k}$ are related through

$$\mathbf{k} = k_x\hat{\mathbf{x}} + k_y\hat{\mathbf{y}} + \gamma\hat{\mathbf{z}} \qquad (7)$$

and

$$\gamma = \pm\sqrt{k^2 - k_x^2 - k_y^2} \qquad (8)$$

where $k = \omega\sqrt{\mu\epsilon}$ as usual and $\hat{x}$, $\hat{y}$, and $\hat{z}$ are unit vectors of the xyz system. The sign is chosen so that γ is positive real in the visible region within the circle $k_x^2 + k_y^2 = k^2$ and negative imaginary outside this circle in the k_xk_y plane. Thus the incident field $\mathbf{E}(\mathbf{r}_0)$ consists of propagating and evanescent waves in the positive-z direction, i.e., from the test antenna toward the probe.

The coefficients $b^s(k_x, k_y)$ can be expressed as

$$b^s(k_x, k_y) = a_0 T^s(k_x, k_y) \qquad (9)$$

where a_0 is the amplitude of a single mode in the feed line to the test antenna and $T^s(k_x, k_y)$ are related to the far-field components of $\mathbf{E}(\mathbf{r})$ through

$$\mathbf{E}(\mathbf{r}) \cong ja_0 2\pi k_{z0}[jT^1(k_{x0}, k_{y0})\hat{\boldsymbol{\phi}} - T^2(k_{x0}, k_{y0})\hat{\boldsymbol{\theta}}]\frac{e^{-jkr}}{r} \qquad (10)$$

This result may be found from (4)–(6) and (9) using the method of steepest descents and introducing a spherical $r\theta\phi$-coordinate system with unit vectors $(\hat{\mathbf{r}}, \hat{\boldsymbol{\theta}}, \hat{\boldsymbol{\phi}})$ and usual orientation with respect to the rectangular xyz-coordinate system. The components (k_{x0}, k_{y0}, k_{z0}) of $\mathbf{k}$ are related to the direction (θ, ϕ) by

$$k_{x0}\hat{\mathbf{x}} + k_{y0}\hat{\mathbf{y}} + k_{z0}\hat{\mathbf{z}} = k\hat{\mathbf{r}} \qquad (11)$$

Because of the relationship (10), $T^s(k_x, k_y)$ is denoted the transmitting characteristic.

The detected signal delivered from the probe is given by the so-called coupling equation

$$b_0'(\mathbf{r}_0) = a_0 \int_{-\infty}^{\infty} \int_{-\infty}^{\infty} \sum_{s=1}^{2} R'^s(k_x, k_y) T^s(k_x, k_y) e^{-j\mathbf{k}\cdot\mathbf{r}_0} dk_x dk_y \quad (12)$$

From a consideration of (4), (9), and (12), it is seen that the $R'^s(k_x, k_y)$ express the sensitivity or the receiving characteristic of the probe to the elementary planar vector functions incident on the probe. Fourier inversion of (12) gives

$$\sum_{s=1}^{2} R'^s(k_x, k_y) T^s(k_x, k_y) = \frac{e^{j\gamma z_0}}{4\pi^2 a_0} \int_{-\infty}^{\infty} \int_{-\infty}^{\infty} b_0'(\mathbf{r}_0) e^{j(k_x x_0 + k_y y_0)} dx_0 dy_0 \quad (13)$$

It can be shown that there exists the following reciprocity relation between $R'^s(k_x, k_y)$ and $T'^s(k_x, k_y)$:

$$\eta_0 R'^s(k_x, k_y) = -\eta \, T'^s(-k_x, -k_y) \quad (14)$$

where η_0 is the admittance in the feed line and η is the free-space admittance. Thus, by measuring the far field of the probe and by making use of (10) and (14), $R'^s(k_x, k_y)$ may be calculated. This means that (13) is one equation in the two unknowns $T^1(k_x, k_y)$ and $T^2(k_x, k_y)$. An additional equation, which is needed in order to find $T^s(k_x, k_y)$, is obtained by repeating the measurements with another probe or the same probe rotated 90°. It is interesting to note that at a finite distance the received signal $b_0'(\mathbf{r}_0)$ in (12) is a Fourier transform of the product of true far-field characteristics of the antennas involved.

When $T^s(k_x, k_y)$ is determined, the gain in a direction characterized by (k_{x0}, k_{y0}, k_{z0}) is

$$G(k_{x0}, k_{y0}, k_{z0}) = 16\pi^3 k_{z0} [|T^1(k_{x0}, k_{y0})|^2 + |T^2(k_{x0}, k_{y0})|^2] \eta/\eta_0 \quad (15)$$

In the above it has been assumed that impedance match exists in the feed lines of the probe and test antenna.

Measurement accuracy may be discussed from a consideration of (13). Evidently integration cannot be made over an infinite range in x_0 and y_0. The scan area has to be truncated. This means that the computed far-field pattern, e.g., in the xz plane as illustrated in Fig. 4, can only be trusted within an angle θ_m given by

$$\tan \theta_m = \frac{L_x - D_x}{2z_0} \quad (16)$$

where L_x and D_x are the scan length and the antenna dimension in the x direction, respectively [17]. Therefore a single scan plane gives coverage only in a conical volume in front of the antenna. However, a combination of scan planes in a polyhedron type of scanning may be arranged to give spherical coverage for the predicted pattern.

Equation 12 shows that $b_0'(\mathbf{r}_0)$ is expressed as an integration in the $k_x k_y$ plane. A consideration of (5)–(8) demonstrates that outside the circle $k_x^2 + k_y^2 = k^2$, evanescent waves contribute to $b_0'(\mathbf{r}_0)$. This means that $b_0'(\mathbf{r}_0)$ is a Fourier transform of a bandlimited function of k_x and k_y. Therefore (13) can be written as a summation and the sampling spacing can be chosen in accordance with the sampling theory. The upper limit

$$\Delta x_0 = \Delta y_0 = \lambda/2 \tag{17}$$

is approached when $z_0 \to \infty$. In general, sampling spacings of $\lambda/3$ are sufficient for test distances z_0 equal to a few wavelengths [18].

5. Cylindrical Scanning

In comparison with a single planar scan giving only coverage in front of the antenna, cylindrical scanning giving azimuthal coverage in the far-field pattern may be preferred. In comparison with spherical scanning, cylindrical scanning may also be preferred when the antenna can only be rotated around a single axis for mechanical reasons. The cylindrical scanning may be implemented as exemplified in Fig. 5. Here, the test antenna is rotated around the z axis of an xyz-coordinate system while the probe is placed on a cylindrical surface at various levels relative to the xy plane.

The field radiated by the test antenna and incident on the probe at $\mathbf{r}_0$ may be expressed as

$$\mathbf{E}(\mathbf{r}_0) = \sum_{n=-\infty}^{\infty} \int_{-\infty}^{\infty} \sum_{s=1}^{2} b_n^s(\gamma) \mathbf{f}_{sn\gamma}^{(4)}(\mathbf{r}_0) d\gamma \tag{18}$$

where, at an arbitrary field point $\mathbf{r}$,

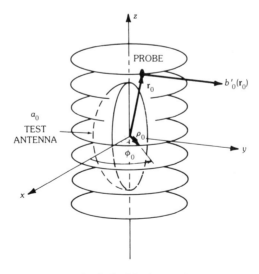

Fig. 5. Cylindrical scanning.

$$\mathbf{f}_{1n\gamma}^{(i)}(\mathbf{r}) = \left[\frac{jn}{\varrho} Z_n^{(i)}(\varkappa\varrho)\hat{\varrho} - \frac{\partial Z_n^{(i)}(\varkappa\varrho)}{\partial\varrho}\hat{\phi}\right] e^{jn\phi} e^{-j\gamma z} \qquad (19)$$

and

$$\mathbf{f}_{2n\gamma}^{(i)}(\mathbf{r}) = \left[-\frac{j\gamma}{k}\frac{\partial Z_n^{(i)}(\varkappa\varrho)}{\partial\varrho}\hat{\varrho} + \frac{n\gamma}{k\varrho} Z_n^{(i)}(\varkappa\varrho)\hat{\phi} + \frac{\varkappa^2}{k}Z_n^{(i)}(\varkappa\varrho)\hat{\mathbf{z}}\right] e^{jn\phi} e^{-j\gamma z} \qquad (20)$$

are the elementary cylindrical vector wave functions. The position vector $\mathbf{r}$ has cylindrical coordinates (ϱ, ϕ, z), n is any integer, $-\infty < \gamma < \infty$, $\varkappa = \sqrt{k^2 - \gamma^2}$ is chosen positive for $\gamma^2 < k^2$ and negative imaginary for $\gamma^2 > k^2$, $Z_n^{(i)}(\varkappa\varrho)$ is the cylinder functions $J_n(\varkappa\varrho)$, $N_n(\varkappa\varrho)$, $H_n^{(1)}(\varkappa\varrho)$, and $H_n^{(2)}(\varkappa\varrho)$ for $i = 1, 2, 3$, and 4, respectively.

The coefficients $b_n^s(\gamma)$ may be expressed by

$$b_n^s(\gamma) = T_n^s(\gamma) a_0 \qquad (21)$$

where the transmitting characteristic $T_n^s(\gamma)$ of the test antenna is related to its far field at a position $\mathbf{r}$ with spherical coordinates (r, θ, ϕ) through

$$E_\phi(\mathbf{r}) \cong -\frac{a_0 2k \sin\theta\, e^{-jkr}}{r} \sum_{n=-\infty}^{\infty} j^n T_n^1(k\cos\theta) e^{jn\phi} \qquad (22)$$

$$E_\theta(\mathbf{r}) \cong -\frac{a_0 2jk \sin\theta\, e^{-jkr}}{r} \sum_{n=-\infty}^{\infty} j^n T_n^2(k\cos\theta) e^{jn\phi} \qquad (23)$$

where $E_\phi(\mathbf{r})$ and $E_\theta(\mathbf{r})$ are the ϕ and θ components of the far field $\mathbf{E}(\mathbf{r})$.

It can be shown that the signal received by the probe is

$$b_0'(\mathbf{r}_0) = a_0 \sum_{n=-\infty}^{\infty} \int_{-\infty}^{\infty} \sum_{s=1}^{2} R_n'^s(\gamma) \left[\sum_{m=-\infty}^{\infty} (-1)^n H_{m-n}^{(2)}(\varkappa\varrho_0) e^{jm\phi_0} e^{-j\gamma z_0} T_m^s(\gamma)\right] d\gamma \qquad (24)$$

Fourier inversion of this equation readily gives

$$\sum_{s=1}^{2} \left[T_m^s(\gamma) \sum_{n=-\infty}^{\infty} (-1)^n H_{m-n}^{(2)}(\varkappa\varrho_0) R_n'^s(\gamma)\right] = \frac{1}{4\pi^2 a_0} \int_{-\infty}^{\infty} \int_{0}^{2\pi} b_0'(\mathbf{r}_0) e^{-jm\phi_0} e^{j\gamma z_0} d\phi_0 dz_0 \qquad (25)$$

which is one equation in the two unknowns $T_m^s(\gamma)$, $s = 1, 2$. An additional equation is obtained as in the planar case.

Using (22) and (23) and the reciprocity relation

$$\eta_0 R_n'^s(\gamma) = (-1)^n \frac{4\pi\varkappa^2}{k} \eta\, T_{-n}'^s(-\gamma) \qquad (26)$$

the receiving characteristic $R_n^{\prime s}(\gamma)$ of the probe can be found from far-field pattern measurements.

The upper limits for the sampling spacings $\Delta\phi_0$ and Δz_0 are determined from

$$\Delta\phi_0 = \lambda/2a \tag{27}$$

and

$$\Delta z_0 = \lambda/2 \tag{28}$$

where a is the radius of the smallest cylinder enclosing the test antenna. These limits are found from considerations analogous to the planar case discussed in the previous section and the spherical case discussed in the following section.

6. Spherical Scanning

While planar scanning requires at least four scans and cylindrical scanning at least two scans, the spherical scanning requires only a single scan over a sphere enclosing the test antenna to give spherical coverage. Mechanically, spherical scanning has the advantage that the test antenna may be mounted on a standard rotator, which can, for example, make a continuous scan in polar angle θ while ϕ is changed in steps as illustrated in Fig. 6. During measurements the probe is kept at a fixed position (see Fig. 9, in Section 8). Electrically, spherical scanning has the advantage that directivity can easily be determined from a single scan.

The field incident on the probe is expressed as

$$\mathbf{E}(\mathbf{r}_0) = \sum_{n=1}^{\infty} \sum_{m=-n}^{n} \sum_{s=1}^{2} b_{mn}^s \mathbf{f}_{smn}^{(4)}(\mathbf{r}_0) \tag{29}$$

where the spherical vector wave functions are

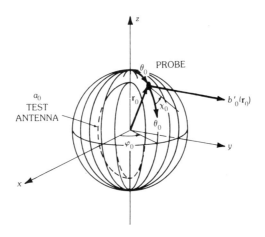

Fig. 6. Spherical scanning.

Near-Field Far-Field Antenna Measurements

$$\mathbf{f}^{(i)}_{1mn}(\mathbf{r}) = z_n^{(i)}(kr)e^{jm\phi}\left[\frac{jm}{\sin\theta}P_n^m(\cos\theta)\hat{\boldsymbol{\theta}} - \frac{\partial P_n^m(\cos\theta)}{\partial\theta}\hat{\boldsymbol{\phi}}\right] \quad (30)$$

and

$$\mathbf{f}^{(i)}_{2mn}(\mathbf{r}) = \frac{n(n+1)}{kr}z_n^{(i)}(kr)P_n^m(\cos\theta)e^{jm\phi}\hat{\mathbf{r}} + \frac{1}{kr}\frac{\partial[rz_n^{(i)}(kr)]}{\partial r}e^{jm\phi}$$
$$\times \left[\frac{\partial P_n^m(\cos\theta)}{\partial\theta}\hat{\boldsymbol{\theta}} + \frac{jm}{\sin\theta}P_n^m(\cos\theta)\hat{\boldsymbol{\phi}}\right] \quad (31)$$

Here, $z_n^{(i)}(kr)$ are the spherical functions $j_n(kr)$, $n_n(kr)$, $h_n^{(1)}(kr)$, and $h_n^{(2)}(kr)$ for $i = 1, 2, 3$, and 4, respectively. The functions $P_n^m(\cos\theta)$ are associated Legendre functions, and n and m are integers, $0 \leqq n \leqq \infty$, $-n \leqq m \leqq n$. The coefficients b_{mn}^s are given by

$$b_{mn}^s = a_0 T_{mn}^s \quad (32)$$

where T_{mn}^s characterizes the test antenna. The relations between T_{mn}^s and the far-field components are

$$E_\phi(\mathbf{r}) \cong j\frac{a_0 e^{-jkr}}{kr}\sum_{n=1}^{\infty}\sum_{m=-n}^{n} j^n e^{jm\phi}\left[-\frac{\partial P_n^m(\cos\theta)}{\partial\theta}T_{mn}^1\right.$$
$$\left. + \frac{m}{\sin\theta}P_n^m(\cos\theta)T_{mn}^2\right] \quad (33)$$

$$E_\phi(\mathbf{r}) \cong \frac{a_0 e^{-jkr}}{kr}\sum_{n=1}^{\infty}\sum_{m=-n}^{n} j^n e^{jm\phi}\left[-\frac{m}{\sin\theta}P_n^m(\cos\theta)T_{mn}^1\right.$$
$$\left. + \frac{\partial P_n^m(\cos\theta)}{\partial\theta}T_{mn}^2\right] \quad (34)$$

By transforming $\mathbf{f}^{(4)}_{smn}(\mathbf{r}_0)$ to a coordinate system associated with the probe, it is found that the coupling equation takes the form

$$b_0'(\mathbf{r}_0) = a_0 \sum_{n=1}^{\infty}\sum_{m=-n}^{n}\sum_{\mu=-n}^{n} D_{\mu m}^{\mu n}(\phi_0, \theta_0, \chi_0) D(m,n,\mu) \quad (35)$$

where

$$D_{\mu m}^{\mu n}(\phi_0, \theta_0, \chi_0) = (-1)^{m+\mu}\sqrt{\frac{(n-\mu)!(n+m)!}{(n+\mu)!(n-m)!}}e^{jm\phi_0}d_{\mu m}^{(n)}(\theta_0)e^{j\mu\chi_0} \quad (36)$$

and $D(m,n,\mu)$ is the so-called coupling product given by

$$D(m,n,\mu) = \sum_{s=1}^{2} T_{mn}^s \sum_{\nu=(1,\mu)}^{\infty}[R_{\mu\nu}'^s(-1)^{n+\nu}A_{\mu\nu}^{\mu n} + R_{\mu\nu}'^{3-s}(-1)^{n+\nu+1}B_{\mu\nu}^{\mu n}] \quad (37)$$

The angle χ_0 is used to characterize the orientation of the probe with respect to the $\hat{\theta}_0$ vector. The coefficients $d^{(n)}_{\mu m}(\theta_0)$, $A^{\mu n}_{\mu \nu}$, and $B^{\mu n}_{\mu \nu}$ are expressed by summations involving the indices μ, ν, m, n, and $\sin(\theta_0/2)$, $\cos(\theta_0/2)$, $h^{(2)}_n(kr_0)$, and the Wigner $3-j$ symbols.

Using the orthogonality integral

$$\int_0^\pi d^{(n)}_{\mu m}(\theta_0) d^{(n')}_{\mu m}(\theta_0) \sin\theta_0 \, d\theta_0 = \frac{2}{2n+1} \delta_{nn'} \tag{38}$$

and (36), Fourier inversion of (35) can be made with respect to ϕ_0, θ_0, and χ_0. This gives

$$D(m,n,\mu) = \frac{(-1)^{\mu+m}(2n+1)}{a_0 \, 8\pi^2} \sqrt{\frac{(n+\mu)!(n-m)!}{(n-\mu)!(n+m)!}} \int_0^{2\pi} \int_0^\pi \int_0^{2\pi}$$
$$\times b'_0(r_0,\phi_0,\theta_0,\chi_0) e^{-jm\phi_0} d^{(n)}_{\mu m}(\theta_0) \sin\theta_0 \, e^{-j\mu\chi_0} d\phi_0 d\theta_0 d\chi_0 \tag{39}$$

When the far field of the probe is measured, its transmitting characteristic T'^s_{mn} can be found from (33) and (34). Then, its receiving characteristic R'^s_{mn} can be found from

$$\eta_0 R'^s_{mn} = (-1)^m \frac{\eta}{k^2} \frac{2\pi n(n+1)}{2n+1} T'^s_{-mn} \tag{40}$$

From using R'^s_{mn} in (37), it is seen that (39) is one equation in the two unknowns T^s_{mn}, with $s = 1, 2$.

It turns out that measurements and calculations are facilitated by choosing a probe with $R'^s_{\mu\nu} \neq 0$ for μ equal to $+1$ and -1 only, e.g., an antenna with rotational symmetry as a linearly polarized circular conical horn fed with the TE_{11} mode. Then, measurements carried out for $\chi_0 = 0°$ and $90°$ are sufficient for making the integration over χ_0. Evaluation of $D(m,n,\mu)$ for μ equal to $+1$ and -1 gives the required two equations for determination of t^s_{mn}.

Due to the fact that the radiated field $\mathbf{E}(\mathbf{r}_0)$ can be truncated for $n = ka$, where a is the radius of the smallest sphere enclosing the antenna, $b'_0(\mathbf{r}_0)$ is a bandlimited function. Therefore the sampling spacings $\Delta\phi_0$ and $\Delta\theta_0$ should be chosen so that

$$\Delta\phi_0 < \lambda/2a \tag{41}$$

and

$$\Delta\theta_0 < \lambda/2a \tag{42}$$

The center of the smallest sphere should be chosen at the intersection between the axes of rotation (see Fig. 9, in Section 8). Therefore the maximum sampling spacing depends on the mounting of the test antenna as well as its size. It should be noted that in case sampling is chosen less than the theoretical maximum, possible modes with $n > ka$, which may be due to noise, may be removed by filtering.

A single computer program based on the coupling equation (35) may be developed to carry out the following major calculations: (*a*) inversion of the coupling equation to find the probe characteristics R'^s_{mn} from spherical scanning of the probe with another probe which can be approximated with a Hertzian dipole (nonprobe correction—this is an alternative to the use of (33) and (34) as described above; (*b*) inversion of the coupling equation to find the unknown characteristics T^s_{mn} from measurements on the unknown antenna with a probe with known R'^s_{mn}, i.e., probe correction; (*c*) the field intensity at any distance from an antenna with known T^s_{mn}; (*d*) the true far-field pattern by specifying the distance as infinite and using asymptotic forms for the mode functions. In the last two types of calculation the coupling equation is used with a probe specified as a Hertzian dipole.

7. Error Budgets for Scanning Ranges

The sources of errors in the determination of the test antenna characteristic T^s can be derived from a consideration of the inverse coupling equations (13), (25), and (39) and their background. Three types of sources exist:

1. Discrepancies between the theoretical model and the actual experiment
2. Inaccuracies in the measurement and digitizing of the quantities in the inverse equations
3. Computations

The first source gives rise to errors due to the neglect of multiple reflections between the test antenna and the probe, multipath reflections from the lining of the test range and the applied instrumentation, and leakage signals. In the derived equations, reflections (mismatches) in the feed lines are also neglected. These, however, may easily be included by multiplying with proper factors. Furthermore, in the planar and cylindrical case there is a practical need for a finite scan area. Nonzero sampling spacing causes neglect of evanescent or higher-order modes. A special source of error is variation in the test antenna structure during measurements. Such variations may be due to temperature variations or deformations in case the antenna is rotated or otherwise moved during measurements.

The second source of error is related to inaccuracies in both the positioning of the probe and electrical characterization of the probe. It also includes inaccurate measurement of probe output signal, input signal to the test antenna, frequency, and reflection coefficients in the receiving system. In some cases determination of the permeability and permittivity of the transmission medium may also influence the determination of T^s.

Because of the development of computers the third source of error is in general insignificant. A particular error may arise in case the determinant of the two equations obtained by measuring two orthogonal components of the radiated field is close to zero.

An error budget for antenna measurements may be carried out in four steps:

1. Specify the test antenna
2. Specify the test range

3. Study the test antenna in the test range
4. Make the error budget

From the first two steps it can be decided whether the test range can handle the test antenna. In this respect, size of antenna and range, frequency, and instrumentation are important factors. When it is decided to start the test, the third step can be initiated. In particular, preliminary experiments, such as those indicating the error signal levels, may be carried out. A systematic analysis and classification of error signals in antenna measurements is given by Appel-Hansen [19]. It is characteristic that all types of error signals are present at the same time. However, each type has its own peculiarities which may be used in the evaluation of the influence of the individual signals. In near-field techniques it is convenient to classify the extraneous error signals into three types. The first type is due to reflections between the test antenna and the probe. These are referred to as *multiple reflections* and have a characteristic $\lambda/2$ period when the probe is moved longitudinally with respect to the test antenna. In particular, multiple reflections may be severe when the two antennas are pointing toward each other. They may be reduced by not measuring too close to the test antenna and by properly matching a probe having a low back-scattering cross section.

The second type of extraneous error signal is due to reflections from the lining of the test range and the instrumentation, which includes the test antenna support. These are referred to as *multipath reflections*. They often cause an irregular variation with change in probe position. They may be reduced by proper choice of probe directivity and of absorbers. Multipath reflections may be measured by moving the probe in various directions for several aspect angles as in the evaluation of the reflectivity level of anechoic chambers [20].

The third type is due to leakage from the transmitting system to the receiving system. Proper shielding and cabling, especially connectors, may reduce this source of error. This type of error signal may be discovered, for example, by probing a possible signal when the transmitting terminals are short-circuited.

To facilitate further insight, major details of error budgets for a planar scanning and a spherical scanning shall be outlined as examples. Additional information may be obtained from references 21 and 22 in the planar case and from reference 23 in the spherical case.

In the first budget the test antenna is a planar array for which the planar near-field technique is a candidate. Some of the details in the four steps of the error budget are the following:

1. The test antenna specifications are:
   Maximum dimension $L_y = 538$ cm
   Minimum dimension $L_x = 202$ cm
   Wavelength $\lambda = 23.5$ cm
2. The range specifications are:
   Maximum scan length $L_{max} = 640$ cm
   Transverse position accuracy $\Delta_x = \Delta_y = \pm 0.005$ cm
   Longitudinal position accuracy $\Delta_z = \pm 0.010$ cm
   Probe gain error $= \pm 0.12$ dB

Near-Field Far-Field Antenna Measurements

The scanner is installed in a hall partly covered with absorbers.

3. The scan length may be obtained by a combination of two or several scan areas. An overlapping of the individual areas may secure proper transition between the patterns obtained from the various scans. The range distance z_0 is chosen to be 37.0 cm, i.e., between 1λ and 2λ. The truncation angle is about 50°. Preliminary transverse movements show a signal level of -35 dB at the edge of the scan plane relative to the maximum signal in the scan area. Longitudinal movement of the probe in front of the test antenna gives a peak-to-peak variation of about 1 dB. Since the distance between two consecutive peaks is $\lambda/2$, the oscillations must be associated with multiple reflections between the probe and test antennas.
4. Using upper-bound expressions [17], the five largest contributions in the error budget for gain measurements are:

Probe gain	±0.12 dB
Receiver nonlinearity	±0.12 dB
Multiple reflections	±0.10 dB
Normalization	±0.05 dB
Mismatch factors	±0.02 dB

It turns out that the position error is less than 0.01 dB. An analysis shows that the root-square sum of all errors is ±0.20 dB in X-band as well as in L-band.

The error budget for side lobe level measurements shows that area truncation error becomes important. It is the third largest component after that of multiple reflections, which is the predominant component, and probe gain. It is concluded that the error budgets give only an indication of expected errors. In particular, this is the case for the side lobe level budgets. However, the budgets probably give good information on the relative magnitudes of the different sources of error. More work seems appropriate in order to obtain agreement between error results derived from theory and from experiment.

1. The test antenna is a pyramidal standard-gain horn. The gain is determined in the frequency range 17.3 GHz to 18.1 GHz.
2. The test range is an anechoic chamber with a precision rotator as described in the next section. Some of the range specifications are as follows:

Positioner accuracy	±0.01°
Distance accuracy	±1 mm
Phase drift	±5°
Amplitude nonlinearity	±0.5 percent

   These values are based on measurements estimated or taken from manufacturers' indications.
3. When the test antenna is placed in the range, a multipath reflection level of -70 dB exists. The level of multiple reflections between the test antenna and the probe is -45 dB.
4. Introducing the individual errors into actually measured data, the five largest contributions for gain measurements are:

Multiple reflections	±0.045 dB
Amplitude nonlinearity	±0.025 dB

Asymmetry of probe	±0.02 dB
Internal loss calculation	$\begin{cases} +0.020 \text{ dB} \\ -0.005 \text{ dB} \end{cases}$
Leakage	< ±0.010 dB

Multipath reflections, phase, positioning, distance, and amplitude drift errors contribute with less than ±0.003 dB each. The root-square-sum value of the upper-bound errors is ±0.06 dB. The value of the theoretically determined internal loss is about 0.020 dB. This is subtracted from the measured directivity values oscillating between 24.647 dB and 24.938 dB. The so-determined gain as a function of frequency is shown in Fig. 7.

Finally, it should be mentioned that the error budgets may be improved by calibrating for systematic inaccuracies such as probe position, probe gain, and, especially, receiver nonlinearity. Part of the influence of the various extraneous reflections may be averaged out from repeated measurements at various range distances as in conventional pattern measurement [20]. The error budgets indicate that the scanning techniques are usually superior to conventional techniques with respect to measurement accuracy.

8. Scanning Ranges

The first commercially available scanner was of the planar type. It used an analog Fourier integral computer [24]. It has turned out, however, that test ranges

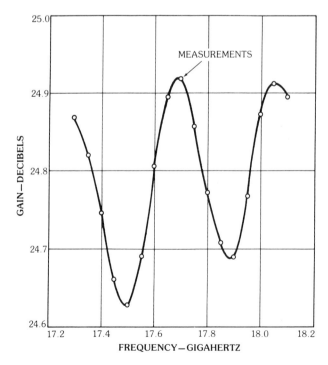

Fig. 7. Gain predicted from spherical near-field measurements. (*After Lemanczyk [23]*)

over the years have usually constructed their own planar scanners in accordance with their specific needs. A recent example is that of Borowick, Holley, Lange, Cummings, and Howard [25]. At present only spherical scanners are readily available. The success of the spherical technique is due to the fact that the mathematical problems have been solved during recent years in such a manner that minicomputers can handle the transformation of measured near-field data to predicted far-field behavior of the test antenna. Furthermore, the spherical-coordinate system is used in conventional far-field measurements. This means that the same instrumentation can be used for near-field work as well as for conventional far-field measurements.

Rectangular XY-scanners have been implemented in several places. They are based on experience obtained in particular at the National Bureau of Standards [26] and at the Georgia Institute of Technology [27]. Vertical as well as horizontal scanners have been built. The horizontal scanner may use existing turntables and rails for movement of the test antenna (see Fig. 8) [28]. Transverse movements from a few meters up to about ten meters have been arranged. Positioning and alignment are the critical works which often require the use of laser systems. When such systems are used, however, the remaining inaccurate positioning is usually an insignificant source of error, as shown in the previous section.

The most often used probe is an open-ended waveguide delivering its signal to a phase-amplitude receiver or a network analyzer. Due to the lengthy measure-

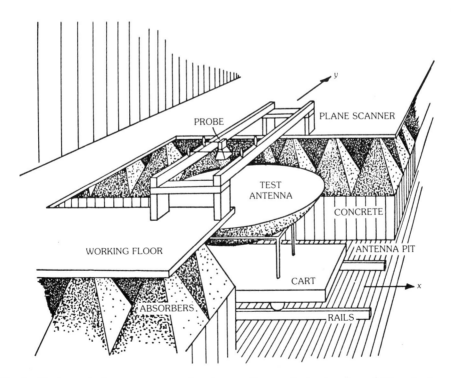

Fig. 8. Horizontal planar scanner setup in a radio anechoic chamber. (*After Anderson, Appel-Hansen, and Larsen [28]*)

ment procedure and in order to keep the mechanical source of error at an insignificant level, measurements in a temperature-stabilized environment may be demanded.

At the Technical University of Denmark a spherical near-field antenna test facility has been developed [29, 30]. In particular, precision measurements of directivity, gain, and copolar and cross-polar patterns can be carried out on microwave antennas of up to 200 wavelengths in electrical diameter and about 6 m in physical diameter. The test distance can be varied from 1.5 m to 5 m. The setup is illustrated in Fig. 9. The test antenna is mounted in a specially designed rotator. The resettability of the intersection between the vertical and horizontal rotation axes is ±0.1 mm. With no load, the intersection accuracy of the axes during rotation is within 0.05 mm. The probe is placed on a special probe tower and aligned accurately to point toward the intersection point. The setup is placed in a large anechoic chamber ($12 \times 14 \times 16$ m^3) lined with 2-m-high pyramidal absorbers.

The probe antennas are conical horns with orthomode transducers for dual linear polarization. This means that during a spherical scan a single probe can be used to measure two orthogonal components simultaneously, i.e., two scans with two different orientations of the probe are not required. The measurements are facilitated by the use of a three-channel receiver, one channel for phase reference and two channels for the two polarizations. In the near-field to far-field transformation, correction is made for the receiving pattern of the probe as a routine.

For antennas with a 10-wavelength diameter the execution time is about 2 hr for measurements and a few minutes for transformation of data. When the diameter is 100 wavelengths, measurements require 10 hr and calculations 5 hr. It should be noted that during one measurement the calculations for the previous measurement can be made on a minicomputer, which together with a microcomputer performs the measurements. When only part of the radiation pattern is needed, truncation of the scanning surface can be made to save time. In case full-sphere scanning is performed, directivity is automatically computed. Typical

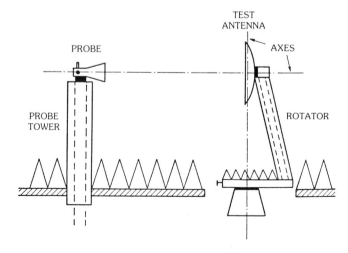

Fig. 9. Spherical near-field scanning.

directivity measurement accuracy is ±0.10 dB (3σ). Determination of gain requires a substitution measurement with a standard-gain horn. The gain of this horn is determined by substracting analytically determined ohmic losses from measured directivity. Cross-polarization levels are measured with an accuracy of a few decibels down to about −55 dB. The use of accurate reference points makes determination of phase centers possible, for example.

9. Plane-Wave Synthesis Techniques

In general, attempts to make far-field measurements may be analyzed from a plane-wave illumination point of view [31]. This may be illustrated by a consideration of the three basic cases shown in Fig. 10. The first case is the simplest and used in the derivation of the conventional far-field criterion (2). Here, a single point source, as shown in Fig. 10a, illuminates the test volume in which the plane-wave quality is specified as an allowable phase error of 22.5°, corresponding to a path length difference of $\lambda/16$ (see Fig. 1).

In the second case, Fig. 10b, a simple array consisting of a few elements that are properly fed is used. In this manner a test volume of sufficient quality may be established at a shorter distance than required by the criterion (2). Lynggaard [32] studied an array of five pyramidal horn antennas theoretically and experimentally. The theory is based on a paper by Hald [33], who has studied the evaluation of the measurement error arising from imperfections in the plane wave. The compact range described in the next section may be considered as a continuous array of elements.

In the third case, Fig. 10c, a complete array of elements creating a perfect plane wave in a certain volume is thought of. It is probably impossible to build such an array. There is another approach, however. Instead of having a radiator at all positions, a single radiator steps from one position to the next. In this manner a

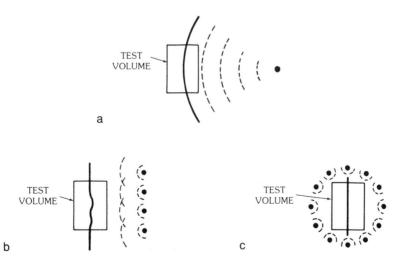

Fig. 10. Plane-wave synthesis. (*a*) Single point source. (*b*) Single array of sources. (*c*) Complete array of sources.

sequence of field contributions, each of which is separated in time, is generated. When a receiving test antenna is inserted in the test volume the response to each field contribution may be sampled. After the sampling, the responses may be multiplied with weighting functions. The result is that the responses added in this manner correspond to the response of the antenna to a synthesized plane-wave illuminator. By proper change of the weighting functions the direction of propagation of the plane wave can be changed and the complete antenna pattern can be obtained from a single scan. In fact, the procedure is analogous to the scanning techniques described in Sections 4 to 8. The approach has been studied by Martsafey [34] and Bennett and Schoessow [35] for the planar and spherical cases, respectively. Ludwig and Larsen [36] have used computer programs for spherical scanning to determine the size of the test volume.

An interesting aspect of the plane-wave synthesis technique is to create a plane wave field in the test volume and at the same time to minimize the illumination outside the test volume [37]. In this manner extraneous reflections may be reduced and the need for absorbers probably eliminated.

10. Compact Range

By *compact range* is usually meant a range which creates a plane wave field at distances which are considerably shorter than those needed under conventional far-field conditions. The required compact range is usually so short that the setup can be arranged indoors. Thus the compact range has all the advantages (controlled conditions, location, etc.) of indoor measurements. The compactness of the range may be obtained by using a set of sources which are properly fed as described in Section 9. Alternatively, the diverging rays from a single-source antenna may be collimated by using a lens or a reflector. Out of these two possibilities, the use of a parabolic reflector is the most widely implemented. Therefore this type of range is here designated "compact range" and will be treated in some detail. In particular, because antenna properties are obtained as in conventional techniques, i.e., much faster than in the scanning techniques which require mathematical transformation of a large number of data sampled in a time-consuming process, the compact range is an important technique.

Fig. 11 shows a reflector illuminated by a feed at its focal point. The principle of operation may be understood by considering the ray tracing in the figure. Let the ray from the feed which passes through the center of the test zone be referred to as the principal ray. Also, let the distance from the feed phase center to the reflector along the principal ray be R_0. Then, the power P_r received by the test antenna with gain G_r is approximately

$$P_r = \left(\frac{\lambda}{4\pi R_0}\right)^2 P_f G_f G_r \qquad (43)$$

where P_f is the power transmitted by the feed with gain G_f [38]. In this derivation use is made of Friis's transmission formula and the fact that the field radiated by the feed is only diverging over a distance R_0 and then collimated in a tube around the principal ray. Thus, in the test zone there is no inverse distance dependence

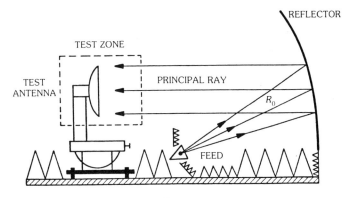

Fig. 11. Compact range.

as in conventional far-field measurements. Also, as opposed to conventional techniques, for fixed antenna dimensions the compact range does not require an increase in distance with an increase in frequency. This is due to the fact that the path lengths of the various rays are the same. As in conventional techniques, it is good practice to probe the field in the test zone to check amplitude and phase taper, and polarization conditions.

In the design and operation of the compact range, specific precautions have to be taken into account. The feed has to illuminate the reflector uniformly. Because the feed does not have a perfect phase center, the positioning of the feed ends up as an optimization problem. Aperture blockage is avoided by using asymmetrical illumination. Side lobes of the feed in the direction of the test antenna have to be minimized or screened with absorbers as illustrated in the figure. In this connection diffraction from the edge of the absorbers may cause adverse effects which also have to be minimized. The test antenna must not be placed too close to the reflector and R_0 must be chosen sufficiently large in order to avoid unwanted coupling between test antenna, probe, and reflector. The edge of the reflector has to be rolled and cut in an irregular manner in order not to give constructive interference of edge reflections. The reflector surface has to be smooth, for example, surface deviations less than $\lambda/100$ over any area of size λ^2. Also, in order to have small depolarization effects, a reflector with a long focal length has to be made.

Proper measurements require that the aperture of the reflector be about three times that of the test antenna. In the frequency range 2 GHz to 40 GHz, test zones with diameters of 1 m and length 1.5 m are typical. In order to increase the test zone diameter without increasing the overall dimensions, a compact range consisting of two crossed parabolic cylindrical reflectors has been suggested [39]. The focal lines of the reflectors are perpendicular to each other, and one reflector acts as a subreflector while the other acts as a main reflector. It is claimed that in the aperture of the main reflector a test zone superior to that of the single paraboloidal reflector with respect to amplitude and phase taper, cross polarization, and extraneous signal level is created. It would be interesting to obtain data for the test zone properties of the two types of compact ranges with the same probing mechanism.

Hess and Tavormina [40] compared results obtained on a compact range with one reflector with results obtained by using non–probe-corrected spherical near-field scanning. In Fig. 12, patterns obtained in the two manners are shown. It is concluded that non–probe-corrected measurements are adequate for many applications. However, it should be noted that, as mentioned in Section 8, ranges exist where probe correction can be made as a routine.

11. Defocusing Techniques

The defocusing techniques are based on path length considerations. For example, in the direction of the major beam of an antenna, the signals radiated from various parts of the antenna structure are usually in phase, i.e., there is no difference in the path lengths over which the signals propagate. It is said that the antenna is focused at infinity. In order to measure the antenna pattern at a finite

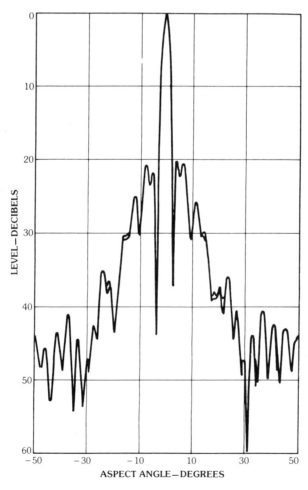

Fig. 12. Comparison of compact range and spherical near-field scanning. (*After Hess [41]*)

distance, the antenna structure is changed so that the path length differences occurring at infinity are established at the finite distance R, which is usually less than the conventional $2D^2/\lambda$ test distance. It is said that the antenna is defocused from infinity or focused at the finite distance. Besides changing the antenna structure physically, for example, bending a linear waveguide array, the defocusing may be done by modifying the feeding network or other components of the antenna.

The problems of the defocusing technique will be discussed in relation to the paraboloidal reflector shown in Fig. 13. The reflector has its focus at F and a focal length f. When a source is placed at F, there are no path length differences at infinity. This situation is approximated at a point F' by displacing the feed a distance ε to a point F'' along the axis and away from the vertex V of the reflector. However, the reflector is only approximately focused at F'. This is due to the fact that a focusing at F' would require an ellipsoidal reflector with foci at F' and F''. The simplest criterion used to find ε is to require that the two path lengths from F'' to F' via the vertex and the edge of the reflector are the same [42]. This is illustratted in the lower part of the figure. Another method is to make an approximate cancellation of the path length differences of all the rays arriving at F'. This gives

$$\varepsilon = \frac{1}{R}\left[f^2 + \left(\frac{D}{4}\right)^2 \right] \qquad [44]$$

where $R = VF'$ and D is the diameter of the reflector. The approximation is valid for f/R small [43]. It should be noted that computations or experience with measurements of particular antenna parameters may be used in the choice of ε [44].

The defocusing is associated with a particular direction. In Fig. 13, defocusing is in the direction of the axis of the paraboloid. Therefore measurements can be trusted only in a certain region around the axis. Range distances should in general be chosen from $D^2/4\lambda$ up to $2D^2/\lambda$, depending on the required accuracy.

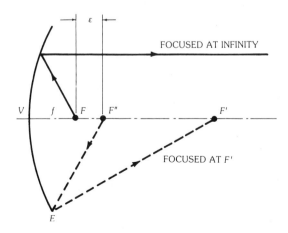

Fig. 13. Focusing at a finite distance.

12. Extrapolation Techniques

In the extrapolation technique two unknown antennas are arranged as in a conventional far-field measurement setup with the feature that the distance d between the antennas can be varied [45]. The theory developed in Section 4 on planar scanning may be used to describe the technique. The antennas are arranged as in Fig. 4 with $z_0 = d$ and $x_0 = y_0 = 0$. Let one of the two antennas be numbered 1 and be used as a transmitting antenna with transmitting characteristic $^1T^s(k_x, k_y)$. Let the other antenna be numbered 2 and be used as a probe with receiving characteristic $^2R'^s(k_x, k_y)$. Using the asymptotic form of the coupling equation (12), the signal received by antenna 2 is given by

$$b_0'(d) \cong j2\pi k a_0 \frac{e^{-jkd}}{d} \sum_{s=1}^{2} {}^2R'^s(0,0) \, {}^1T^s(0,0) \tag{45}$$

By theoretically extrapolating d to infinity we obtain one equation

$$\sum_{s=1}^{2} {}^2R'^s(0,0) \, {}^1T^s(0,0) = \lim_{d\to\infty} \left[\frac{b_0'(d) d \, e^{jkd}}{j2\pi k a_0} \right] \tag{46}$$

in the four unknowns $^2R'^s(0,0)$ and $^1T^s(0,0)$. By rotating antenna 2 about the direction to antenna 1 by 90°, an additional equation is obtained. Introducing an antenna number 3 and carrying out analogous experiments with the antenna pair 1 and 3 and the pair 2 and 3, it is understood [by making use of the reciprocity relation (14)] that six equations with the six unknown characteristics $^nT^s(0,0)$, for $s = 1, 2$ and $n = 1, 2, 3$, are obtained. The equations can in general be solved when no two antennas are circularly polarized. It should be noted that only those antenna characteristics along the axis between the antennas, usually in the main lobe directions, i.e., $(k_x, k_y) = (0,0)$, are measured.

To facilitate the solution, use is made of the series expression [46] for the received field:

$$b_0'(d) = a_0 \sum_{p=0}^{\infty} \frac{e^{-j(2p+1)kd}}{d^{2p+1}} \sum_{q=0}^{\infty} \frac{A_{pq}}{d^q} \tag{47}$$

For $p = 0$, $p = 1$, and $p > 1$ series representing the direct signal, the first-order multiple reflections, and the higher-order multiple reflections, respectively, are obtained. The magnitude A_{00} is in the dominant term for the far field. In fact, insertion of (47) into (46) gives

$$\sum_{s=1}^{2} {}^2R'^s(0,0) \, {}^1T^s(0,0) = \frac{A_{00}}{j2\pi k} \tag{48}$$

Determination of A_{00} is made by fitting a truncated part of the series (47) to the measured values of $b_0'(d)$ in the interval over which d is varied. Statistical tests of significance may be used as a criterion to find a satisfactory number of terms in the polynomial fitting. In actual measurements the ratio between the received and transmitted powers is measured as illustrated in Fig. 14. This means that $|A_{00}|^2$ is determined and not A_{00}. Therefore, in order to solve the system of six equations, phase differences are measured between the signals obtained for the two orientations of each pair of antennas.

The oscillations in Fig. 14 are due to multiple reflections, which may be averaged out by fitting processes. The same may be the case for ground reflections which have a longer period than the $\lambda/2$ characteristic for the multiple reflections. Repjar, Newell, and Baird [47] demonstrated a technique which overcomes unwanted oscillations due to ground reflections by including phase versus distance information. Hunter and Morgan [48] have combined the series expression (47) with theoretical correction factors in an extrapolation technique. The extrapolation technique is usually carried out for distances from $0.2D^2/\lambda$ and up to the conventional far-field criterion $2D^2/\lambda$. Newell, Baird, and Wacker [45] give further details on the extrapolation technique. In Fig. 15 their setup is shown.

In some recent investigations, Panicali and Nakamura [49] and Hollmann [50] have demonstrated that accurate results can be obtained by using the concept of an amplitude center, suggested by Appel-Hansen [51]. Scanning techniques and efficient methods of computation are discussed by Yaghjian [52]. Additional references to near-field far-field antenna measurements may be found in Hansen [53], which deals with spherical near-field measurements.

Acknowledgments

The author wishes to thank Johnny Vang Dahlberg and Steen Vedsegaard for reading and commenting on the manuscript. Thanks are also due to Flemming Holm Larsen and Jesper E. Hansen for many fruitful discussions on near-field techniques; to Mette Flagstad for creating most of the drawings; and to Ellinor Barfoed for typing the manuscript.

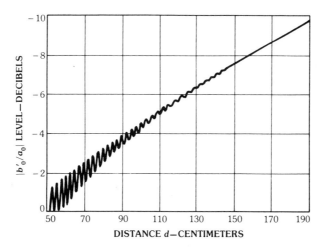

Fig. 14. Data for extrapolation to true far field. (*After Newell, Baird, and Wacker,* © *1973 IEEE*)

Fig. 15. Outdoor extrapolation range. (*After Newell, Baird, and Wacker [45]*, © 1973 *IEEE*)

13. References

[1] J. Appel-Hansen, ed., *Antenna Measurements*, FN 36, vols. I and II, Electromagnetics Institute, Technical Univ. of Denmark, Lyngby, October 1980.

[2] A. W. Rudge, K. Milne, A. D. Olver, and P. Knight, eds., *The Handbook of Antenna Design*, chapter 8, Stevenage, Herts., UK: Peter Peregrinus, 1982.

[3] S. Silver, ed., *Microwave Antenna Theory and Design*, MIT Radiation Lab Series, vol. 12, New York: McGraw-Hill Book Co., 1949, pp. 92–95.

[4] "IEEE Standard Test Procedure for Antennas," *IEEE Std. 149-1979*, pp. 18–20, Institute of Electrical and Electronics Engineers, 1979.

[5] J. D. Dyson, "Measurement of near fields of antennas and scatterers," *IEEE Trans. Antennas Propag.*, vol. AP-21, no. 4, pp. 446–460, July 1973.

[6] D. M. Kerns and E. S. Dayhoff, "Theory of diffraction in microwave interferometry," *J. Res. NBS B*, Mathematics and Mathematical Physics, vol. 64B, no. 1, pp. 1–13, January–March 1960.

[7] D. M. Kerns, "Plane-wave scattering-matrix theory of antennas and antenna-antenna interactions," *NBS Monograph 162*, National Bureau of Standards, Boulder, June 1981.

[8] J. Brown and E. V. Jull, "The prediction of aerial radiation patterns from near-field measurements," *Proc. IEE*, vol. 109B, pap. no. 3649 E, pp. 635–644, November 1961.

[9] F. Jensen, "Electromagnetic near-field–far-field correlations," *LD 15*, Electromagnetics Institute, Technical Univ. of Denmark, July 1970.

[10] W. M. Leach, Jr., and D. T. Paris, "Probe compensated near-field measurements on a cylinder," *IEEE Trans. Antennas Propag.*, vol. AP-21, no. 4, pp. 435–445, July 1973.

[11] A. D. Yaghjian, "Near-field antenna measurements on a cylindrical surface: a source

scattering matrix formulation," *NBS Tech. Note 696*, National Bureau of Standards, Boulder, September 1977.

[12] P. F. Wacker, "Non-planar near-field measurements: spherical scanning," *NBSIR 75-809*, National Bureau of Standards, Boulder, June 1975.

[13] F. H. Larsen, "Probe correction of spherical near-field measurements," *Electron. Lett.*, vol. 13, no. 14, pp. 393–395, July 1977.

[14] J. Appel-Hansen, "On cylindrical near-field scanning techniques," *IEEE Trans. Antennas Propag.*, vol. AP-28, no. 2, pp. 231–234, March 1980.

[15] P. F. Wacker, "Plane-radial scanning techniques with probe correction; natural orthogonalities with respect to summation on planar measurement lattices," *1979 Intl. Symp. Dig. IEEE Antennas Propag. Soc.*, vol. 2, pp. 561–564, Seattle, June 1979.

[16] Y. Rahmat-Samii, V. Galindo-Israel, and R. Mittra, "A plane-polar approach for far-field construction from near-field measurements," *IEEE Trans. Antennas Propag.*, vol. AP-28, no. 2, pp. 216–230, March 1980.

[17] A. D. Yaghjian, "Upper-bound errors in far-field antenna parameters determined from planar near-field measurements," *NBS Tech. Note 667, Part 1: Analysis*, National Bureau of Standards, Boulder, October 1975.

[18] E. B. Joy and D. T. Paris, "Spatial sampling and filtering in near-field measurements," *IEEE Trans. Antennas Propag.*, vol. AP-20, no. 3, pp. 253–261, May 1972.

[19] J. Appel-Hansen, "Error signals in antenna and scattering measurements," *CPEM Dig. 1982*, pp. F4–6, Conference on Precision Electromagnetic Measurements, Univ. of Colorado, Boulder, June 28–July 1, 1982.

[20] J. Appel-Hansen, "Reflectivity level of radio anechoic chambers," *IEEE Trans. Antennas Propag.*, vol. AP-21, no. 4, pp. 490–498, July 1973.

[21] K. R. Carver and A. C. Newell, "A comparison of independent far-field and near-field measurements of large spaceborne planar arrays with controlled surface deformations," *1979 Intl. Symp. Dig. IEEE Antennas Propag. Soc.*, vol. 2, pp. 494–497, Seattle, June 1979.

[22] J. Appel-Hansen, F. Jensen, and A. C. Ludwig, "SAR antenna test techniques," *TICRA ApS Rep. S-117-02*, Copenhagen, January 1980.

[23] J. H. Lemanczyk, "Calibration report P-band standard gain horn Scientific-Atlanta Model 12-12 SN215," *R 265*, Electromagnetics Institute, Technical Univ. of Denmark, Lyngby, September 1982.

[24] L. Clayton, Jr., and J. S. Hollis, "Calculation of microwave antenna radiation patterns by the Fourier integral method," *The Essay*, no. 1, Scientific-Atlanta, Inc., March 1960.

[25] J. Borowick, A. E. Holley, W. Lange, R. Cummings, and R. W. Howard, "A near field antenna measurement system," *CPEM Dig. 1982*, p. F-10, Conference on Precision Electromagnetic Measurements, Univ. of Colorado, Boulder, June 28–July 1, 1982.

[26] P. F. Wacker and A. C. Newell, "Advantages and disadvantages of planar, circular cylindrical, and spherical scanning and description of the NBS antenna scanning facilities," *ESA Preprint SP127*, pp. 115–121, Workshop on Antenna Testing Techniques held at ESTEC, European Space Agency, Noordwijk, The Netherlands, June 6–8, 1977.

[27] E. B. Joy, W. M. Leach, Jr., G. P. Rodrigue, and D. T. Paris, "Application of probe-compensated near-field measurements," *IEEE Trans. Antennas Propag.*, vol. AP-26, no. 3, pp. 379–389, May 1978.

[28] J. Anderson, J. Appel-Hansen, and F. H. Larsen, "Horizontal planar scanner for near-field measurements," *Conf. Dig. IX Nat. Conv. Radio Sci. and the Nordic Antenna Seminar*, pp. 100–103, URSI Finnish National Committee, Otaniemi, Finland, October 27–29, 1976.

[29] J. Lemanczyk, F. H. Larsen, J. E. Hansen, and J. Aasted, "Calibration of standard gain antennas using a spherical near-field technique," *IEE Conf. Pub. No. 195*, pt. 1, pp. 241–245, Second International Conference on Antennas and Propagation, Univ. of York, Heslington, York, UK, April 13–16, 1981.

[30] J. E. Hansen et al., "Introduction to the TUD-ESA spherical near-field far-field antenna test facility," draft report, Electromagnetics Institute, Technical Univ. of Denmark, July 1, 1983.

[31] J. E. Hansen, "Spherical near-field testing of spacecraft antennas," *ESA Journal*, vol. 4, pp. 89–102, 1980.

[32] S. K. Lynggaard, "Antenna measurements using plane wave synthesis," MSc thesis (in Danish), Electromagnetics Institute, Technical Univ. of Denmark, Lyngby, August 1982.

[33] J. Hald, "Synthesis and evaluation of spherical quasi plane wave regions for antenna pattern measurements," *1981 Intl. Symp. Dig. IEEE Antennas Propag. Soc.*, vol. II, pp. 573–576, Los Angeles, June 16–19, 1981.

[34] V. V. Martsafey, "Measurement of electrodynamic antenna parameters by the method of synthesized aperture," *Radio Eng. Electron. Phys.*, vol. 13, no. 12, pp. 1869–1873, 1968.

[35] J. C. Bennett and E. P. Schoessow, "Antenna near-field/far-field transformation using a plane-wave synthesis technique," *Proc. IEE*, vol. 125, no. 3, pp. 179–184, March 1978.

[36] A. C. Ludwig and F. H. Larsen, "Spherical near-field measurements from a 'compact range' viewpoint," *IEE Conf. Pub. No. 195*, pt. 1, pp. 274–277, Second International Conference on Antennas and Propagation, Univ. of York, Heslington, York, UK, April 13–16, 1981.

[37] J. F. R. Pereira, A. P. Anderson, and J. C. Bennett, "A procedure for near field measurement of microwave antennas without anechoic environments," *IEE Conf. Pub. No. 219*, pt. 1, pp. 219–223, Third International Conference on Antennas and Propagation, Univ. of East Anglia, Norwich, UK, April 12–15, 1983.

[38] D. W. Hess and R. C. Johnson, "Compact ranges provide accurate measurement of radar cross-section," *Microwave Systems News*, vol. 12, no. 9, pp. 150–160, September 1982.

[39] V. J. Vokurka, "Compact-antenna range performance at 70 GHz," *1980 Intl. Symp. Dig. IEEE Antennas Propag. Symp.*, pp. 260–263, Université Laval, Quebec, vol. I, June 2–6, 1980.

[40] D. W. Hess and J. J. Tavormina, "Verification testing of a spherical near-field algorithm and comparison to compact range measurements," *1981 Intl. Symp. Dig. IEEE Antennas Propag. Soc.*, vol. I, pp. 242–271, Los Angeles, June 16–19, 1981.

[41] D. W. Hess, "Comparison of spherical near-field vs. compact range," private communication, Scientific-Atlanta, Inc., Atlanta, August 1981.

[42] S. T. Moseley, "On-axis defocus characteristics of the paraboloidal reflector," final report, USAF Contract AF30(602)-925, Department of Electrical Engineering, Syracuse Univ., August 1, 1954.

[43] D. K. Cheng and S. T. Moseley, "On-axis defocus characteristics of the paraboloidal reflector," *IRE Trans. Antennas Propag.*, vol. AP-3, pp. 214–216, October 1955.

[44] R. C. Johnson, H. A. Ecker, and J. S. Hollis, "Determination of far-field antenna patterns from near-field measurements," *Proc. IEEE*, vol. 61, no. 12, pp. 1668–1694, December 1973.

[45] A. C. Newell, R. C. Baird, and P. F. Wacker, "Accurate measurement of antenna gain and polarization at reduced distances by an extrapolation technique," *IEEE Trans. Antennas Propag.*, vol. AP-21, no. 4, pp. 418–431, July 1973.

[46] P. F. Wacker, "Theory and numerical techniques for accurate extrapolation of near-zone antenna and scattering measurements," unpublished report, National Bureau of Standards, Boulder, April 1972.

[47] A. Repjar, A. Newell, and R. Baird, "Antenna gain measurements by an extended version of the NBS extrapolation method," *CPEM Dig. 1982*, pp. F7–9, Conference on Precision Electromagnetic Measurements, Univ. of Colorado, Boulder, June 28–July 1, 1982.

[48] J. D. Hunter and I. G. Morgan, "Gain measurements in the National Measurement Laboratory anechoic chamber," *Dig. Seminar on Electromagnetic Antenna and*

Scattering Measurements, CSIRO Division of Applied Physics, pt. II, pp. 197–204, National Measurement Laboratory, Sydney, November 1982.

[49] A. R. Panicali and M. M. Nakamura, "On the amplitude center of radiating apertures," *IEEE Trans. Antennas Propag.*, vol. AP-33, no. 3, pp. 330–335, March 1985.

[50] H. Hollmann, "Gain measurements for pyramidal horns," a summary of "Gewinnmessung von Hornstrahlern," *Tech. Rep. 454 TB 42*, Forschungsinstitut beim FTZ, Deutsche Bundespost, Darmstadt, Germany, August 1987.

[51] J. Appel-Hansen, "Centers of structures in electromagnetism: a critical analysis," *IEEE Trans. Antennas Propag.*, vol. AP-30, no. 4, pp. 606–610, July 1982.

[52] A. D. Yaghjian, "An overview of near-field antenna measurements," *IEEE Trans. Antennas Propag.*, vol. AP-34, no. 1, pp. 30–45, January 1986.

[53] J. E. Hansen, ed., *Spherical Near-Field Antenna Measurements*, Stevenage, Herts., UK: Peter Peregrinus Ltd., to be published.

Appendixes

CONTENTS

A. Physical Constants, International Units, Conversion of Units, and Metric Prefixes	A-3
B. The Frequency Spectrum	B-1
C. Electromagnetic Properties of Materials	C-1
D. Vector Analysis	D-1
E. VSWR Versus Reflection Coefficient and Mismatch Loss	E-1
F. Decibels Versus Voltage and Power Ratios	F-1

Appendix A

Physical Constants, International Units, Conversion of Units, and Metric Prefixes

Yi-Lin Chen
*University of Illinois**

Physical Constants

Quantity	Symbol	Value
Speed of light in vacuum	c	$2.997\,925 \times 10^8\,\text{ms}^{-1}$
Electron charge	e	$1.602\,192 \times 10^{-19}\,\text{C}$
Electron rest mass	m_e	$9.109\,558 \times 10^{-31}\,\text{kg}$
Boltzmann constant	k	$1.380\,622 \times 10^{-23}\,\text{JK}^{-1}$
Dielectric constant in vacuum	ϵ_0	$8.854\,185 \times 10^{-12}\,\text{Fm}^{-1}$ $\cong (36\pi \times 10^9)^{-1}\,\text{Fm}^{-1}$
Permeability in vacuum	μ_0	$4\pi \times 10^{-7}\,\text{Hm}^{-1}$

International System of Units (SI Units): Basic Units

Quantity	Symbol	Units
Length	ℓ	meters (m)
Mass	m	kilograms (kg)
Time	t	seconds (s)
Electric current	I	amperes (A)
Temperature	T	kelvins (K)
Luminous intensity	I	candelas (cd)

---

*On leave from the Chinese Aeronautical Laboratory, Beijing, China, during 1983.

Derived Units in Electromagnetics

Quantity	Symbol	Units
Electric-field strength	**E**	volts per meter (V/m)
Magnetic-field strength	**H**	amperes per meter (A/m)
Electric-flux density	**D**	coulombs per meter squared (C/m^2)
Magnetic-flux density	**B**	teslas (T) = Wb/m^2
Electric-current density	**J**	amperes per meter squared (A/m^2)
Magnetic-current density	**K**	volts per meter squared (V/m^2)
Electric-charge density	ϱ	coulombs per meter cubed (C/m^3)
Magnetic-charge density	ϱ_m	webers per meter cubed (Wb/m^3)
Voltage	V	volts (V)
Electric current	I	amperes (A)
Dielectric constant (permittivity)	ϵ	farads/meter (F/m)
Permeability	μ	henrys/meter (H/m)
Conductivity	σ	siemens per meter (S/m) = ℧/m
Resistance	R	ohms (Ω)
Inductance	L	henrys (H)
Capacitance	C	farads (F)
Impedance	Z	ohms (Ω)
Admittance	Y	siemens (S) or mhos (℧)
Power	P	watts (W)
Energy	W	joules (J)
Radiation intensity	I	watts per steradian (W/sr)
Frequency	f	hertz (Hz)
Angular frequency	ω	radians per second (rad/s)
Wavelength	λ	meters (m)
Wave number	k	1 per meter (m^{-1})
Phase shift constant	β	radians per meter (rad/m)
Attenuation factor	α	nepers per meter (Np/m)

Conversions of Units

Quantity	Symbol	SI Unit	Equivalent Number of CGS Electromagnetic Unit	Equivalent Number of CGS Electrostatic Unit
Electric charge	q	coulombs	10^{-1} abcoulomb	3×10^9 statcoulombs
Current	I	amperes	10^{-1} abampere	3×10^9 statamperes
Volume current density	$\mathbf{J}$	amperes/meter^2	10^{-5} abampere/centimeter^2	3×10^5 statamperes/centimeter^2
Voltage	V	volts	10^8 abvolts	$\frac{1}{3} \times 10^{-2}$ statvolt
Electric-field intensity	$\mathbf{E}$	volts/meter	10^6 abvolts/cm	$\frac{1}{3} \times 10^{-4}$ statvolt/centimeter
Electric-flux density	$\mathbf{D}$	coulombs/meter^2	$4\pi \times 10^{-5}$ abcoulomb/centimeter^2	$12\pi \times 10^5$ statcoulombs/centimeter^2
Magnetic-field intensity	$\mathbf{H}$	amperes/meter	$4\pi \times 10^{-3}$ oersted	$12\pi \times 10^7$ oersteds
Magnetic-flux intensity	$\mathbf{B}$	webers/meter^2	10^4 gausses	$\frac{1}{3} \times 10^{-6}$ gauss
Permittivity	ϵ	farads/meter	$4\pi \times 10^{-11}$ abfarad/centimeter	$36\pi \times 10^9$ statfarads/centimeter
Permeability	μ	henrys/meter	$\frac{1}{4\pi} \times 10^7$ gauss/oersted	$\frac{1}{36\pi} \times 10^{-13}$ gauss/oersted
Magnetic flux	Φ	webers	10^8 gilberts	$\frac{1}{3} \times 10^{-2}$ gilbert
Resistance	R	ohms	10^9 abohms	$\frac{1}{9} \times 10^{-11}$ statohm
Inductance	L	henrys	10^9 abhenrys	$\frac{1}{9} \times 10^{-11}$ stathenry
Capacitance	C	farads	10^{-9} abfarad	9×10^{11} statfarads
Conductivity	σ	siemens/meter	10^{-11} absiemen/centimeter	9×10^9 statsiemens/centimeter
Work	W	joules	10^7 ergs	10^7 ergs
Power	P	watts	10^7 ergs/second	10^7 ergs/second

Conversion of Length Units

Meters	Centimeters	Inches	Feet	Miles
1	100	39.37	3.281	6.214×10^{-4}
0.01	1	0.3937	3.281×10^{-2}	
0.0254	2.540	1	8.333×10^{-2}	
0.3048	30.48	12	1	1.894×10^{-4}
1609			5279	1

Metric Prefixes and Symbols*

Multiplication Factor	Prefix	Symbol
10^{18}	exa	E
10^{15}	peta	P
10^{12}	tera	T
10^{9}	giga	G
10^{6}	mega	M
10^{3}	kilo	k
10^{2}	hecto	h
10	deka	da
10^{-1}	deci	d
10^{-2}	centi	c
10^{-3}	milli	m
10^{-6}	micro	μ
10^{-9}	nano	n
10^{-12}	pico	p
10^{-15}	femto	f
10^{-18}	atto	a

*From *IEEE Standard Dictionary of Electrical and Electronics Terms*, p. 682, The Institute of Electrical and Electronics Engineers, Inc., 1984.

Appendix B

The Frequency Spectrum

Li-Yin Chen
*University of Illinois**

The wavelength of an electromagnetic wave in free space is $\lambda_0 = c/f$.

$$\lambda_0 = \frac{300\,000}{f(\text{kHz})}\,\text{m} = \frac{300}{f(\text{MHz})}\,\text{m} = \frac{30}{f(\text{GHz})}\,\text{cm}$$

$$= \frac{9.843 \times 10^5}{f(\text{kHz})}\,\text{ft} = \frac{9.843 \times 10^2}{f(\text{MHz})}\,\text{ft} = \frac{11.81}{f(\text{GHz})}\,\text{in}$$

The wave number of an electromagnetic wave in free space: $k_0 = \omega\sqrt{\mu_0\epsilon_0} = 2\pi f/c$.

$$\begin{aligned}
k_0 &= f(\text{Hz}) \times 2.0944 \times 10^{-8}\,\text{m}^{-1} = f(\text{kHz}) \times 2.0944 \times 10^{-5}\,\text{m}^{-1} \\
&= f(\text{MHz}) \times 2.0944 \times 10^{-2}\,\text{m}^{-1} = f(\text{GHz}) \times 20.944\,\text{m}^{-1} \\
&= f(\text{Hz}) \times 6.383 \times 10^{-9}\,\text{ft}^{-1} = f(\text{kHz}) \times 6.383 \times 10^{-6}\,\text{ft}^{-1} \\
&= f(\text{MHz}) \times 6.383 \times 10^{-3}\,\text{ft}^{-1} = f(\text{GHz}) \times 6.383\,\text{ft}^{-1} \\
&= f(\text{GHz}) \times 0.532\,\text{in}^{-1}
\end{aligned}$$

---

*On leave from the Chinese Aeronautical Laboratory, Beijing, China, during 1983.

Nomenclature of Frequency Bands

Adjectival Designation	Frequency Range	Metric Subdivision	Wavelength Range
elf: Extremely low frequency	30 to 300 Hz	Megametric waves	10 000 to 1000 km
vf: Voice frequency	300 to 3000 Hz		1000 to 100 km
vlf: Very low frequency	3 to 30 kHz	Myriametric waves	100 to 10 km
lf: Low frequency	30 to 300 kHz	Kilometric waves	10 to 1 km
mf: Medium frequency	300 to 3000 kHz	Hectrometric waves	1000 to 100 m
hf: High frequency	3 to 30 MHz	Decametric waves	100 to 10 m
vhf: Very high frequency	30 to 300 MHz	Metric waves	10 to 1 m
uhf: Ultrahigh frequency	300 to 3000 MHz	Decimetric waves	100 to 10 cm
shf: Superhigh frequency	3 to 30 GHz	Centimetric waves	10 to 1 cm
ehf: Extremely high frequency	30 to 300 GHz	Millimetric waves	10 to 1 mm
	300 to 3000 GHz	Decimillimetric waves	1 to 0.1 mm

Standard Radar-Frequency Letter Bands*

Band Designation	Nominal Frequency Range
hf	3–30 MHz
vhf	30–300 MHz
uhf	300–1000 MHz
L	1000–2000 MHz
S	2000–4000 MHz
C	4000–8000 MHz
X	8000–12 000 MHz
K_u	12.0–18 GHz
K	18–27 GHz
K_a	27–40 GHz
Millimeter	40–300 GHz

*Reprinted from ANSI/IEEE Std. 100-1984, *IEEE Standard Dictionary of Electrical and Electronics Terms*, © 1984 by The Institute of Electrical and Electronics Engineers, Inc., by permission of the IEEE Standards Department.

Television Channel Frequencies*

Channel Number[†]	Band (MHz)	Channel Number[†]	Band (MHz)	Channel Number[†]	Band (MHz)
2	54–60	29	560–566	57	728–734
3	60–66	30	566–572	58	734–740
4	66–72	31	572–578	59	740–746
5	76–82	32	578–584	60	746–752
6	82–88	33	584–590	61	752–758
7	174–180	34	590–596	62	758–764
8	180–186	35	596–602	63	764–770
9	186–192	36	602–608	64	770–776
10	192–198	37	608–614	65	776–782
11	198–204	38	614–620	66	782–788
12	204–210	39	620–626	67	788–794
13	210–216	40	626–632	68	794–800
14	470–476	41	632–638	69	800–806
15	476–482	42	638–644	70	806–812
16	482–488	43	644–650	71	812–818
17	488–494	44	650–656	72	818–824
18	494–500	45	656–662	73	824–830
19	500–506	46	662–668	74	830–836
20	506–512	47	668–674	75	836–842
21	512–518	48	674–680	76	842–848
22	518–524	49	680–686	77	848–854
23	524–530	50	686–692	78	854–860
24	530–536	51	692–698	79	860–866
25	536–542	52	698–704	80	866–872
26	542–548	53	704–710	81	872–878
27	548–554	54	710–716	82	878–884
28	554–560	55	716–722	83	884–890
		56	722–728		

Note: The carrier frequency for the video portion is the lower frequency plus 1.25 MHz. The audio carrier frequency is the upper frequency minus 0.25 MHz. All channels have a 6-MHz bandwidth. For example, channel 2 video carrier is at 55.25 MHz and the audio carrier is at 59.75 MHz.

[†]Channels 2 through 13 are vhf; channels 14 through 83 are uhf. Channels 70 through 83 were withdrawn and reassigned to tv translator stations until licenses expire.

Appendix C

Electromagnetic Properties of Materials

Yi-Lin Chen
*University of Illinois**

Resistivities and Skin Depth of Metals and Alloys

Material	Resistivity* ($\mu\Omega$-cm)	Skin Depth† (μm at 1 GHz)
Aluminum	2.62	2.576
Brass	7.5	4.3586
(66% Cu, 34% Zn)		
Copper	1.7241	2.0898
Gold	2.44	2.4861
Iron	9.71	4.9594
Nickel	6.9	4.1807
Silver	1.62	2.0257
Steel	13–22	5.7384–7.465
(0.4–0.5% C, balance Fe)		
Steel, stainless	90	15.0988
(0.1% C, 18% Cr, 8% Ni, balance Fe)		
Tin	11.4	5.3737
Titanium	47.8	11.0036

*In solid form at 20°C; resistivity = (conductivity)$^{-1}$.
†Skin depth $\delta = (\pi\mu\sigma f)^{-1/2} = (20\pi)^{-1} [\sigma f(\text{GHz})]^{-1/2}$ m. The δs in the column are calculated at $f = 1$ GHz. For other frequencies, multiply them by $[f(\text{GHz})]^{-1/2}$.

---

*On leave from the Chinese Aeronautical Laboratory, Beijing, China, during 1983.

Characteristics of Insulating Materials*

Material Composition	T(°C)	Dielectric Constant[‡] at (Frequency in Hertz)				Dissipation Factor[‡] at (Frequency in Hertz)				Dielectric Strength in Volts/Mil at 25°C	DC Volume Resistivity in Ohm-cm at 25°C	Thermal Expansion (Linear) in Parts/°C	Softening Point in °C	Moisture Absorption in Percent
		10^4	10^6	3×10^9	2.5×10^{10}	10^4	10^6	3×10^9	2.5×10^{10}					
Ceramics:														
Aluminum oxide	25	8.80	8.80	8.79	—	0.00033	0.00030	0.0010	—	—	10^{12}–10^{13}	—	1400–1430	—
Barium titanate[†]	26	1143	—	600	100	0.0105	—	0.30	0.60	75	10^{12}–10^{14}	—	1510	0.1
Calcium titanate	25	167.7	167.7	165	—	0.0002	—	0.0023	—	100	—	—	—	<0.1
Magnesium oxide	25	9.65	9.65	—	—	<0.0003	<0.0003	—	—	—	—	—	—	—
Magnesium silicate	25	5.97	5.96	5.90	—	0.0005	0.0004	0.0012	—	—	>10^{14}	9.2×10^{-6}	1350	0.1–1
Magnesium titanate	25	13.9	13.9	13.8	13.7	0.0004	0.0005	0.0017	0.0065	—	—	—	—	—
Oxides of aluminum, silicon, magnesium, calcium, barium	24	6.04	—	—	—	0.0011	—	—	—	—	—	—	1325	—
Porcelain (dry process)	25	5.08	5.04	5.90	—	0.0075	0.0078	0.0024	—	—	—	7.7×10^{-6}	—	—
Steatite 410	25	5.77	5.77	5.7	—	0.0007	0.0006	0.00089	—	—	—	—	—	—
Strontium titanate	25	232	232	—	—	0.0002	0.0001	—	—	100	10^{12}–10^{14}	—	1510	0.1
Titanium dioxide (rutile)	26	100	100	—	—	0.0003	0.00025	—	—	—	—	—	1667	—
Glasses:														
Iron-sealing glass	24	8.30	8.20	7.99	7.84	0.0005	0.0009	0.00199	0.0112	—	10^{10} at 250°	132×10^{-7}	484	poor
Soda-borosilicate	25	4.84	4.84	4.82	4.65	0.0036	0.0030	0.0054	0.0090	—	7×10^7 at 250°	50×10^{-7}	693	—
100% silicon dioxide (fused quartz)	25	3.78	3.78	3.78	3.78	0.0001	0.0002	0.00006	0.00025	410 (0.25″)	>10^{18}	5.7×10^{-7}	1667	—
Plastics:														
Alkyd resin	25	4.76	4.55	4.50	—	0.0149	0.0138	0.0108	—	—	—	—	—	—
Cellulose acetate-butyrate, plasticized	26	3.30	3.08	2.91	—	0.018	0.017	0.028	—	250–400 (0.125″)	—	$11–17 \times 10^{-5}$	60–121	2.3
Cresylic acid–formaldehyde, 50% α-cellulose	25	4.51	3.85	3.43	3.21	0.036	0.055	0.051	0.038	1020 (0.033″)	3×10^{12}	—	>125	1.2
Cross-linked polystyrene	25	2.58	2.58	2.58	—	0.0016	0.0020	0.0019	—	—	—	3×10^{-5}	—	—
Epoxy resin (Araldite CN-501)	25	3.62	3.35	3.09	—	0.019	0.034	0.027	—	405 (0.125″)	>3.8×10^7	4.77×10^{-5}	109 (distortion)	0.14
Epoxy resin (Epon resin RN-48)	25	3.52	3.32	3.04	—	0.0142	0.0264	0.021	—	—	—	—	—	—
Foamed polystyrene, 0.25% filler	25	1.03	1.03	1.03	1.03	<0.0002	—	0.0001	—	—	—	—	85	low
Melamine—formaldehyde, α-cellulose	24	7.00	6.0	4.93	—	0.041	0.085	0.103	—	300–400	—	—	99 (stable)	0.4–0.6
Melamine—formaldehyde, 55% filler	26	5.75	5.5	—	—	0.0115	0.020	—	—	—	—	1.7×10^{-5}	—	0.6
Phenol—formaldehyde (Bakelite BM 120)	25	4.36	3.95	3.70	3.55	0.0280	0.0380	0.0438	0.0390	300 (0.125″)	10^{11}	$30–40 \times 10^{-6}$	<135 (distortion)	<0.6
Phenol—formaldehyde, 50% paper laminate	26	4.60	4.04	3.57	—	0.034	0.057	0.060	—	—	—	—	—	—
Phenol—formaldehyde, 65% mica, 4% lubricants	24	4.78	4.72	4.71	—	0.0082	0.0115	0.0126	—	—	—	—	—	—
Polycarbonate	—	2.96	—	—	—	0.010	—	—	—	—	2×10^{16}	7×10^{-5}	—	—
Polychlorotrifluoroethylene	25	2.42	2.32	2.29	2.28	0.0082	0.0028	0.0028	0.0053	364 (0.125″)	10^{13}	—	135 (deflection)	—

Material	Temp (°C)	ε (60 Hz)	ε (1 kHz)	ε (1 MHz)	ε (high freq)	tan δ (60 Hz)	tan δ (1 kHz)	tan δ (1 MHz)	tan δ (high freq)	Dielectric strength (V/mil)	Volume resistivity (Ω·cm)	Coeff. expansion	Heat distortion (°C)	Water absorption (%)
Polyethylene	25	2.26	2.26	2.26	2.26	<0.0002	0.0002	0.00031	0.0006	1200 (0.033″)	10^{17}	19×10^{-5} (varies)	95–105 (distortion)	0.03
Polyethylene-terephthalate	—	2.98	—	—	—	0.016	—	—	—	4000 (0.002″)	—	—	—	—
Polyethylmethacrylate	22	2.55	2.52	2.51	2.5	0.0090	—	0.0075	0.0083	—	—	—	60 (distortion)	low
Polyhexamethylene-adipamide (nylon)	25	3.14	3.0	2.84	2.73	0.0218	0.0200	0.0117	0.0105	400 (0.125″)	8×10^{14}	10.3×10^{-5}	65 (distortion)	1.5
Polyimide	—	3.4	—	—	—	0.003	—	—	—	570	—	—	—	—
Polyisobutylene	25	2.23	2.23	2.23	—	0.0001	0.0003	0.00047	—	600 (0.010″)	—	—	25 (distortion)	low
Polymer of 95% vinyl-chloride, 5% vinyl-acetate	20	2.90	2.8	2.74	—	0.0150	0.0080	0.0059	—	—	—	$8\text{–}9 \times 10^{-5}$	70–75 (distortion)	0.3–0.6
Polymethyl methacrylate	27	2.76	—	2.60	—	0.0140	—	0.0057	—	990 (0.030″)	$>5 \times 10^{16}$	—	—	—
Polyphenylene oxide	—	2.55	—	2.55	—	0.0007	—	0.0011	—	500 (0.125″)	10^{17}	5.3×10^{-5}	195 (deflection)	—
Polypropylene	—	2.55	—	—	—	<0.0005	—	—	—	650 (0.125″)	6×10^{16}	$6\text{–}8.5 \times 10^{-5}$	99–116 (deflection)	—
Polystyrene	25	2.56	2.55	2.55	2.54	0.000007	<0.0001	0.00033	0.0012	500–700 (0.125″)	10^{18}	$6\text{–}8 \times 10^{-5}$	82 (distortion)	0.05
Polytetrafluoroethylene (Teflon)	22	2.1	2.1	2.1	2.08	<0.0002	<0.0002	0.00015	0.0006	1000–2000 (0.005″–0.012″)	10^{17}	9.0×10^{-5}	66 (distortion) (stable to 300)	0.00
Polyvinylcyclohexane	24	2.25	2.25	2.25	—	<0.0002	<0.0002	0.00018	—	860 (0.034″)	$>5 \times 10^{16}$	7.7×10^{-5}	190	—
Polyvinyl formal	26	2.92	2.80	2.76	2.7	0.019	0.013	0.0113	0.0115	260 (0.125″)	2×10^{14}	12×10^{-5}	148 (deflection)	1.3
Polyvinylidene fluoride	—	6.6	5.1	—	—	0.17	0.050	0.0555	—	375 (0.085″)	—	2.6×10^{-5}	152 (distortion)	—
Urea-formaldehyde, cellulose	27	5.65	—	4.57	—	0.027	—	—	—	450–500 (0.125″)	—	$10\text{–}20 \times 10^{-5}$	—	2
Urethane elastomer	—	6.5–7.1	—	—	—	—	—	—	—	—	—	—	—	—
Vinylidene–vinyl chloride copolymer	23	3.18	2.82	2.71	—	0.057	0.0180	0.0072	—	300 (0.125″)	$10^{14}\text{–}10^{16}$	15.8×10^{-5}	150	<0.1
100% aniline-formaldehyde (Dilectane-100)	25	3.58	3.50	3.44	—	0.0061	0.0033	0.0026	—	810 (0.068″)	10^{16}	5.4×10^{-5}	125	0.06–0.08
100% phenol-formaldehyde	24	5.4	4.4	3.64	—	0.060	0.077	0.052	—	277 (0.125″)	—	$8.3\text{–}13 \times 10^{-5}$	50 (distortion)	0.42
100% polyvinyl-chloride	20	2.88	2.85	2.84	—	0.0160	0.0081	0.0055	—	400 (0.125″)	10^{14}	6.9×10^{-5}	54 (distortion)	0.05–0.15
Organic Liquids:														
Aviation gasoline (100 octane)	25	1.94	1.94	1.92	—	—	0.0001	0.0014	—	—	—	—	—	—
Benzene (pure, dried)	25	2.28	2.28	2.28	2.28	<0.0001	<0.0001	<0.0001	<0.0001	—	—	—	—	—
Carbon tetrachloride	25	2.17	2.17	2.17	—	<0.0004	<0.0002	0.0004	—	—	—	—	—	—
Ethyl alcohol (absolute)	25	24.5	23.7	6.5	—	0.090	0.062	0.250	—	—	—	—	—	—
Ethylene glycol	25	41	41	12	—	0.030	0.045	1.00	—	—	—	—	—	—
Jet fuel (JP-3)	25	2.08	2.08	2.04	—	0.0001	—	0.0055	—	—	—	—	—	—
Methyl alcohol (absolute analytical grade)	25	31	31.0	23.9	—	0.20	0.038	0.64	—	—	—	—	—	—
Methyl or ethyl siloxane polymer (1000 cs)	22	2.78	—	2.74	—	<0.0003	—	0.0096	—	—	—	—	—	—
Monomeric styrene	22	2.40	2.40	2.40	—	<0.0003	—	0.0020	—	300 (0.100″)	3×10^{12}	—	—	0.06
Transil oil	26	2.22	2.20	2.18	—	<0.0005	0.0048	0.0028	—	300 (0.100″)	—	—	−40 (pour point)	—
Vaseline	25	2.16	2.16	2.16	—	<0.0001	<0.0004	0.00066	—	—	—	—	—	—

Characteristics of Insulating Materials* (cont'd.)

Material Composition	T(°C)	Dielectric Constant† at (Frequency in Hertz)				Dissipation Factor† at (Frequency in Hertz)				Dielectric Strength in Volts/Mil at 25°C	DC Volume Resistivity in Ohm-cm at 25°C	Thermal Expansion (Linear) in Parts/°C	Softening Point in °C	Moisture Absorption in Percent
		10^4	10^6	3×10^9	2.5×10^{10}	10^4	10^6	3×10^9	2.5×10^{10}					
Waxes:														
Beeswax, yellow	23	2.53	2.45	2.39	—	0.0092	0.0090	0.0075	—	—	—	—	45–64 (melts)	—
Dichloronaphthalenes	23	2.98	2.93	2.89	—	0.0003	0.0017	0.0037	—	—	—	—	35–63 (melts)	nil
Polybutene	25	2.34	2.30	2.27	—	0.00133	0.00133	0.0009	—	—	—	—	—	—
Vegetable and mineral waxes	25	2.3	2.3	2.25	—	0.0004	0.0004	0.00046	—	—	—	—	57	—
Rubbers:														
Butyl rubber	25	2.35	2.35	2.35	—	0.0010	0.0010	0.0009	—	—	—	—	—	—
GR-S rubber	25	2.50	2.82	2.75	—	0.0120	0.0080	0.0057	—	870 (0.040″)	2×10^{15}	—	—	—
Gutta-percha	25	2.53	2.47	2.40	—	0.0042	0.0120	0.0060	—	—	10^{15}	—	—	—
Hevea rubber (pale crepe)	25	2.4	2.4	2.15	—	0.0018	0.0050	0.0030	—	—	—	—	—	—
Hevea rubber, vulcanized (100 pts pale crepe. 6 pts sulfur)	27	2.74	2.42	2.36	—	0.0446	0.0180	0.0047	0.025	—	—	—	—	—
Neoprene rubber	24	6.26	4.5	4.00	4.0	0.038	0.090	0.034	0.10	300 (0.125″)	8×10^{12}	—	—	nil
Organic polysulfide. fillers	23	110	30	16	13.6	0.39	0.28	0.22	—	—	—	—	—	—
Silicone-rubber compound	25	3.20	3.16	3.13	—	0.0030	0.0032	0.0097	—	—	—	—	—	—
Woods:‡														
Balsa wood	26	1.37	1.30	1.22	—	0.0120	0.0135	0.100	—	—	—	—	—	—
Douglas fir	25	1.93	1.88	1.82	1.78	0.026	0.033	0.027	0.032	—	—	—	—	—
Douglas fir, plywood	25	1.90	—	—	1.6	0.0230	—	—	0.0220	—	—	—	—	—
Mahogany	25	2.25	2.07	1.88	1.6	0.025	0.032	0.025	0.020	—	—	—	—	—
Yellow birch	25	2.70	2.47	2.13	1.87	0.029	0.040	0.033	0.026	—	—	—	—	—
Yellow poplar	25	1.75	—	1.50	1.4	0.019	—	0.015	0.017	—	—	—	—	—
Miscellaneous:														
Amber (fossil resin)	25	2.65	—	2.6	—	0.0056	—	0.0090	—	2300 (0.125″)	Very high	—	200	—
DeKhotinsky cement	23	3.23	—	2.96	—	0.024	—	0.021	—	—	—	—	80–85	—
Gilsonite (99.9% natural bitumen)	26	2.58	2.56	—	—	0.0016	0.0011	—	—	—	—	9.8×10^{-5}	155 (melts)	—
Shellac (natural XL)	28	3.47	3.10	2.86	—	0.031	0.030	0.0254	—	—	10^{16}	—	80	low after baking
Mica, glass-bonded	25	7.39	—	—	—	0.0013	—	—	—	—	—	—	—	—
Mica, glass, titanium dioxide	24	9.0	—	—	—	0.0026	—	0.0040	—	—	—	—	—	—
Ruby mica	26	5.4	5.4	5.4	—	0.0003	0.0002	0.0003	—	3800–5600 (0.040″)	5×10^{13}	—	400	<0.5
Paper, royalgrey	25	2.99	2.77	2.70	—	0.038	0.066	0.056	—	202 (0.125″)	—	—	—	—
Selenium (amorphous)	25	6.00	6.00	6.00	6.00	<0.0003	<0.0002	0.00018	0.0013	—	—	—	—	—
Asbestos fiber–chrysotile paper	25	3.1	—	—	—	0.025	—	—	—	—	—	—	—	—
Sodium chloride (fresh crystals)	25	5.90	—	—	5.90	0.0002	0.0002	—	<0.0005	—	—	—	—	—

Material								
Soil, sandy dry	25	2.59	2.55	2.55	—	0.017	0.0062	—
Soil, loamy dry	25	2.53	2.48	2.44	—	0.018	0.0011	—
Ice (from pure distilled water)	−12	4.15	3.45	3.20	—	0.12	0.0009	—
Freshly fallen snow	−20	1.20	1.20	1.20	—	0.0215	0.00029	—
Hard-packed snow followed by light rain	−6	1.55	—	1.5	—	0.29	0.0009	—
Water (distilled)	25	78.2	78	76.7	34	0.040	0.157	0.2650	10^4	—

*Reproduced with permission of the publisher, Howard W. Sams & Company, Indianapolis, *Reference Data for Engineers: Radio, Electronics, Computer, and Communications*, 7th ed., by E. C. Jordan, ed., © 1985.

†The dissipation factor is defined as the ratio of the energy dissipated to the energy stored in the dielectric, or as the tangent of the loss angle. Dielectric constant and dissipation factor depend on electrical field strength.

‡Field perpendicular to grain.

Properties of Soft Magnetic Metals*

Name	Composition (%)	Permeability Initial	Permeability Maximum	Coercivity H_c (A/m)	Retentivity B_r (T)	B_{max} (T)	Resistivity (μΩ-cm)
Ingot iron	99.8 Fe	150	5 000	80	0.77	2.14	10
Low carbon steel	99.5 Fe	200	4 000	100	—	2.14	12
Silicon iron, unoriented	3 Si, bal Fe	270	8 000	60	—	2.01	47
Silicon iron, grain oriented	3 Si, bal Fe	1 400	50 000	7	1.20	2.01	50
4750 alloy	48 Ni, bal Fe	11 000	80 000	2	—	1.55	48
4-79 Permalloy	4 Mo, 79 Ni, bal Fe	40 000	200 000	1	—	0.80	58
Supermalloy	5 Mo, 80 Ni, bal Fe	80 000	450 000	0.4	—	0.78	65
2V-Permendur	2V, 49 Co, bal Fe	800	8 000	160	—	2.30	40
Supermendur	2V, 49 Co, bal Fe	—	100 000	16	2.00	2.30	26
Metglas[†] 2605SC	$Fe_{81}B_{13.5}Si_{3.5}C_2$	—	210 000	14	1.46	1.60	125
Metglas[†] 2605S-3	$Fe_{79}B_{16}Si_5$	—	30 000	8	0.30	1.58	125

*Reproduced with permission of the publisher, Howard W. Sams & Company, Indianapolis, *Reference Data for Engineers: Radio, Electronics, Computer, and Communications*, 7th ed., by E. C. Jordan, ed., © 1985.
[†]Metglas is Allied Corporation's registered trademark for amorphous alloys.

Appendix D

Vector Analysis

Yi-Lin Chen
University of Illinois[*]

1. Change of Coordinate Systems

The transformations of the coordinate components of a vector **A** among the rectangular (x, y, z), cylindrical (θ, ϕ, z), and spherical (r, θ, ϕ) coordinates are given by the following relations (see Fig. 1):

$$A_x = A_\rho \cos\phi - A_\phi \sin\phi = A_r \sin\theta \cos\phi + A_\theta \cos\theta \cos\phi - A_\phi \sin\phi$$
$$A_y = A_\rho \sin\phi + A_\phi \cos\phi = A_r \sin\theta \sin\phi + A_\theta \cos\theta \sin\phi + A_\phi \cos\phi$$

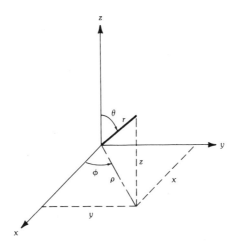

Fig. 1. Rectangular, cylindrical, and spherical coordinate systems.

---

[*]On leave from the Chinese Aeronautical Laboratory, Beijing, China, during 1983.

$$A_z = A_r \cos\theta - A_\theta \sin\theta$$
$$A_\varrho = A_x \cos\phi + A_y \sin\phi = A_r \sin\theta + A_\theta \cos\theta$$
$$A_\phi = -A_x \sin\phi + A_y \cos\phi$$
$$A_r = A_x \sin\theta \cos\phi + A_y \sin\theta \sin\phi + A_z \cos\theta = A_\varrho \sin\theta + A_z \cos\theta$$
$$A_\theta = A_x \cos\theta \cos\phi + A_y \cos\theta \sin\phi - A_z \sin\theta = A_\varrho \cos\theta - A_z \sin\theta$$

Differential element of volume:
$$dV = dx\,dy\,dz = \varrho\,d\varrho\,d\phi\,dz = r^2 \sin\theta\,dr\,d\theta\,d\phi$$

Differential element of vector area:
$$\mathbf{dS} = \hat{\mathbf{x}}\,dy\,dz + \hat{\mathbf{y}}\,dx\,dz + \hat{\mathbf{z}}\,dx\,dy$$
$$= \hat{\boldsymbol{\varrho}}\varrho\,d\phi\,dz + \hat{\boldsymbol{\phi}}\,d\varrho\,dz + \hat{\mathbf{z}}\varrho\,d\varrho\,d\phi$$
$$= \hat{\mathbf{r}}r^2 \sin\theta\,d\theta\,d\phi + \hat{\boldsymbol{\theta}}r\sin\theta\,dr\,d\phi + \hat{\boldsymbol{\phi}}r\,dr\,d\theta$$

Differential element of vector length:
$$d\boldsymbol{\ell} = \hat{\mathbf{x}}\,dx + \hat{\mathbf{y}}\,dy + \hat{\mathbf{z}}\,dz$$
$$= \hat{\boldsymbol{\varrho}}\,d\varrho + \hat{\boldsymbol{\phi}}\varrho\,d\phi + \hat{\mathbf{z}}\,dz$$
$$= \hat{\mathbf{r}}\,dr + \hat{\boldsymbol{\theta}}r\,d\theta + \hat{\boldsymbol{\phi}}r\sin\theta\,d\phi$$

2. ∇ Operator

In rectangular coordinates (x, y, z):

$$\nabla\Phi = \left(\hat{\mathbf{x}}\frac{\partial}{\partial x} + \hat{\mathbf{y}}\frac{\partial}{\partial y} + \hat{\mathbf{z}}\frac{\partial}{\partial z}\right)\Phi$$

$$\nabla\cdot\mathbf{A} = \frac{\partial A_x}{\partial x} + \frac{\partial A_y}{\partial y} + \frac{\partial A_z}{\partial z}$$

$$\nabla\times\mathbf{A} = \begin{vmatrix} \hat{\mathbf{x}} & \hat{\mathbf{y}} & \hat{\mathbf{z}} \\ \frac{\partial}{\partial x} & \frac{\partial}{\partial y} & \frac{\partial}{\partial z} \\ A_x & A_y & A_z \end{vmatrix}$$

$$\nabla^2\Phi = \nabla\cdot\nabla\Phi = \left(\frac{\partial^2}{\partial x^2} + \frac{\partial^2}{\partial y^2} + \frac{\partial^2}{\partial z^2}\right)\Phi$$

$$\nabla^2\mathbf{A} = \hat{\mathbf{x}}\nabla^2 A_x + \hat{\mathbf{y}}\nabla^2 A_y + \hat{\mathbf{z}}\nabla^2 A_z$$

In cylindrical coordinates (ϱ, ϕ, z):

$$\nabla\Phi = \left(\hat{\boldsymbol{\varrho}}\frac{\partial}{\partial \varrho} + \hat{\boldsymbol{\phi}}\frac{\partial}{\varrho\,\partial\phi} + \hat{\mathbf{z}}\frac{\partial}{\partial z}\right)\Phi$$

$$\nabla \cdot \mathbf{A} = \frac{1}{\varrho}\frac{\partial}{\partial \varrho}(\varrho A_\varrho) + \frac{1}{\varrho}\frac{\partial A_\phi}{\partial \phi} + \frac{\partial A_z}{\partial z}$$

$$\nabla \times \mathbf{A} = \frac{1}{\varrho}\begin{vmatrix} \hat{\varrho} & \varrho\hat{\phi} & \hat{z} \\ \frac{\partial}{\partial \varrho} & \frac{\partial}{\partial \phi} & \frac{\partial}{\partial z} \\ A_\varrho & \varrho A_\phi & A_z \end{vmatrix}$$

$$= \hat{\varrho}\left(\frac{1}{\varrho}\frac{\partial A_z}{\partial \phi} - \frac{\partial A_\phi}{\partial z}\right) + \hat{\phi}\left(\frac{\partial A_\varrho}{\partial z} - \frac{\partial A_z}{\partial \varrho}\right) + \hat{z}\left[\frac{1}{\varrho}\frac{\partial}{\partial \varrho}(\varrho A_\phi) - \frac{1}{\varrho}\frac{\partial A_\varrho}{\partial \phi}\right]$$

$$\nabla^2 \Phi = \frac{1}{\varrho}\frac{\partial}{\partial \varrho}\left(\varrho\frac{\partial \Phi}{\partial \varrho}\right) + \frac{1}{\varrho^2}\frac{\partial^2 \Phi}{\partial \phi^2} + \frac{\partial^2 \Phi}{\partial z^2}$$

$$\nabla^2 \mathbf{A} = \nabla\nabla\cdot\mathbf{A} - \nabla\times\nabla\times\mathbf{A} \neq \hat{\varrho}\nabla^2 A_\varrho + \hat{\phi}\nabla^2 A_\phi + \hat{z}\nabla^2 A_z$$

In spherical coordinates (r, θ, ϕ):

$$\nabla\Phi = \left(\hat{r}\frac{\partial}{\partial r} + \hat{\theta}\frac{1}{r}\frac{\partial}{\partial \theta} + \hat{\phi}\frac{1}{r\sin\theta}\frac{\partial}{\partial \phi}\right)\Phi$$

$$\nabla \cdot \mathbf{A} = \frac{1}{r^2}\frac{\partial}{\partial r}(r^2 A_r) + \frac{1}{r\sin\theta}\frac{\partial}{\partial \theta}(A_\theta \sin\theta) + \frac{1}{r\sin\theta}\frac{\partial A_\phi}{\partial \phi}$$

$$\nabla \times \mathbf{A} = \frac{1}{r^2 \sin\theta}\begin{vmatrix} \hat{r} & r\hat{\theta} & (r\sin\theta)\hat{\phi} \\ \frac{\partial}{\partial r} & \frac{\partial}{\partial \theta} & \frac{\partial}{\partial \phi} \\ A_r & rA_\theta & (r\sin\theta)A_\phi \end{vmatrix}$$

$$= \hat{r}\frac{1}{r\sin\theta}\left[\frac{\partial}{\partial \theta}(A_\phi \sin\theta) - \frac{\partial A_\theta}{\partial \phi}\right] + \hat{\theta}\frac{1}{r}\left[\frac{1}{\sin\theta}\frac{\partial A_r}{\partial \phi} - \frac{\partial}{\partial r}(rA_\phi)\right]$$

$$+ \hat{\phi}\frac{1}{r}\left[\frac{\partial}{\partial r}(rA_\phi) - \frac{\partial A_r}{\partial \theta}\right]$$

$$\nabla^2 \Phi = \frac{1}{r^2}\frac{\partial}{\partial r}\left(r^2\frac{\partial \Phi}{\partial r}\right) + \frac{1}{r^2 \sin\theta}\frac{\partial}{\partial \theta}\left(\sin\theta\frac{\partial \Phi}{\partial \theta}\right) + \frac{1}{r^2 \sin^2\theta}\frac{\partial^2 \Phi}{\partial \phi^2}$$

$$\nabla^2 \mathbf{A} = \nabla\nabla\cdot A - \nabla\times\nabla\times A \neq \hat{r}\nabla^2 A_r + \hat{\theta}\nabla^2 A_\theta + \hat{\phi}\nabla^2 A_\phi$$

3. Identities

$$\mathbf{a}\cdot\mathbf{b}\times\mathbf{c} = \mathbf{a}\times\mathbf{b}\cdot\mathbf{c} = \mathbf{b}\cdot\mathbf{c}\times\mathbf{a}$$

$$\mathbf{a}\times(\mathbf{b}\times\mathbf{c}) = (\mathbf{a}\cdot\mathbf{c})\mathbf{b} - (\mathbf{a}\cdot\mathbf{b})\mathbf{c}$$

$$(\mathbf{a}\times\mathbf{b})\cdot(\mathbf{c}\times\mathbf{d}) = \mathbf{a}\cdot\mathbf{b}\times(\mathbf{c}\times\mathbf{d}) = \mathbf{a}\cdot[(\mathbf{b}\cdot\mathbf{d})\mathbf{c} - (\mathbf{b}\cdot\mathbf{c})\mathbf{d}] = (\mathbf{a}\cdot\mathbf{c})(\mathbf{b}\cdot\mathbf{d}) - (\mathbf{a}\cdot\mathbf{d})(\mathbf{b}\cdot\mathbf{c})$$

$$(\mathbf{a}\times\mathbf{b})\times(\mathbf{c}\times\mathbf{d}) = (\mathbf{a}\times\mathbf{b}\cdot\mathbf{d})\mathbf{c} - (\mathbf{a}\times\mathbf{b}\cdot\mathbf{c})\mathbf{d}$$

$$\nabla(\Phi + \psi) = \nabla\Phi + \nabla\psi$$

$$\nabla(\Phi\psi) = \Phi\nabla\psi + \psi\nabla\Phi$$

$$\nabla\cdot(\mathbf{a} + \mathbf{b}) = \nabla\cdot\mathbf{a} + \nabla\cdot\mathbf{b}$$

$$\nabla \times (\mathbf{a} + \mathbf{b}) = \nabla \times \mathbf{a} + \nabla \times \mathbf{b}$$

$$\nabla \cdot (\Phi \mathbf{a}) = \mathbf{a} \cdot \nabla \Phi + \Phi \nabla \cdot \mathbf{a}$$

$$\nabla \times (\Phi \mathbf{a}) = \nabla \Phi \times \mathbf{a} + \Phi \nabla \times \mathbf{a}$$

$$\nabla (\mathbf{a} \cdot \mathbf{b}) = (\mathbf{a} \cdot \nabla)\mathbf{b} + (\mathbf{b} \cdot \nabla)\mathbf{a} + \mathbf{a} \times (\nabla \times \mathbf{b}) + \mathbf{b} \times (\nabla \times \mathbf{a})$$

$$\nabla \times (\mathbf{a} \times \mathbf{b}) = \mathbf{a}\nabla \cdot \mathbf{b} - \mathbf{b}\nabla \cdot \mathbf{a} + (\mathbf{b} \cdot \nabla)\mathbf{a} - (\mathbf{a} \cdot \nabla)\mathbf{b}$$

$$\nabla \cdot (\mathbf{a} \times \mathbf{b}) = \mathbf{b} \cdot \nabla \times \mathbf{a} - \mathbf{a} \cdot \nabla \times \mathbf{b}$$

$$\nabla \times \nabla \times \mathbf{a} = \nabla \nabla \cdot \mathbf{a} - \nabla^2 \mathbf{a}$$

$$\nabla \times \nabla \Phi \equiv 0$$

$$\nabla \cdot \nabla \times \mathbf{a} \equiv 0$$

$$\iiint_V \nabla \cdot \mathbf{a} \, dV = \oiint_S \mathbf{a} \cdot d\mathbf{S} \quad \text{(Gauss's theorem)}$$

$$\iint_S \nabla \times \mathbf{a} \cdot d\mathbf{S} = \oint_C \mathbf{a} \cdot d\boldsymbol{\ell} \quad \text{(Stokes's theorem)}$$

Green's first and second identities:

$$\iiint_V (\nabla \psi \cdot \nabla \Phi + \Phi \nabla^2 \psi) \, dV = \oint_S \Phi \nabla \psi \cdot d\mathbf{S}$$

$$\iiint_V (\Phi \nabla^2 \psi - \psi \nabla^2 \Phi) \, dV = \oint_S (\Phi \nabla \psi - \psi \nabla \Phi) \cdot d\mathbf{S}$$

$$\iiint_V (\nabla \times \mathbf{A} \cdot \nabla \times \mathbf{B} - \mathbf{A} \cdot \nabla \times \nabla \times \mathbf{B}) \, dV = \iiint_V (\nabla \cdot \mathbf{A} \times \nabla \times \mathbf{B}) \, dV$$

$$= \oint_S \mathbf{A} \times \nabla \times \mathbf{B} \cdot d\mathbf{S}$$

$$\iiint_V (\mathbf{B} \cdot \nabla \times \nabla \times \mathbf{A} - \mathbf{A} \cdot \nabla \times \nabla \times \mathbf{B}) \, dV = \oint_S (\mathbf{A} \times \nabla \times \mathbf{B} - \mathbf{B} \times \nabla \times \mathbf{A}) \cdot d\mathbf{S}$$

Appendix E

VSWR Versus Reflection Coefficient and Mismatch Loss

Yi-Lin Chen
*University of Illinois**

The following relations are used in the construction of the vswr table below.

$$\text{vswr} = \frac{1 + |\Gamma|}{1 - |\Gamma|}, \quad |\Gamma| = \frac{\text{vswr} - 1}{\text{vswr} + 1}$$

$$\text{mismatch loss (dB)} = -10\log_{10}(1 - |\Gamma|^2)$$

VSWR Versus Reflection Coefficient (Γ) and Mismatch Loss

VSWR	$|\Gamma|$	Mismatch Loss (dB)	VSWR	$|\Gamma|$	Mismatch Loss (dB)
1.01	.0050	.0001	1.12	.0566	.0139
1.02	.0099	.0004	1.13	.0610	.0162
1.03	.0148	.0009	1.14	.0654	.0186
1.04	.0196	.0017	1.15	.0698	.0212
1.05	.0244	.0026	1.16	.0741	.0239
1.06	.0291	.0037	1.17	.0783	.0267
1.07	.0338	.0050	1.18	.0826	.0297
1.08	.0385	.0064	1.19	.0868	.0328
1.09	.0431	.0081	1.20	.0909	.0360
1.10	.0476	.0099	1.21	.0950	.0394
1.11	.0521	.0118	1.22	.0991	.0429

*On leave from the Chinese Aeronautical Laboratory, Beijing, China, during 1983.

VSWR Versus Reflection Coefficient (Γ) and Mismatch Loss (cont'd.)

VSWR	$|\Gamma|$	Mismatch Loss (dB)	VSWR	$|\Gamma|$	Mismatch Loss (dB)
1.23	.1031	.0464	1.73	.2674	.3222
1.24	.1071	.0501	1.74	.2701	.3289
1.25	.1111	.0540	1.75	.2727	.3357
1.26	.1150	.0579	1.76	.2754	.3425
1.27	.1189	.0619	1.77	.2780	.3493
1.28	.1228	.0660	1.78	.2806	.3561
1.29	.1266	.0702	1.79	.2832	.3630
1.30	.1304	.0745	1.80	.2857	.3698
1.31	.1342	.0789	1.81	.2883	.3767
1.32	.1379	.0834	1.82	.2908	.3837
1.33	.1416	.0880	1.83	.2933	.3906
1.34	.1453	.0927	1.84	.2958	.3976
1.35	.1489	.0974	1.85	.2982	.4046
1.36	.1525	.1023	1.86	.3007	.4116
1.37	.1561	.1072	1.87	.3031	.4186
1.38	.1597	.1121	1.88	.3056	.4257
1.39	.1632	.1172	1.89	.3080	.4327
1.40	.1667	.1223	1.90	.3103	.4398
1.41	.1701	.1275	1.91	.3127	.4469
1.42	.1736	.1328	1.92	.3151	.4540
1.43	.1770	.1382	1.93	.3174	.4612
1.44	.1803	.1436	1.94	.3197	.4683
1.45	.1837	.1490	1.95	.3220	.4755
1.46	.1870	.1546	1.96	.3243	.4827
1.47	.1903	.1602	1.97	.3266	.4899
1.48	.1935	.1658	1.98	.3289	.4971
1.49	.1968	.1715	1.99	.3311	.5043
1.50	.2000	.1773	2.00	.3333	.5115
1.51	.2032	.1831	2.05	.3443	.5479
1.52	.2063	.1890	2.10	.3548	.5844
1.53	.2095	.1949	2.15	.3651	.6212
1.54	.2126	.2009	2.20	.3750	.6582
1.55	.2157	.2069	2.25	.3846	.6952
1.56	.2188	.2130	2.30	.3939	.7324
1.57	.2218	.2191	2.35	.4030	.7696
1.58	.2248	.2252	2.40	.4118	.8069
1.59	.2278	.2314	2.45	.4203	.8441
1.60	.2308	.2377	2.50	.4286	.8814
1.61	.2337	.2440	2.55	.4366	.9186
1.62	.2366	.2503	2.60	.4444	.9557
1.63	.2395	.2566	2.65	.4521	.9928
1.64	.2424	.2630	2.70	.4595	1.0298
1.65	.2453	.2695	2.75	.4667	1.0667
1.66	.2481	.2760	2.80	.4737	1.1035
1.67	.2509	.2825	2.85	.4805	1.1402
1.68	.2537	.2890	2.90	.4872	1.1767
1.69	.2565	.2956	2.95	.4937	1.2131
1.70	.2593	.3022	3.00	.5000	1.2494
1.71	.2620	.3088	3.10	.5122	1.3215
1.72	.2647	.3155	3.20	.5238	1.3929

VSWR Versus Reflection Coefficient (Γ) and Mismatch Loss

VSWR	$|\Gamma|$	Mismatch Loss (dB)	VSWR	$|\Gamma|$	Mismatch Loss (dB)
3.30	.5349	1.4636	8.10	.7802	4.0754
3.40	.5455	1.5337	8.20	.7826	4.1170
3.50	.5556	1.6030	8.30	.7849	4.1583
3.60	.5652	1.6715	8.40	.7872	4.1992
3.70	.5745	1.7393	8.50	.7895	4.2397
3.80	.5833	1.8064	8.60	.7917	4.2798
3.90	.5918	1.8727	8.70	.7938	4.3196
4.00	.6000	1.9382	8.80	.7959	4.3591
4.10	.6078	2.0030	8.90	.7980	4.3982
4.20	.6154	2.0670	9.00	.8000	4.4370
4.30	.6226	2.1302	9.10	.8020	4.4754
4.40	.6296	2.1927	9.20	.8039	4.5135
4.50	.6364	2.2545	9.30	.8058	4.5513
4.60	.6429	2.3156	9.40	.8077	4.5888
4.70	.6491	2.3759	9.50	.8095	4.6260
4.80	.6552	2.4355	9.60	.8113	4.6628
4.90	.6610	2.4945	9.70	.8131	4.6994
5.00	.6667	2.5527	9.80	.8148	4.7356
5.10	.6721	2.6103	9.90	.8165	4.7716
5.20	.6774	2.6672	10.00	.8182	4.8073
5.30	.6825	2.7235	11.00	.8333	5.1491
5.40	.6875	2.7791	12.00	.8462	5.4665
5.50	.6923	2.8340	13.00	.8571	4.7625
5.60	.6970	2.8884	14.00	.8667	6.0399
5.70	.7015	2.9421	15.00	.8750	6.3009
5.80	.7059	2.9953	16.00	.8824	6.5472
5.90	.7101	3.0479	17.00	.8889	6.7804
6.00	.7143	3.0998	18.00	.8947	7.0017
6.10	.7183	3.1513	19.00	.9000	7.2125
6.20	.7222	3.2021	20.00	.9048	7.4135
6.30	.7260	3.2525	30.00	.9355	9.0354
6.40	.7297	3.3022	40.00	.9512	10.2145
6.50	.7333	3.3515	50.00	.9608	11.1411
6.60	.7368	3.4002	60.00	.9672	11.9045
6.70	.7403	3.4485	70.00	.9718	12.5536
6.80	.7436	3.4962	80.00	.9753	13.1182
6.90	.7468	3.5435	90.00	.9780	13.6178
7.00	.7500	3.5902	100.00	.9802	14.0658
7.10	.7531	3.6365	200.00	.9900	17.0330
7.20	.7561	3.6824	300.00	.9934	18.7795
7.30	.7590	3.7277	400.00	.9950	20.0217
7.40	.7619	3.7727	500.00	.9960	20.9865
7.50	.7647	3.8172	600.00	.9967	21.7754
7.60	.7674	3.8612	700.00	.9971	22.4428
7.70	.7701	3.9049	800.00	.9975	23.0212
7.80	.7727	3.9481	900.00	.9978	23.5315
7.90	.7753	3.9909	1000.00	.9980	23.9881
8.00	.7778	4.0334			

Appendix F

Decibels Versus Voltage and Power Ratios*

Yi-Lin Chen
University of Illinois[†]

The decibel chart below indicates decibels for any ratio of voltage or power up to 100 dB. For voltage ratios greater than 10 (or power ratios greater than 100) the ratio can be broken down into two products, the decibels found for each separately, the two results then added. For example, to convert a voltage ratio of 200:1 to dB, a 200:1 voltage ratio equals the product of 100:1 and 2:1. Now, 100:1 equals 40 dB; 21:1 equals 6 dB. Therefore a 200:1 voltage ratio equals 40 dB + 6 dB, or 46 dB.

$$dB = 20\log_{10}(\text{voltage ratio}) = 10\log_{10}(\text{power ratio})$$

---

*Reprinted with permission of *Microwave Journal*, from *The Microwave Engineer's Handbook and Buyer's Guide*, 1966 issue, © 1966 Horizon House–Microwave, Inc.
[†]On leave from the Chinese Aeronautical Laboratory, Beijing, China, during 1983.

Decibels Versus Voltage and Power Ratios

Voltage Ratio	Power Ratio	−dB +	Voltage Ratio	Power Ratio	Voltage Ratio	Power Ratio	−dB +	Voltage Ratio	Power Ratio	Voltage Ratio	Power Ratio	−dB +	Voltage Ratio	Power Ratio
1.0000	1.0000	0	1.000	1.000	.5309	.2818	5.5	1.884	3.548	.2818	.07943	11.0	3.548	12.59
.9886	.9772	.1	1.012	1.023	.5248	.2754	5.6	1.905	3.631	.2876	.07762	11.1	3.589	12.88
.9772	.9550	.2	1.023	1.047	.5188	.2692	5.7	1.928	3.715	.2754	.07586	11.2	3.631	13.18
.9661	.9333	.3	1.035	1.072	.5129	.2630	5.8	1.950	3.802	.2723	.07413	11.3	3.673	13.49
.9550	.9120	.4	1.047	1.096	.5070	.2570	5.9	1.972	3.890	.2692	.07244	11.4	3.715	13.80
.9441	.8913	.5	1.059	1.122	.5012	.2512	6.0	1.995	3.981	.2661	.07079	11.5	3.758	14.13
.9333	.8710	.6	1.072	1.148	.4955	.2455	6.1	2.018	4.074	.2630	.06918	11.6	3.802	14.45
.9226	.8511	.7	1.084	1.175	.4898	.2399	6.2	2.042	4.169	.2600	.06761	11.7	3.846	14.79
.9120	.8318	.8	1.096	1.202	.4842	.2344	6.3	2.065	4.266	.2570	.06607	11.8	3.890	15.14
.9016	.8128	.9	1.109	1.230	.4786	.2291	6.4	2.089	4.365	.2541	.06457	11.9	3.936	15.49
.8913	.7943	1.0	1.122	1.259	.4732	.2239	6.5	2.113	4.467	.2512	.06310	12.0	3.981	15.85
.8810	.7762	1.1	1.135	1.288	.4677	.2188	6.6	2.138	4.571	.2483	.06166	12.1	4.027	16.22
.8710	.7586	1.2	1.148	1.318	.4624	.2138	6.7	2.163	4.677	.2455	.06026	12.2	4.074	16.60
.8610	.7413	1.3	1.161	1.349	.4571	.2089	6.8	2.188	4.786	.2427	.05888	12.3	4.121	16.98
.8511	.7244	1.4	1.175	1.380	.4519	.2042	6.9	2.213	4.898	.2399	.05754	12.4	4.169	17.38
.8414	.7079	1.5	1.189	1.413	.4467	.1995	7.0	2.239	5.012	.2371	.05623	12.5	4.217	17.78
.8318	.6918	1.6	1.202	1.445	.4416	.1950	7.1	2.265	5.129	.2344	.05495	12.6	4.266	18.20
.8222	.6761	1.7	1.216	1.479	.4365	.1905	7.2	2.291	5.248	.2317	.05370	12.7	4.315	18.62
.8128	.6607	1.8	1.230	1.514	.4315	.1862	7.3	2.317	5.370	.2291	.05248	12.8	4.365	19.05
.8035	.6457	1.9	1.245	1.549	.4266	.1820	7.4	2.344	5.495	.2265	.05129	12.9	4.416	19.50
.7943	.6310	2.0	1.259	1.585	.4217	.1778	7.5	2.371	5.623	.2239	.05012	13.0	4.467	19.95
.7852	.6166	2.1	1.274	1.622	.4169	.1738	7.6	2.399	5.754	.2213	.04898	13.1	4.519	20.42
.7762	.6026	2.2	1.288	1.660	.4121	.1698	7.7	2.427	5.888	.2188	.04786	13.2	4.571	20.89
.7674	.5888	2.3	1.303	1.698	.4074	.1660	7.8	2.455	6.026	.2163	.04677	13.3	4.624	21.38
.7586	.5754	2.4	1.318	1.738	.4027	.1622	7.9	2.483	6.166	.2138	.04571	13.4	4.677	21.88

.7499	.5623	2.5	1.334	1.778	.3981	.1585	8.0	2.512	6.310	.2113	.04467	13.5	4.732	22.39
.7413	.5495	2.6	1.349	1.820	.3936	.1549	8.1	2.541	6.457	.2089	.04365	13.6	4.786	22.91
.7328	.5370	2.7	1.365	1.862	.3890	.1514	8.2	2.570	6.607	.2065	.04266	13.7	4.842	23.44
.7244	.5248	2.8	1.380	1.905	.3846	.1479	8.3	2.600	6.761	.2042	.04169	13.8	4.898	23.99
.7161	.5129	2.9	1.396	1.950	.3802	.1445	8.4	2.630	6.918	.2018	.04074	13.9	4.955	24.55
.7079	.5012	3.0	1.413	1.995	.3758	.1413	8.5	2.661	7.079	.1995	.03981	14.0	5.012	25.12
.6998	.4898	3.1	1.429	2.042	.3715	.1380	8.6	2.692	7.244	.1972	.03890	14.1	5.070	25.70
.6918	.4786	3.2	1.445	2.089	.3673	.1349	8.7	2.723	7.413	.1950	.03802	14.2	5.129	26.30
.6839	.4677	3.3	1.462	2.138	.3631	.1318	8.8	2.754	7.586	.1928	.03715	14.3	5.188	26.92
.6761	.4571	3.4	1.479	2.188	.3589	.1288	8.9	2.786	7.762	.1905	.03631	14.4	5.248	27.54
.6683	.4467	3.5	1.496	2.239	.3548	.1259	9.0	2.818	7.943	.1884	.03548	14.5	5.309	28.18
.6607	.4365	3.6	1.514	2.291	.3508	.1230	9.1	2.851	8.128	.1862	.03467	14.6	5.370	28.84
.6531	.4266	3.7	1.531	2.344	.3467	.1202	9.2	2.884	8.318	.1841	.03388	14.7	5.433	29.51
.6457	.4169	3.8	1.549	2.399	.3428	.1175	9.3	2.917	8.511	.1820	.03311	14.8	5.495	30.20
.6383	.4074	3.9	1.567	2.455	.3388	.1148	9.4	2.951	8.710	.1799	.03236	14.9	5.559	30.90
.6310	.3981	4.0	1.585	2.512	.3350	.1122	9.5	2.985	8.913	.1778	.03162	15.0	5.623	31.62
.6237	.3890	4.1	1.603	2.570	.3311	.1096	9.6	3.020	9.120	.1758	.03090	15.1	5.689	32.36
.6166	.3802	4.2	1.622	2.630	.3273	.1072	9.7	3.055	9.333	.1738	.03020	15.2	5.754	33.11
.6095	.3715	4.3	1.641	2.692	.3236	.1047	9.8	3.090	9.550	.1718	.02951	15.3	5.821	33.88
.6026	.3631	4.4	1.660	2.754	.3199	.1023	9.9	3.126	9.772	.1698	.02884	15.4	5.888	34.67
.5957	.3548	4.5	1.679	2.818	.3162	.1000	10.0	3.162	10.000	.1679	.02818	15.5	5.957	35.48
.5888	.3467	4.6	1.698	2.884	.3126	.09772	10.1	3.199	10.23	.1660	.02754	15.6	6.026	36.31
.5821	.3388	4.7	1.718	2.951	.3090	.09550	10.2	3.236	10.47	.1641	.02692	15.7	6.095	37.15
.5754	.3311	4.8	1.738	3.020	.3055	.09333	10.3	3.273	10.72	.1622	.02630	15.8	6.166	38.02
.5689	.3236	4.9	1.758	3.090	.3020	.09120	10.4	3.311	10.96	.1603	.02570	15.9	6.237	38.90
.5623	.3162	5.0	1.778	3.162	.2985	.08913	10.5	3.350	11.22	.1585	.02512	16.0	6.310	39.81
.5559	.3090	5.1	1.799	3.236	.2951	.08710	10.6	3.388	11.48	.1567	.02455	16.1	6.383	40.74
.5495	.3020	5.2	1.820	3.311	.2917	.08511	10.7	3.428	11.75	.1549	.02399	16.2	6.457	41.69
.5433	.2951	5.3	1.841	3.388	.2884	.08318	10.8	3.467	12.02	.1531	.02344	16.3	6.531	42.66
.5370	.2884	5.4	1.862	3.467	.2851	.08128	10.9	3.508	12.30	.1514	.02291	16.4	6.607	43.65

Decibels Versus Voltage and Power Ratios, (cont'd.)

Voltage Ratio	Power Ratio	−dB +	Voltage Ratio	Power Ratio	Voltage Ratio	Power Ratio	−dB +	Voltage Ratio	Power Ratio	Voltage Ratio	Power Ratio	−dB +	Voltage Ratio	Power Ratio
.1496	.02239	16.5	6.683	44.67	.1259	.01585	18.0	7.943	63.10	.1059	.01122	19.5	9.441	89.13
.1479	.02188	16.6	6.761	45.71	.1245	.01549	18.1	8.035	64.57	.1047	.01096	19.6	9.550	91.20
.1462	.02138	16.7	6.839	46.77	.1230	.01514	18.2	8.128	66.07	.1035	.01072	19.7	9.661	93.33
.1445	.02089	16.8	6.918	47.86	.1216	.01479	18.3	8.222	67.61	.1023	.01047	19.8	9.772	95.50
.1429	.02042	16.9	6.998	48.98	.1202	.01445	18.4	8.318	69.18	.1012	.01023	19.9	9.886	97.72
.1413	.01995	17.0	7.079	50.12	.1189	.01413	18.5	8.414	70.79	.1000	.01000	20.0	10.000	100.00
.1396	.01950	17.1	7.161	51.29	.1175	.01380	18.6	8.511	72.44					
.1380	.01905	17.2	7.244	52.48	.1161	.01349	18.7	8.610	74.13		10^{-3}	30		10^{3}
.1365	.01862	17.3	7.328	53.70	.1148	.01318	18.8	8.710	75.86	10^{-2}	10^{-4}	40	10^{2}	10^{4}
.1349	.01820	17.4	7.413	54.95	.1135	.01288	18.9	8.811	77.62			50		10^{5}
.1334	.01778	17.5	7.499	56.23	.1122	.01259	19.0	8.913	79.43		10^{-5}	60		10^{6}
.1318	.01738	17.6	7.586	57.54	.1109	.01230	19.1	9.016	81.28	10^{-3}	10^{-6}	70	10^{3}	10^{7}
.1303	.01698	17.7	7.674	58.88	.1096	.01202	19.2	9.120	83.18		10^{-7}	80		10^{8}
.1288	.01660	17.8	7.762	60.26	.1084	.01175	19.3	9.226	85.11	10^{-4}	10^{-8}	90	10^{4}	10^{9}
.1274	.01622	17.9	7.852	61.66	.1072	.01148	19.4	9.333	87.10	10^{-5}	10^{-10}	100	10^{5}	10^{10}

Index

Index

Abbe sine condition
  with lenses, 16-19 to 16-23, 16-29, 16-56, 21-18
  radiation pattern of lenses obeying, 19-97 to 19-99
Aberrations with lenses, 16-12 to 16-19, 21-23. *See also* Coma lens aberrations
Absolute gain measurements, 32-42 to 32-49
Absolute models, 32-69
Absolute polarization measurements, 32-60 to 32-61
Absorbers with UTD solutions, 20-17
Absorption
  of ionospheric propagation, 29-24
  of satellite-earth propagation, 29-8
ACS (attitude control systems), 22-46
Active element patterns
  of periodic arrays, 13-6
  with phased arrays, 21-8
Active region
  with log-periodic antennas
    dipole, 9-17 to 9-21, 9-24
    mono arrays, 9-65
    phasing of, 9-56
    zigzag wire, 9-35 to 9-36
  with log-spiral antennas, 9-75
    conical, 9-83, 9-84, 9-98 to 9-99
Adcock rotating antennas systems, 25-4, 25-9 to 25-10
  for in-rotating null patterns, 25-11 to 25-12
Admittance
  of edge slot arrays, 12-26
  of medium, 1-18
  mutual, between slots, 4-91
  of ridged TE/TM waveguides, 28-47
  of thin-wire antennas, 7-7 to 7-8, 7-14 to 7-15
  of transmission lines, 9-47 to 9-49, 28-5
Admittance matrices, 3-61
  and symmetries, 3-65, 3-67 to 3-68
Admittivity and dielectric modeling, 32-72
Advanced microwave sounding unit, 22-25, 22-48
Advanced Multifrequency Scanning Radiometer (AMSR), 22-48, 22-50, 22-51 to 22-52
AF method. *See* Aperture field method
Airborne Radiation Pattern code, 20-4, 20-70 to 20-85
Air coaxial transmission lines, 28-16
Aircraft and aircraft antennas
  blade antennas, 30-12, 30-16 to 30-19
  EMP responses for, 30-4, 30-6, 30-17 to 30-31
  interferometers mounted on, 25-21 to 25-23
  military, simulation of, 20-78 to 20-88
  modeling of, 32-74 to 32-76, 32-79 to 32-86
  monopoles, 20-62 to 20-68, 20-70, 20-73 to 20-75, 20-87 to 20-88
  numerical solutions for, 20-70 to 20-90
  radomes on, 31-3 to 31-4
  simulation of, 20-30, 20-37 to 20-42, 20-63 to 20-65
  structure of, as antenna, 20-61
Air gap with leaky-wave antennas, 17-97 to 17-98
Airport radar, antennas for, 19-19, 19-56
Air striplines, 21-58, 21-60
  for power divider elements, 21-50 to 21-51
  with satellite antennas, 21-7
Air traffic control antennas, 19-19, 19-56
Alignment of antenna ranges, 32-33 to 32-34
Aluminum alloys for satellite antennas, 21-28 to 21-29
AM antennas
  directional feeders for, 26-37 to 26-47
  ground systems for, 26-37

AM antennas (*cont.*)
  horizontal plane field strength in, 26-33 to 26-34
  patterns for
    augmented, 26-36
    size determination of, 26-22 to 26-33
    standard, 26-35 to 26-36
    theoretical, 26-34 to 26-35
    two-tower, 26-22
  power for system losses in, 26-36 to 26-37
  sky-wave propagation with, 29-25 to 29-27
  standard reference, 26-3 to 26-22
Ammeters, 2-30 to 2-32
Amplitude meters, 2-30 to 2-32
Amplitude patterns, measurement of, 32-39 to 32-41
Amplitude quantization errors, 21-12
Amplitude source, ideal, 2-29 to 2-30
AMSR, 22-48, 22-50, 22-51 to 22-52
AMSU (advanced microwave sounding unit), 22-25, 22-50, 22-51 to 22-52
Analytical formulations of modeling codes, 3-89
Analytical validation of computer code, 3-76
Anechoic chambers for indoor ranges, 32-25 to 32-28
  evaluation of, 32-38 to 32-39
Angles with log-periodic antennas, 9-12, 9-14, 9-15, 9-23
Annular patches for microstrip antennas, 10-43 to 10-45
  characteristics of, 10-17
  resonant frequency for, 10-47, 10-48
Annular phased array applicator for medical applications, 24-16 to 24-18
Annular-sector patches for microstrip antennas
  characteristics of, 10-17
  resonant frequency of, 10-47, 10-49
Annular slot antennas, 30-13
Antarctica, brightness temperature image of, 22-15
Antenna arrays. *See* Arrays
Antenna ranges. *See* Ranges, antenna
Antenna sampling systems for feeder systems, 26-41 to 26-44
Antipodal fin lines, 28-55
Aperiodic arrays, 14-3 to 14-6
  linear arrays as, 11-41
  optically fed, 19-57 to 19-59
  probabilistic approach to, 14-8 to 14-35
  space-tapered, 14-6 to 14-8
Aperture antennas, 5-5
  and discrete arrays, 11-29 to 11-30
  optimization of, 11-74
Aperture blockage, 5-16
  with Cassegrain feed systems, 19-62
  in compact ranges, 32-29, 33-23
  with HIHAT antennas, 19-87
  and lens antennas, 16-5, 21-16
  with millimeter-wave antennas, 17-28
  and offset parabolic antennas, 15-80
    dual-reflector, 15-61
    with off-focus feeds, 15-58
  with phased array feeds, 19-60 to 19-61
  with reflector antennas, 15-17 to 15-18
  and space-fed arrays, 19-54 to 19-55
Aperture couplers power division elements, 21-51 to 21-52
Aperture distributions. *See* Apertures and aperture distributions
Aperture efficiency
  of dielectric-loaded horn antennas, 8-73
  and effective area, 5-27 to 5-28
  of geodesic antennas, 17-32
  with lenses, 16-19, 16-20
  with offset phased-array feeds, 19-61
  and parallel feed networks, 19-6
  with reflector antennas, 21-24 to 21-25
  and reflector surface errors, 15-105
Aperture field method, 5-5
  for pyramidal horn fields, 15-92 to 15-93
  with reflector antennas, 8-72, 15-7, 15-13 to 15-15
Aperture fields, 15-88
  for circular arrays, 21-92
  for circumferential slots, 21-92
  of horn antennas
    conical, 15-93
    corrugated, 8-53
    E-plane, 8-5 to 8-7
    H-plane, 8-20
    pyramidal, 8-34, 8-36
Aperture-matched horn antennas, 8-50, 8-64
  bandwidths of, 8-65 to 8-66
  radiated fields of, 8-58, 8-66 to 8-67
  vswr of, 8-68
Apertures and aperture distributions, 11-13. *See also* Circular apertures; Planar apertures; Tapered aperture distributions; Uniform aperture distributions

with Butler matrix, 19-8
on curved surfaces, 13-50 to 13-51
efficiency of. *See* Aperture efficiency
electric and magnetic fields in, 5-8
with feed systems
   radial transmission line, 19-28, 19-31, 19-33
   semiconstrained, 19-17
   unconstrained, 19-51
and gain, 5-26 to 5-27, 17-7, 17-8
illumination of, 13-30 to 13-32, 13-36 to 13-37
with lens antennas, 21-14, 21-16, 21-17
   phase distributions of, 16-7, 16-36
   power distributions of, 16-33, 16-36
of longitudinal array slots, 12-24
and parallel feed networks, 19-6
in perfectly conducting ground, 3-22
radiation from, 1-28
   circular, 5-20 to 5-25
   and equivalent currents, 5-8 to 5-10
   near-field, 5-24 to 5-26
   planar aperture distributions, 5-10 to 5-11
   and plane-wave spectra, 5-5 to 5-7
   rectangular aperture, 5-11 to 5-20
reflections of, 8-67
with reflector antennas, 15-15 to 15-23, 21-20
phase error of, 15-107 to 15-108
for satellite antennas, 21-5, 21-7, 21-14, 21-16, 21-17
size of
   and conical horn beamwidth, 8-61
   of E-plane horn antenna directivity, 8-16
   of lenses, 16-6
square, 5-32 to 5-33
with Taylor line source synthesis, 13-26
Aperture taper
   and radiometer beam efficiency, 22-32, 22-33
   with reflector antennas, 21-24
Apparent phase center, 8-75
Appleton-Hartee formula for ionospheric refractive index, 29-21
Applied excitation of periodic arrays, 13-6 to 13-7
Arabsat satellite antenna, 21-77
Arbitrary ray optical field, 4-8
Archimedean-spiral curves, 9-107 to 9-108, 9-110

Array blindness, 13-45 to 13-49
   with flared-notch antennas, 13-59
Array collimations, 13-7 to 13-10
Array geometry, linear transformations in, 11-23 to 11-25
Array pattern functions, 3-35, 11-8
   aperiodic, 14-4 to 14-5, 14-8 to 14-9, 14-15, 14-22, 14-35
   Dolph-Chebyshev, 11-17 to 11-18
   linear, 11-8, 11-9, 11-11 to 11-14
   with longitudinal slots, 12-17
   phased, 21-10
   transformation of, 11-23 to 11-48
   with UTD solutions, 20-6
Arrays, 3-33, 3-35 to 3-36, 3-38 to 3-43. *See also* Aperiodic arrays; Array theory; Broadside arrays; Circular arrays; Log-periodic arrays; Log-spiral antennas; Periodic arrays; Planar arrays; Slot arrays
element patterns with, 14-25
errors in, from phase quantization, 13-52 to 13-57
horn antennas as elements in, 8-4
large, design of, 12-28 to 12-34
of microstrip antennas, for medical applications, 24-25
millimeter-wave, 17-23 to 17-27
phase control of, 13-62 to 13-64
scan characteristics of, 14-24 to 14-25
of tapered dielectric-rod antennas, 17-47
thinning of, with aperiodic arrays, 14-3
Array theory
   directivity in, 11-63 to 11-76
   general formulation of, 11-5 to 11-8
   for linear arrays, 11-8 to 11-23
   linear transformations in, 11-23 to 11-48
   pattern synthesis in, 11-76 to 11-86
   for planar arrays, 11-48 to 11-58
   SNR in, 11-58 to 11-63, 11-70 to 11-76
Artificial-dielectric plates, 21-19
A-sandwich panels, 31-17 to 31-19
Aspect ratio
   of dielectric grating antennas, 17-77 to 17-78
   surface corrugations for, 17-64
   and leakage constants, 17-62, 17-63
   with log-periodic antennas, 9-12, 9-14
   and directivity, 9-21 to 9-24
Assistance with computer models, 3-68
Astigmatic lens aberrations, 16-15, 16-20, 21-13

Astigmatic ray tube, 4-8
Asymmetric aperture distributions, 21-17
Asymmetric strips, 17-84 to 17-87, 17-97 to 17-98
AT-536/ARN marker beacon antenna, 30-27 to 30-28
AT-1076 uhf antenna, 30-21
Atmosphere
  absorption by
    of millimeter-wave antennas, 17-5 to 17-6
    and satellite-earth propagation, 29-8
  noise from, 6-25, 29-50 to 29-51
  refraction of, and line-of-sight propagation, 29-32 to 29-33
  refractive index of, 29-30 to 29-32
  remote sensing of, 22-6 to 22-7, 22-13 to 22-14, 22-16, 22-21 to 22-22
  standard radio, 29-32
Attachment coefficients, 4-96
  for conducting cylinder, 21-97
  of microstrip line, 17-118
Attenuation and attenuation constants
  of atmosphere, 17-5 to 17-6, 29-8
  with biconical horns, 21-104
  for conducting cylinder, 21-97
  ground wave, 29-45
  with log-periodic antennas, 9-4 to 9-5
  with microstrip patch antennas, 17-110
  with periodic loads, 9-49 to 9-54
  from radiation, 7-12, 9-5
  by rain, 29-10 to 29-13
  of rf cables, 28-26
  of standard waveguides, 28-40 to 28-41
    circular, 1-49 to 1-50
    rectangular, 1-39, 1-43
    TE/TM, 28-37
  for transmission lines, 28-7
  with uniform leaky-wave antennas, 17-89, 17-92
Attitude control systems, 22-46
Augmented am antenna pattern, 26-36 to 26-37
Auroral blackout and ionospheric propagation, 29-25
Automated antenna ranges, 32-30 to 32-31
Axial current on thin-wire antennas, 7-5 to 7-6
Axial feed displacements, 15-49 to 15-55
Axial gain, near-field, 5-29 to 5-33
Axial power density, 5-32 to 5-33

Axial ratios
  with lens antenna feeds, 21-79, 21-80
  and polarization, 32-51
    with log-spiral antennas, 9-77 to 9-78, 9-90 to 9-93
  with polarization ellipse, 1-15 to 1-16
Axis slot on elliptical cylinder, 4-75, 4-77, 21-92, 21-95
Azimuth-over-elevation positioners, 32-12
Azimuthal symmetry, 1-33

Babinet principle, 2-13 to 2-16
  and EMP and slot antennas, 30-14 to 30-16
Backfire arrays, 9-64
Back lobes and horn antennas
  aperture-matched, 8-67
  corrugated, 8-55, 8-58
Baklanov-Tseng-Cheng design, 11-55 to 11-56
Balanced antennas
  dipole, 9-72 to 9-73
  log-periodic zigzag, 9-45
  slot, 9-72 to 9-73, 9-78 to 9-79
  spiral, 9-78
Balanced bifilar helix, 9-83
Balanced two- and four-wire transmission lines, 28-12, 28-13
Baluns
  for log-spiral antennas, 9-75
    conical, 9-95, 9-103, 9-105
  microstrip, 18-8
Bandwidth, 3-28 to 3-31
  of beam-forming networks, 18-21, 18-24, 18-26, 21-32
  and diode phase shifters, 13-62 to 13-64, 18-16
  of feed circuits
    array feeds, 13-61 to 13-62
    broadband array, 13-39 to 13-41
    parallel feed, 19-5 to 19-6
    space-fed beam-forming feeds, 18-18
    true-time-delay, 19-77 to 19-78
  of horn antennas
    aperture-matched, 8-64 to 8-65
    corrugated, 8-64
    hybrid-mode, 15-99
    millimeter-wave, 17-23
  of leaky-wave antennas, 17-98
  of lens antennas
    constrained, 16-55
    dielectric, 16-54 to 16-56

equal group delay, 16-45 to 16-46
millimeter-wave, 17-14
Rinehart-Luneberg, 19-44, 19-46 to
19-49
zoning of, 16-39, 16-43, 16-54 to
16-55, 21-16
of log-periodic dipole antennas, 9-17, 9-21
of log-spiral conical antennas, 9-98 to
9-99
of longitudinal-shunt-slot array, 17-33
of microstrip antennas, 10-6, 10-46,
17-106, 17-118, 17-122
with circular polarization, 10-59, 10-61
dipole, 17-118 to 17-120, 17-125 to
17-127
and impedance matching, 10-50 to
10-52
of millimeter-wave antennas, 17-5
biconical, 17-32
dielectric grating, 17-74 to 17-75
holographic, 17-130
horn, 17-13
lens, 17-14
microstrip arrays, 17-106, 17-114,
17-116
tapered dielectric-rod, 17-39 to 17-40,
17-44
of phased arrays, 13-19 to 13-20, 18-8,
21-12
and Q-factor, 6-22, 10-6
of receiving antennas, 6-22, 6-24
of satellite antennas, 21-4 to 21-5, 21-7
of self-complementary antennas, 9-10
with side-mount tv antennas, 27-21
of subarrays, 13-32 to 13-35, 13-39
and substrate height, 17-107
and temperature sounder sensitivity,
22-22
of TEM waveguides, 21-57
Bar-line feed network, 21-74, 21-78
Bar line transmission types, 21-58
Barn antennas, 21-84 to 21-85
Base impedance of am antennas, 26-8 to
26-17
Basic Scattering Code, 20-4
for antennas on noncurved surfaces,
20-68
for numerical simulations
of aircraft, 20-70, 20-85 to 20-86, 20-90
of ships, 20-90 to 20-97
Basis functions in modeling codes, 3-81,
3-83

Batwing tv antennas, 27-23 to 27-24
Bayliss line source pattern synthesis,
13-27 to 13-29
BCS (bistatic cross section), 2-22 to 2-23
BDF. *See* Beam deviation factor
Beacon antennas
EMP responses for, 30-23, 30-27 to
30-28
with satellite antennas, 21-84
Beads with coaxial lines, 28-21 to 28-22
Beam-broadening factor
with linear arrays, 11-19 to 11-21
with periodic arrays, 13-17 to 13-19
with phased arrays, 21-10
Beam deviation factor
with offset antennas, 15-81, 21-24
with scanned beams, 15-51, 15-53, 15-55
to 15-59, 15-62
Beam dithering, 13-56
Beam efficiency
in arrays, 11-74, 11-76
for conical horn antennas, 8-62
for microwave radiometers, 22-28 to
22-30, 22-32 to 22-33
Beam-forming feed networks
constrained, 19-3 to 19-15
cylindrical arrays, 19-98 to 19-119
optical transform, 19-91 to 19-98
for phased arrays, 18-17 to 18-26
for satellite antennas, 21-5 to 21-6, 21-29
to 21-57
semiconstrained, 19-15 to 19-49
transmission lines for, 21-63
unconstrained, 19-49 to 19-91
Beam-pointing errors, 18-27 to 18-28
Beam scanning, 14-25 to 14-26
of arrays
aperiodic, 14-32 to 14-35
Dolph-Chebyshev, 11-52
millimeter-wave, 17-25 to 17-26
beam-forming feed networks for, 18-21,
18-24, 21-27 to 21-34, 21-37 to
21-49
conical, 22-40 to 22-41
with microwave radiometers, 22-33 to
22-43
with millimeter-wave antennas, 17-9,
17-10, 17-24 to 17-26
array, 17-25 to 17-26
dielectric grating, 17-69 to 17-72
with offset parabolic antennas, 15-51,
15-53

Beam scanning (*cont.*)
　with satellite antennas, 21-5
　　phased-array, 21-71
　　polar orbiting, 22-35 to 22-38
Beam-shaping efficiency, 21-5 to 21-6
Beam squint
　with constrained feeds, 18-19, 18-21 to 18-22
　with offset antennas, 15-48
　with phase-steered arrays, 13-20
　with reflector satellite antennas, 21-19
Beam steering
　computer for, in phased array design, 18-3 to 18-4, 18-26
　with lens antennas, 17-22
　with Wheeler Lab approach, 19-107
Beam tilt with tv antennas, 27-7, 27-26
Beamwidth, 1-20. *See also* Half-power beamwidths
　of arrays
　　aperiodic, 14-3, 14-12
　　linear, 11-15
　　periodic, 13-10 to 13-12, 13-17 to 13-19
　　phased, 19-61, 21-11 to 21-12
　　planar, 11-53 to 11-55
　　subarrays, 13-32
　and Dolph-Chebyshev pattern synthesis, 13-24 to 13-25
　of feed circuits
　　limited scan, 19-57
　　series feed networks, 19-5
　and gain, with tv antennas, 27-7
　of horn antennas
　　diagonal, 8-70
　　stepped-horn, 21-82 to 21-84
　and lens F/D ratio, 21-16
　of log-periodic antennas
　　planar, 9-9 to 9-10
　　zigzag, 9-39 to 9-40, 9-45 to 9-46
　of millimeter-wave antennas
　　dielectric grating, 17-69 to 17-70, 17-74, 17-81 to 17-82
　　geodesic, 17-32
　　integrated, 17-134
　　lens, 17-20
　　metal grating, 17-69
　　microstrip dipole, 17-119
　　pillbox, 17-29
　　reflector, 17-11, 17-13, 17-14
　　slotted-shunt-slot array, 17-33
　　spiral, 17-28
　　tapered rod, 17-47
　　uniform-waveguide leaky-wave, 17-85, 17-90, 17-98
　of multibeam antennas, 21-58
　and optimum rectangular aperture patterns, 5-18 to 5-20
　of reflector antennas
　　offset, 15-37, 15-82
　　tapered-aperture, 15-16 to 15-17
　of subarrays, 13-32
　and Taylor line source pattern synthesis, 13-26 to 13-27
　of tv antennas, 27-7
　　side-mount tv antennas, 27-20
　of uhf antennas, 27-27
Beamwidths scanned and gain loss with reflector antennas
　dual-reflector, 15-73
　offset parabolic, 15-55, 15-58, 15-62 to 15-66
Benelux Cross antenna, 14-17 to 14-18
Beyond-the-horizon transmission, 29-21 to 29-30
Bhaskara-I and -II satellites, 22-17
Biconical antennas, 3-24 to 3-26, 3-29
　coaxial dipole, 3-30
　horn, 3-30
　millimeter-wave, 17-32
　for satellite antennas, 21-89, 21-100 to 21-109
Bifilar helix, 9-83
Bifilar zigzag wire antennas, 9-35, 9-37
Bifocal lenses, 16-30 to 16-33
Bilateral fin lines, 28-55
Binomial linear arrays, 11-10 to 11-11
Bistatic cross section, 2-22 to 2-23
Bistatic radar, 2-36 to 2-39
Blackbody radiation, 22-3 to 22-4
Blade antennas
　EMP responses for, 30-17 to 30-19
　equivalent circuits for, 30-16
　equivalent receiving area of, 30-12
Blass matrices
　with beam-forming feed networks, 18-25, 19-10 to 19-11
　with space-fed subarray systems, 19-68
　for switch beam circuits, 21-31 to 21-33
Blass tilt traveling-wave arrays, 19-80
Blind angles with arrays, 14-25 to 14-27, 14-32 to 14-35
Blindness, array, 13-45 to 13-49
　with flared-notch antennas, 13-59

Index

Blind spots with horn antennas, 8-4
Blockage. *See* Aperture blockage
Block 5D satellite, 22-18
Body-stabilized satellites, 21-90
Boeing 707 aircraft and EMP, 30-4, 30-6
Boeing 737 aircraft
   simulation of, 20-64, 20-70 to 20-78
   slot antennas on, 20-85 to 20-86, 20-89, 20-90
Boeing 747 aircraft and EMP, 30-4, 30-6
Bombs, aircraft, simulation of, 20-79
Bone, human, phantom models for, 24-55
Booker-Gordon formula for scattering, 29-28
Boom length of log-periodic antennas
   dipole, 9-21, 9-23, 9-24, 9-26
   zigzag, 9-39
Bootlace lenses, 16-46 to 16-48
Boresight beam
   gain of, with reflector antennas, 15-5
   with offset parabolic antennas, 15-81
   of phased-array satellite antennas, 21-71
   and radomes, 31-3 to 31-4, 31-7 to 31-9, 31-27 to 31-29
Born-Rytov and computer solutions, 3-54
Boundary conditions, 1-8 to 1-9
   with corrugated horns, 8-50, 15-96 to 15-97
   and current distributions, 3-38 to 3-39
   and lens zoning, 16-39
   in longitudinal slots, 12-6
   numerical implementation of, 3-55
   numerical validation of, 3-79 to 3-80
   with patches, 10-16 to 10-17
   and perfect ground planes, 3-18 to 3-19
   for thin wires, 3-44 to 3-45
Bow-tie dipoles, 17-137 to 17-138
Bragg condition for diffraction, 29-28
Brain, human, phantom models for, 24-55
Bra-Ket notation and reciprocity, 2-16
Branch guide coupler power division elements, 21-51 to 21-52
Branch line couplers
   for array feeds, 13-61
   directional power division element, 21-50
Branch line guide for planar arrays, 12-21 to 12-23
Brewster angle phenomena
   modeling of, 32-79
   and radome design, 31-12 to 31-13

Brightness temperature
   of emitters, 2-40 to 2-41
   and humidity, 22-6 to 22-7
   of microwave radiometers, 22-28
   vs. physical temperature, 22-10
   and surface emissivity, 22-7 to 22-9, 22-15
Broadband array feeds, 13-39 to 13-41
Broadside arrays, 3-40, 11-11, 11-15
   maximum directivity of, 11-64 to 11-67
   radiation pattern of, 13-12 to 13-13
Broadside coupled transmission lines, 28-16, 28-17
Broadside radiation of microstrip arrays, 17-115
Broadwall millimeter-wave array antennas, 17-26
Broadwall shunt-slot radiators, 18-8
Broadwall slots
   for center-inclined arrays, 12-24
   longitudinal, 12-4
BSC. *See* Basic Scattering Code
Bulge, earth, and line-of-sight propagation 29-34
Butler matrices
   with arrays
      multimode circular, 25-18 to 25-19
      multiple-beam, 19-80
   with beam-forming feed networks, 18-23 to 18-24, 19-8 to 19-10, 19-12
   with cylindrical array feeds, 19-101 to 19-103
   as Fourier transformer, 19-95
   with reflector-lens limited-scan feed concept, 19-68
   with space-fed subarray systems, 19-68
   for switch beam circuits, 21-31, 21-33
   in transform feeds, 19-92 to 19-93

Cable effect with modeling, 32-84
Cabling with scanning ranges, 33-16
CAD (computer-aided design), 17-120 to 17-122
Calibration
   accuracy of, with microwave radiometers, 22-23
   of AMSU, 22-50
   of field probes, 24-54 to 24-57
   horn antennas as standard for, 8-3, 8-43, 8-69
   of microwave radiometers, 22-25 to 22-26, 22-48

Cameras, thermographic, 24-51
Cancer therapy. *See* Hyperthermia
Candelabras for multiple-antenna installations, 27-37
Capacitance of in-vivo probes, 24-33 to 24-34
Capacitor-plate hyperthermia applicators, 24-46 to 24-47
Capture area of receiving antenna, 6-6
Cardioid patterns
 with loop antennas, 25-6 to 25-7
 with slot antennas
  coaxial, 27-28
  waveguide, 27-30
 from slotted cylinder reflectors, 21-110
 for T T & C, 21-89 to 21-90
Carrier-to-noise ratio, 29-48
Cartesian coordinate systems, 15-115 to 15-120
Cassegrain feed systems
 near-field, 19-62 to 19-63
 for reflector antennas
  millimeter-wave, 17-9, 17-13 to 17-14
  satellite, 21-21 to 21-22
 with wide-angle systems, 19-84, 19-88
Cassegrain offset antennas
 cross polarization with, 15-69 to 15-70, 15-72 to 15-74
 dual-mode horn antennas for, 8-72
 parameters of, 15-67 to 15-68
 performance evaluation of, 15-69 to 15-72
 scan performance of, 15-71 to 15-73, 15-78 to 15-79
Caustic distances, 4-10
 edge-diffracted, 4-27
 and GO reflected field, 4-18
Caustics, 4-6, 8-76
 with diffracted rays, 4-5, 4-6
  surface-diffracted, 4-83
 of diffracted waves, 4-96 to 4-102
 and GO representation, 4-16
 and lens antennas, 17-16
 matching functions for, 4-103
Cavity-backed radiators
 in panel fm antennas, 27-35 to 27-36
 in side-mount tv antennas, 27-20 to 27-23
Cavity-backed slots with log-periodic arrays, 9-68 to 9-71
Cavity model for microstrip antennas, 10-10 to 10-21
Centered broad wall slots, 12-4

Center-fed antennas
 dipole
  microstrip, 17-123 to 17-124
  transient response of, 7-22
 slot, for tv, 27-29 to 27-30
Center-fed arrays, 17-119 to 17-120
Center-fed dual series feeds, 18-20
Center-fed reflectarrays, 19-55
Center-inclined slot arrays, 12-24
Channel-diplexing, 22-46
Channel guide mode for leaky-wave antennas, 17-98
Characteristic impedance, 3-28 to 3-31
 of dipoles, 7-6, 9-28, 9-30
 of fin lines, 28-57, 28-58
 of holographic antennas, 17-130
 of log-periodic zigzag antennas, 9-42 to 9-44, 9-48
 of microstrip arrays, 17-109 to 17-111
 of slot lines, 28-53 to 28-54
 of transmission lines, 28-7
  biconical, 3-24 to 3-25
  coaxial, 28-20
  microstrip planar, 28-30 to 28-31
  TEM, 28-10 to 28-18
  triplate stripline, 28-22, 28-28 to 28-29
  two-wire, 7-39, 28-11, 28-19
  waveguide planar, 28-33
 of waveguides
  rectangular, 2-28
  ridged, 24-14 to 24-16
  TE/TM, 28-36, 28-37
Chebyshev polynomials, 11-15 to 11-17. *See also* Dolph-Chebyshev arrays
Check cases for validating computer code, 3-80
Chi-square distribution with aperiodic arrays, 14-10 to 14-12
Chokes
 with coaxial sleeve antennas, 7-31 to 7-33, 7-36
 with microstrip antennas, 10-68
Circles, 15-25 to 15-26
 and intersection curve, 15-29
Circuit characteristics
 measurements for, 32-4
 models for, 10-26 to 10-28
Circular apertures
 Bayliss line source pattern synthesis for, 13-29
 distribution of
  axial field, 5-32
  with reflector antennas, 15-17, 15-23

efficiency of, 5-29, 5-30
   radiation patterns of, 5-20 to 5-24
      near-field, 5-25
Circular arrays
   for direction-finding antennas, 25-17 to 25-20
   of probes, with transmission line feeds, 19-26 to 19-30
   sections of, 13-8 to 13-10
   transformations with, 11-33 to 11-36, 11-41 to 11-47
   for T T & C, 21-89 to 21-100
Circular coaxial lines, 28-20 to 28-22
Circular cones
   geometry for, 4-60
   intersection of reflector surface with, 15-26 to 15-29
Circular corrugated horns, 15-97, 15-99
Circular cylinders
   and intersection curve, 15-30
   radiation patterns of, 4-75, 4-77 to 4-79
Circular-disk microstrip patches
   characteristics of, 10-16, 10-25
   with CP microstrip antennas, 10-61, 10-62
   elements for, 13-60
   principal-plane pattern of, 10-22
   resonant frequency of, 10-47
Circularity, of multiple antennas, 27-38 to 27-40
Circular log-periodic antenna, 9-3, 9-4
Circular loop antennas
   for geophysical applications, 23-19 to 23-21
   as magnetic-field probes, 24-40 to 24-43
   radiation resistance of, 6-19
   thin-wire, 7-42
Circular open-ended waveguides
   radiators for, with phased arrays, 18-6
   with reflector antennas, 21-74
   for reflector feeds, 15-90 to 15-93
Circular patches, 10-41 to 10-44
Circular-phase fields, 9-109
Circular pillbox feeds, 19-19, 19-21 to 19-22
Circular polarization, 1-15 to 1-16, 1-20, 1-28 to 1-29
   of arrays, 11-7
   with constrained lenses, 16-43
   conversion of linear to, 13-60 to 13-61
   with dual-mode converters, 21-78, 21-80
   with fm antennas, 27-33 to 27-34
   gain measurements for, 32-49 to 32-51
   with horn antennas
      biconical, 21-107 to 21-108
      corrugated, 8-51, 8-64
      diagonal, 8-70
   with image-guide-fed slot array antennas, 17-50
   of log-spiral conical antennas, 9-90 to 9-91, 9-99
   measurements of, 32-52, 32-58, 32-60
   with microstrip antennas, 10-57 to 10-63, 10-69 to 10-70
   microstrip radiator types for, 13-60
   with phased arrays, 18-6
   for power dividers, 21-78, 21-80
   with reflector antennas
      offset parabolic, 15-36, 15-48 to 15-50
      for satellites, 21-19, 21-73, 21-74, 21-78 to 21-80
   for satellite antennas, 21-7, 21-29, 21-89
   reflector, 21-19, 21-73, 21-74, 21-78 to 21-80
   for shipboard simulation, 20-91 to 20-93
   with spiral antennas, 17-27
   for T T & C, 21-89
   with tv antennas, 27-8 to 27-9, 27-38 to 27-40
      helical, 27-15 to 27-19
      side-mount, 27-20 to 27-23
      skewed-dipole, 27-11 to 27-12, 27-18
      vee-dipole, 27-9 to 27-11
Circular-sector patches, 10-17, 10-44
   resonant frequency of, 10-47, 10-48
Circular tapered dielectric-rod antennas, 17-37 to 17-38
Circular waveguides, 1-44 to 1-50
   array of, with infinite array solutions, 13-48 to 13-49
   for conical horn antennas, 8-46, 15-93
   dielectric, 28-50 to 28-51
   open-ended, 15-90 to 15-93, 15-105
   TE/TM, 28-42 to 28-45
   transmission type, 21-60
Circumferential slots
   on cones, 4-92
   on cylinders, 4-75, 4-77, 21-92 to 21-93, 21-96
CLAD (controlled liquid artificial dielectrics), 17-22
Classical antennas, 3-26 to 3-31
Cluttered environments, simulation of, 20-90 to 20-97
Cmn, explicit expressions for, 2-11

Coaxial cable. *See also* Sleeve antennas
  with implantable antennas, 24-26 to 24-28
  for in-vivo measurements, 24-32 to 24-35
  for log-periodic zigzag antennas, 9-44 to 9-45
  with log-spiral antennas, 9-75, 9-80
  with medical applications, 24-24 to 24-25
  for microstrip antennas, 10-51, 17-120
  with TEM waveguides, 21-57
Coaxial current loops for hyperthermia, 24-45 to 24-46
Coaxial dipole antennas, 3-30
Coaxial hybrids as baluns, 9-95
Coaxial-line corporate feed networks, 13-61
Coaxial line transmission type, 21-57, 21-60
  for TEM transmission lines, 28-23 to 28-25
    circular, 28-20 to 28-22
    impedance of, 28-15, 28-16, 28-18
Coaxial-probe feeds, 10-28, 10-29
Coaxial radiators
  dipole, for phased arrays, 18-7
  excited-disk, 18-8 to 18-9
Coaxial slot antennas, 27-28 to 27-29
COBRA modeling code, 3-88 to 3-91
Codes, computer modeling, 3-80 to 3-96
Collimations, array, 13-7 to 13-10
Collinear arrays, 3-43
Collocation moments method, 3-60
Column arrays
  collimation of, 13-8
  of dipoles, coupling between, 13-44
Coma lens aberrations, 16-15, 16-17, 21-13 to 21-14
  and Abbe sine condition, 16-19 to 16-21
  and millimeter-wave antennas, 17-19 to 17-21
  with Ruze lenses, 19-39
  with spherical thin lenses, 16-29
  and zoning, 16-38 to 16-41
Communication satellite antennas, 21-4 to 21-6
  beam-forming networks for, 21-29 to 21-57
  design of, 21-6 to 21-8, 21-68 to 21-80
  feed arrays for, 21-27 to 21-29
  multibeam, 21-57 to 21-68
  types of, 21-8 to 21-26

Compact antenna ranges, 32-28 to 32-30
  for field measurements, 33-22 to 33-24
Compensation theorem and computer solutions, 3-54
Complementary planes, 2-14 to 2-15
Completely overlapped space-fed subarray system, 19-68 to 19-76
Complex environments
  airborne antenna patterns, 20-70 to 20-90
  antennas for, 20-3 to 20-4
  far fields with, 20-53 to 20-54, 20-61 to 20-70
  numerical simulations
    for antennas, 20-5 to 20-7
    for environment, 20-7 to 20-53
  shipboard antenna patterns, 20-90 to 20-97
Complex feeds for reflectors, 15-94 to 15-99
Complex pattern function for arrays, 11-7
Complex polarization ratios, 32-52
Complex poles of antennas, 10-11
Complex power, 1-7
Component errors and phased-array performance, 18-26 to 18-28
Compound distributions for rectangular apertures, 5-16 to 5-17
Computation with computer models, 3-54 to 3-55, 3-68 to 3-72
Computer-aided design, 17-120 to 17-122
Computers and computer programs
  for beam-forming network topology, 21-57
  for beam steering, 18-3 to 18-4, 18-26
  for complex environment simulations, 20-3 to 20-4, 20-18
  for dipole-dipole arrays, 23-7
  for Gregorian feed systems, 19-65
  for integral equations, 3-52 to 3-55
    codes for, 3-80 to 3-96
    computation with, 3-68 to 3-72
    numerical implementation of, 3-55 to 3-68
    validation of, 3-72 to 3-80
  for ionospheric propagation, 29-25
  modeling with, 32-65
  for rf personnel dosimeters, 24-43
  for spherical scanning, 33-15
Computer storage and time
  with frequency-domain solutions, 3-61 to 3-62, 3-65

using NEC and TWTD modeling codes, 3-94
and symmetry, 3-65, 3-67
with time-domain solutions, 3-63 to 3-65
and unknowns, 3-57
Concentric lenses, 19-80
Concentric ring arrays, 19-57 to 19-58
Conceptualization for computer models, 3-53 to 3-54
Conductance
  in inclined narrow wall slot arrays, 12-25 to 12-26
  with microstrip antennas, 10-7 to 10-8
  of resonant-length slots, 12-14
Conducting cones, fields of, 1-10
Conducting wedges, fields of, 1-9 to 1-10
Conduction losses, 1-7
  of microstrip patch antennas, 17-111
  of transmission lines
    circular coaxial, 28-21
    coplanar waveguide planar, 28-34
    microstrip planar, 28-32
    triplate stripline, 28-28
  of waveguides, 1-39
    rectangular TE/TM, 28-38, 28-42
Conductivity of troposphere, 29-31
Conductors, modeling of, 32-70 to 32-71
Conductor surfaces, linear current density in, 12-3
C-141 aircraft, simulation of, 20-82 to 20-84
Cone-tip diffraction, 4-93 to 4-94
Cones
  circular, 4-60
  conducting, 1-10
  coupling coefficient of slots on, 4-92
  radial slots in, 4-78 to 4-80
  surface-ray paths on, 4-75 to 4-76
Conformal arrays, 13-49 to 13-52
  nonplanar, 13-41
  probabilistic approach to, 14-33 to 14-34
  for T T & C, 21-90
Conical cuts, 32-6
Conical horn antennas, 8-4. *See also* Corrugated horn antennas, conical
  design procedures for, 8-48 to 8-49
  directivity of, 8-46 to 8-47
  dual-mode, 8-71 to 8-72, 15-95, 15-100 to 15-101
  for earth coverage satellite antennas, 21-82
  for feeds for reflectors, 15-93 to 15-94, 15-99 to 15-101
  gain of, 8-48
  millimeter-wave, 17-23, 17-25
  phase center of, 8-76, 8-78 to 8-80
  radiated fields of, 8-46
Conical log-spiral antennas, 9-3, 9-4
  active region of, 9-83, 9-84, 9-98 to 9-99
  axial ratios with, 9-90 to 9-93
  current wave of, 9-83 to 9-84
  directivity of, 9-87 to 9-90
  feeds for, 9-95 to 9-96, 9-99 to 9-104, 9-111 to 9-112
  front-to-back ratios with, 9-89, 9-92
  half-power beamwidth of, 9-86, 9-87, 9-89, 9-97 to 9-98
  impedance of, 9-95 to 9-97, 9-99, 9-107
  phase center of, 9-83, 9-91 to 9-92, 9-94
  polarization of, 9-86, 9-90 to 9-91, 9-99
  propagation constant with, 9-80 to 9-84
  radiation fields for, 9-85 to 9-86, 9-89 to 9-92, 9-99 to 9-109
  transmission lines with, 9-95
  truncation with, 9-79, 9-84, 9-89 to 9-90, 9-95 to 9-97
Conical scanning, 22-38 to 22-43, 22-48
Conic-section-generated reflector antennas, 15-23 to 15-31
Constant-K lenses, 17-15 to 17-16, 17-19
Constituent waves with radomes, 31-5
Constrained feeds, 18-19 to 18-26, 21-34
  multimode element array technique, 19-12 to 19-15
  multiple-beam matrix, 19-8 to 19-12
  for overlapped subarrays, 13-36
  parallel feed networks, 19-5 to 19-6
  series feed networks, 19-3 to 19-5
  true time-delay, 19-7 to 19-8
Constrained lenses, 16-5 to 16-6, 16-41 to 16-48
  analog of dielectric lenses for, 16-49 to 16-51
  bandwidth of, 16-55
  wide-angle multiple-beam, 19-80 to 19-85
Constrained variables and radome performance, 31-4
Contiguous subarrays, 13-32 to 13-35
Continuity equations, 1-6. *See also* Discontinuities
  for electromagnetic fields, 3-6
  for field distributions, 5-18 to 5-19

Continuous metal strips for leaky-wave antennas, 17-84 to 17-86
Continuous scans, 22-35
  compared to step scans, 22-43
Contour beam reflectors, 15-5, 15-80 to 15-87
Contours and tv antenna strength, 27-6
Controlled liquid artificial dielectrics, 17-22
Convergence
  in large arrays, 12-28
  measures of, and modeling errors, 3-72 to 3-75
Convex cylinder, surface-ray paths on, 4-75 to 4-76
Convex surfaces
  GTD for, 4-28 to 4-30
  UTD for
    and mutual coupling, 4-84 to 4-96
    and radiation, 4-63 to 4-84
    and scattering, 4-50 to 4-63
Coordinate systems
  for antenna ranges, 32-8 to 32-14
  feed vs. reflector, 15-88 to 15-89
  for lenses, 16-7 to 16-9
  for numerical aircraft solutions, 20-76
  for probe antennas, 32-6 to 32-7
  with reflector antennas, 15-7
  transformations of, 15-115 to 15-120
Cophasal arrays
  Dolph-Chebyshev, 11-19
  end-fire, 11-69, 11-70
  uniform, 11-64 to 11-65
  circular, linear transformations with, 11-46 to 11-47
Coplanar waveguide transmission types, 21-59, 21-61
  applicators for medical applications, 24-24 to 24-25
  planar quasi-TEM, 28-33 to 28-34
  strip planar quasi-TEM, 28-35
Copolarization. *See* Reference polarization
Copper for satellite antenna feeds, 21-28 to 21-29
Corner-diffracted fields
  coefficient for, 4-44 to 4-48
  and UTD solutions, 20-12 to 20-13
Corporate feed networks, 19-5 to 19-6
  for arrays, 13-61
  for beam-forming, 18-24
Correction factors
  for propagation, 29-5

with reflector surface errors, 15-106 to 15-107, 15-113
Corrugated horn antennas, 8-4, 8-50 to 8-51
  circular, 15-97, 15-99
  conical, 8-59 to 8-63
    cross-polarized pattern for, 8-69
    millimeter-wave, 17-23
    radiated fields of, 8-64, 8-65
  efficiency of, 8-50
  millimeter-wave, 17-23
  pyramidal, 8-51 to 8-52
    aperture fields of, 8-53
    half-power beamwidth of, 8-55
    radiated fields of, 8-53 to 8-59
  for reflector antennas, 15-96, 15-102 to 15-104
  for satellite antennas, 21-82
Corrugations
  depth of, and EDC, 17-55 to 17-56
  with periodic dielectric antennas, 17-48
Cosine lenses, hyperbolic, 16-53 to 16-54
Cosine representations of far fields, 1-28 to 1-29
Cosinusoidal aperture distributions
  and corrugated horns, 8-53
  near-field reduction factors for, 5-30 to 5-32
Cosmos-243 satellite, 22-17
Cosmos-384 satellite, 22-17
Cos type patterns in reflector feeds, 15-99, 15-102 to 15-105
Coupler/phase-shifter variable power dividers, 21-43 to 21-45
Couplers
  for array feeds, 13-61
  for series feed networks, 19-4 to 19-5
Coupling, 6-9. *See also* Mutual coupling
  of beam-forming feed networks, 18-25, 18-26
  with dipole arrays, 13-44
  microstrip antenna to waveguide, 17-119 to 17-121
  and phase shift, 9-58
  of transmitting and receiving antennas, 20-68
  with UTD solutions, 20-6 to 20-7
    for curved surfaces, 20-18
  of waveguides for planar arrays, 12-21 to 12-22
Coupling coefficient, 4-92
  with aperiodic arrays, 14-24, 14-26

Coupling equation
  for planar scanning, 33-8 to 33-9
  for spherical scanning, 33-13
Coverage area. *See also* Footprints
  of satellite antennas, 21-4 to 21-6, 21-73, 21-77
    maps for, 15-80 to 15-81
  of tv antennas, 27-4 to 27-6
Creeping waves
  with conformal arrays, 13-50 to 13-51
  with UTD solutions
    for curved surfaces, 20-20, 20-22
    for noncurved surfaces, 20-45
Critical frequencies of ionospheric propagation, 29-23
Crops, remote sensing for, 22-15
Cross dipole antennas, 1-24
Crossed yagis phased arrays, 21-12
Cross-flux of fields, 2-16, 2-18
Crossover
  of beam-forming feed networks, 18-25
  with pillbox feeds, 19-22 to 19-23
Cross polarization, 1-23 to 1-24
  with arrays
    center-inclined broad wall slot, 12-24
    conformal, 13-50
    inclined narrow wall slot, 12-24
    microstrip patch resonator, 17-108
    phased, 18-8
  with beam-forming networks, 21-57
  with edge-slot array antennas, 17-27
  with horn antennas
    aperture-matched, 8-68
    corrugated, 8-64, 8-69
    millimeter-wave, 17-23
  and matched feeds, 15-99
  with microwave antennas, 33-20 to 33-21
  with microwave radiometers, 22-30, 22-45
  with millimeter-wave antennas
    horn, 17-23
    reflector, 17-9
  with probe antennas, 32-8
  with reflector antennas, 15-5 to 15-6
    dual-reflector, 15-69 to 15-70, 15-72 to 15-74
    millimeter-wave, 17-9
    offset parabolic, 15-42, 15-46 to 15-48
    satellite, 21-19 to 21-20, 21-23
  with satellite antennas, 6-26, 6-27, 21-5, 21-7
  of side lobes, slots for, 8-63

Cross sections
  for leaky-wave antennas, 17-88
  receiving. *See* Receiving cross section
  with scattering, 2-22 to 2-25
  of tapered dielectric-rod antennas, 17-38 to 17-40
  of transmission lines, 21-63
Cross-track scanning, 33-7 to 33-10
Cubic phase errors with lenses, 16-14, 16-16 to 16-19, 16-38
  and coma aberrations, 17-19
Current and current distribution. *See also* Magnetic current distributions; Surface current
  with arrays, 3-33, 3-35
  in biconical transmission lines, 3-24
  and boundary-condition mismatch, 3-80
  with cavities, 10-12
  with coaxial sleeve antennas, 7-33, 7-35 to 7-37
  density of
    with microstrip antennas, 10-12
    PO, 4-19 to 4-21
    for thin wires, 3-44 to 3-45
  with dipole antennas
    arrays, 13-43
    folded, 7-37 to 7-39
    linear, 7-11
    loaded, 7-9, 7-19
    short, 6-14 to 6-15
    unloaded linear, 7-12 to 7-14, 7-18 to 7-19
  edge, 10-55 to 10-57
  of far fields, 1-21 to 1-22
  and imperfect ground planes, 3-27, 3-48
  with Kirchhoff approximations, 2-21
  measurement of, 2-31, 33-5 to 33-6
  of perfect conductors, 3-38 to 3-39
  in rectangular waveguides, 2-28 to 2-29, 12-3 to 12-4
  of reflected waves, 7-14
  short-circuit, 2-32, 6-5
  sinusoidal, 3-14, 3-17
  with sleeve antennas, 7-23 to 7-33
  with straight wire antennas, 3-49 to 3-50
  with thin-wire antennas, 7-5 to 7-6
    loops, 7-42 to 7-48
  for traveling-wave antennas, 3-16
  for Tseng-Cheng pattern, 11-56
  for UTD antenna solutions, 20-5
Current elements, 26-8
Current reflection coefficient, 2-28

Current source, ideal, 2-29 to 2-30
Current waves
 with log-spiral antennas, 9-74 to 9-75
  conical, 9-83 to 9-84
 in transmission lines, 28-6, 28-7
Curvature of field with lenses, 16-15, 16-20
Curved surfaces
 effects of, on conformal arrays, 13-50
 for diffraction, 8-64
 UTD solutions with, 20-4 to 20-5, 20-18 to 20-37
Cutoff frequency of waveguides, 28-40 to 28-41
 circular dielectric, 1-49, 28-50
 rectangular, 1-42
Cutoff wavelength
 of transmission lines, circular coaxial TEM, 28-20
 of waveguides, 28-40 to 28-41
  circular TE/TM, 28-42, 28-44
  rectangular TE/TM, 28-38
  ridged, 24-14 to 24-15
Cylinders and cylindrical antennas
 circular, 4-75, 4-77 to 4-79
 conducting, 21-92 to 21-100
 current distribution of, 3-14, 3-16
 dipole, 3-28
 dipole arrays on, 13-50 to 13-52
 elliptical, 4-75, 4-77
 illumination of, 4-12 to 4-13
 for intersection curve, 15-30
 for multiple-antenna analyses, 27-37 to 27-38
 radiation pattern for, 4-62 to 4-63
 strips mounted on, 20-23 to 20-24
 thin-wire, 7-10 to 7-11
Cylinder-to-cylinder interactions, 20-37, 20-44 to 20-53
Cylindrical array feeds, 19-98 to 19-101
 matrix-fed, 19-101 to 19-119
Cylindrical scanning for field measurements, 33-10 to 33-12

Data transmission, millimeter-wave antennas for, 17-5
Decoupling with coaxial sleeve antennas, 7-31 to 7-32
Defense meteorological satellites, 22-14
Defocusing techniques for field measurements, 33-24 to 33-26
Deicing of tv antennas, 27-9, 27-26
Delay lenses, equal group, 16-45 to 16-46

Design procedures and parameters
 for antenna ranges, 32-14 to 32-19
 for arrays
  Dolph-Chebyshev, 11-15 to 11-23
  feed, for satellite antennas, 21-27 to 21-29
  large, 12-28 to 12-34
  of longitudinal slots, 12-10 to 12-24
  phased, 18-3 to 18-28
  space-tapered, 14-6 to 14-8
  of wall slots, 12-24 to 12-28
 for dielectric grating antennas, 17-62 to 17-69
 for dielectric lenses, 16-8 to 16-9
 for horn antennas
  conical, 8-48 to 8-49
  E-plane, 8-19 to 8-20
  H-plane, 8-33 to 8-34
  pyramidal, 8-43 to 8-45
 for log-periodic antennas
  dipole, 9-20 to 9-32
  zigzag, 9-37 to 9-46
 for offset parabolic reflectors, 15-80 to 15-84
 and radome performance, 31-4, 31-8 to 31-10
 for satellite antennas, 21-6 to 21-8, 21-20 to 21-26
 feed arrays for, 21-27 to 21-29
Detour parameters, 4-42
Diagnostics, medical. See Medical applications
Diagonal horn antennas, 8-70 to 8-71
Diameter of offset parabolic reflectors, 15-31
Diathermy applicators, 24-46 to 24-48
Dicke radiometers, 22-24 to 22-25
Dielectric antennas
 bifocal lens, 16-30 to 16-33
 in integrated antennas, 17-132
 periodic, 17-34 to 17-35, 17-48 to 17-82
Dielectric constants, 1-6 to 1-7, 10-5. See also Effective dielectric constants
 and Abbe sine condition, 16-20
 and cross sections, 17-39
 of microstrip antennas, 10-7, 10-23 to 10-24
 vs. thickness, with lenses, 16-12 to 16-13, 16-23
Dielectric grating antennas, 17-50 to 17-53, 17-80 to 17-82

Dielectric horn antennas, 17-23, 17-25
Dielectric lenses, 16-5, 21-16
   constrained analog of, 16-49 to 16-51
   design principles of, 16-7 to 16-9
   and Snell's law, 16-6
   taper-control, 16-33 to 16-38
   wide-angle, 16-19 to 16-33
   zoning of, 16-38 to 16-41
Dielectric loading
   with horn antennas, 8-73
      for satellite antennas, 21-82, 21-84 to 21-86
   with lenses for cylindrical array feeds, 19-111, 19-114
   with pillbox feeds, 19-21
   with waveguides, 13-48 to 13-49
      for medical applications, 24-9
Dielectric logging for rock conductivity, 23-25
Dielectric loss
   in constant-K lenses, 17-19
   tangent for, 1-6
   with transmission lines, 28-9
      coplanar waveguide planar quasi-TEM, 28-34
      microstrip planar quasi-TEM, 28-31
   with waveguides
      rectangular, 1-39
      TE/TM, 28-37
Dielectric matching sheets, 16-56
Dielectric materials
   with cavities, 10-21
   and feed reactance, with microstrip antennas, 10-33
   and phase shift, 9-64
   properties of, 28-10
Dielectric millimeter-wave antennas, 17-48 to 17-82
Dielectric probes
   open-ended coaxial cable as, 24-32 to 24-34
   short monopoles as, 24-30 to 24-32
Dielectric resonator antennas, 17-35 to 17-36
Dielectric-rod antennas, 17-34 to 17-48, 17-133 to 17-134
Dielectrics
   and modeling, 32-70 to 32-75
   and radiation patterns, 32-81 to 32-82
Dielectric sheets, plane wave propagation through, 31-10 to 31-18, 31-20 to 31-25
Dielectric-slab polarizers, 21-30
Dielectric substrate in microstrip antennas, 10-5, 10-6
Dielectric waveguide transmission types, 21-59, 21-61
   circular, 28-50 to 28-51
   for fiber optics, 28-47
   rectangular, 28-51, 28-53
Dielectric weight of lens antennas, 17-15
Dielguides, 8-73
Difference beams with beam-forming feed networks, 18-25 to 18-26
Differential formulations, compared to integral, 3-55 to 3-58, 3-82
Diffracted rays, 4-3 to 4-5
   caustic regions with, 4-5
Diffractions and diffracted fields. *See also* Edges and edge-diffracted fields; High-frequency techniques
   coefficients of
      with reflector antennas, 15-13
      for UTD solutions, 20-13, 20-43, 20-44
   and curved surfaces, 8-64
   diffracted-diffracted, 20-28, 20-29, 20-34
   diffracted-reflected
      for curved surfaces, 20-24, 20-29, 20-34
      for noncurved surfaces, 20-37, 20-45, 20-46, 20-56
   and Fresnel ellipsoid, 29-36
   horn antennas to reduce, 8-4, 8-50
   propagation by, 29-30, 29-36, 29-41 to 29-44
   and radomes, 31-8, 31-10
   with UTD solutions, 20-13, 20-16, 20-43, 20-44
      correction of, 4-105 to 4-106
      for curved surfaces, 20-26, 20-29, 20-32, 20-34
      for noncurved surfaces, 20-49 to 20-50, 20-52, 20-59 to 20-60
Digital communications and rain attenuation, 29-13
Digital phase shifters
   for periodic arrays, 13-63 to 13-64
   for phased arrays, 18-12 to 18-13, 18-15, 18-18
   quantization errors wtih, 21-12
   in true-time-delay systems, 19-80
Dimensions of standard waveguides, 28-40 to 28-41

Diode detectors for rf radiation, 24-37
Diode-loaded circular loops, 24-40 to 24-43
Diode phase shifters
  for periodic arrays, 13-62 to 13-64
  for phased arrays, 18-12 to 18-17
  variable, for beam-forming networks, 21-38, 21-40 to 21-42
Diodes. *See* Pin diodes; Schottky diodes; Varactor diodes
Diode switches for beam-forming networks, 21-7, 21-37 to 21-38
Diode variable power dividers, 21-46 to 21-48
Diplexer/circulators, 19-110 to 19-113
Diplexers
  with fixed beam-forming networks, 21-7, 21-53, 21-54
  orthomode transducers as, 22-46
Dipole antennas, 2-27, 3-30. *See also* Dipole arrays; Electric dipole antennas; Log-periodic dipole antennas; Short antennas, dipole
  balanced, 9-72 to 9-73
  bow-tie, with integrated antennas, 17-137 to 17-138
  coaxial, 3-30
  compared to monopole, 3-20
  coupling between, 20-68
  for direction-finding antennas, 25-9 to 25-10
  as E-field probes, 24-38 to 24-39
  and EMP, 30-9 to 30-14
  in feed guides, for circular polarization, 8-64
  folded, 3-29, 7-36 to 7-40
  with implantable antennas, 24-27 to 24-28
  linear, 3-13 to 3-14, 7-6, 7-11 to 7-23
  magnetic, 4-59, 30-9
  microstrip, 17-106, 17-118 to 17-119, 17-122 to 17-126
  millimeter-wave, 17-32 to 17-33
  near-field radiation pattern of, 20-69
  with side-mount tv antennas, 27-20 to 27-21
  skewed, 27-11 to 27-18, 27-34 to 27-35
  sleeve, 7-26 to 7-29
  small, 3-28, 30-9
  spherical, 30-14
  thick, 3-28
  for vehicular-mounted interferometers, 25-23

Dipole arrays, 13-43 to 13-44, 13-58
  blindness in, 13-47 to 13-48
  on cylinders, 13-50 to 13-52
  mutual coupling with, 13-41 to 13-42
Dipole-dipole arrays
  for earth resistivity, 23-6 to 23-7
  for E-field probes, 24-51
Dipole radiators for phased arrays, 18-6 to 18-7, 18-10
Direct-contact waveguide applicators, 24-9
Direct-current mode with microstrip antennas, 10-25
Direct-current resistivity of earth, 23-4 to 23-8
Direct-ray method with radome analysis, 31-20 to 31-24
Direction-finding antennas and systems, 25-3
  with am antennas, 26-37 to 26-47
  with conical log-spiral antennas, 9-109 to 9-110
  interferometers, 25-21 to 25-23
  multimode circular arrays, 25-17 to 25-20
  multiple-signal, 25-24 to 25-25
  rotating antenna patterns, 25-4 to 25-17
Directive gain, 1-25, 3-12, 5-26, 5-27
  of arrays, 3-37
  of biconical antennas, 3-25
  of lens satellite antennas, 21-74
  of millimeter-wave antennas
    dielectric grating, 17-73 to 17-74, 17-76
    geodesic, 17-32
    integrated, 17-135 to 17-136
    lens, 17-18
    reflector, 17-12
    tapered dielectric-rod, 17-41, 17-44, 17-46 to 17-47
  of receiving antennas, 6-7
Directivity, 1-29, 3-12. *See also* Directive gain
  and aperture area, 17-7, 17-8
  of arrays, 11-5, 11-6, 11-58 to 11-76, 11-78
    aperiodic, 14-3, 14-15, 14-17 to 14-19, 14-32
    backfire, 9-64
    Dolph-Chebyshev, 11-19 to 11-21
    millimeter-wave, 17-33
    multimode element, 19-12

Index

periodic, 13-10 to 13-11, 13-54 to 13-55
planar, 9-10, 11-63 to 11-65
scanned in one plane, 13-12
uniform linear, 11-11 to 11-13
of batwing tv antennas, 27-24
of beam-forming feed networks, 18-25
of conical log-spiral antennas, 9-87 to 9-90
for dielectric grating antennas, 17-79
and Dolph-Chebyshev pattern synthesis, 13-24 to 13-25
with DuFort-Uyeda lenses, 19-49
of ferrite loop antennas, 6-18, 6-20
and gain, 1-25
of horn antennas
conical, 8-46 to 8-47
E-plane, 8-14 to 8-18
H-plane, 8-29 to 8-33
pyramidal, 8-37, 8-39 to 8-42
of lens antennas, 17-18
of log-periodic antennas, 9-10
dipole arrays, 9-21 to 9-24, 9-27, 9-61, 9-64, 9-66
zigzag, 9-39 to 9-40
measurement of, 32-39 to 32-41
with modeling, 32-85
of microwave antennas, 33-20
of millimeter-wave antennas, 17-5, 17-33
lens, 17-18
reflector, 17-12 to 17-14
tapered dielectric-rod, 17-38 to 17-39
of monopole on aircraft, 20-76
of offset parabolic reflectors, 15-36
of open-ended circular waveguide feeds, 15-105
of short dipoles, 6-10 to 6-11
of side-mount tv antennas, 27-22
with spherical scanning, 33-12
Disc-cone antennas, 3-30
millimeter-wave, 17-32
Discontinuities and UTD solutions, 20-10, 20-12
with curved surfaces, 20-23 to 20-24, 20-30
with noncurved surfaces, 20-37, 20-48, 20-52 to 20-53
Discrete arrays and aperture antennas, 11-29 to 11-30
Disk radiators for phased arrays, 18-6
Dispersion characteristics
of dielectric grating antenna, 17-66
of dielectric rod, 17-39
of rectangular dielectric waveguides, 28-54
Displaced rectangular aperture distributions, 5-17 to 5-18
Dissipation factor of triplate stripline TEM transmission lines, 28-28, 28-29
Distortion
from beam-forming networks, 21-57
of integrated antennas, 17-135
and lateral feed displacement, 21-20
from lens aberrations, 16-15, 16-20, 21-13
with microstrip patch resonator arrays, 17-109
with offset parabolic antennas, 15-49
from pin diodes, 21-38
from reflector surface errors, 15-105 to 15-114
with satellite antennas, 21-5
scanning, 11-36 to 11-38
with tapered dielectric-rod antennas, 17-44
thermal, 21-28 to 21-29, 21-57
from transformations, 11-57
of wavefronts, from radomes, 31-7, 31-9
Distribution function with aperiodic arrays, 14-6, 14-10 to 14-12
Divergence factor and ocean reflection coefficient, 29-39 to 29-40
DMS (defense meteorological satellites), 22-14
DMSP Block 5D satellite, 22-17
DNA Wideband Satellite Experiment, 29-19
Documentation of computer models, 3-68, 3-70
Dolph-Chebyshev arrays
linear, 11-13 to 11-23
planar, two-dimensional, 11-49 to 11-52
probabilistic approach to, 14-8
Dolph-Chebyshev pattern synthesis, 13-24 to 13-25
Doppler direction-finding antennas, 25-16 to 25-17
Doppler effects and mobile communication, 29-30
Dosimeters, personnel, rf, 24-43
Double-braid rf cable, 28-23
Double-ridged waveguides
for medical applications, 24-12 to 24-16
TE/TM waveguides, 28-45 to 28-47, 28-49

Doubly tuned waveguide array elements, 13-57
Downlink signals for satellite tv, 6-26
Driven-ferrite variable polarizer power dividers, 21-43 to 21-44
Driving-point impedance
 with directional antenna feeder systems, 26-43 to 26-45
 of microstrip antennas, 10-20, 10-34, 10-53
Drought conditions, remote sensing of, 22-15
Dual-band microstrip elements, 10-63 to 10-67
Dual fields, 2-5 to 2-6
 with Maxwell equations, 3-9, 3-11
Dual-frequency array with infinite array solutions, 13-48 to 13-49
Dual-hybrid variable power dividers, 21-43
Dual-lens limited-scan feed concept, 19-66 to 19-67
Dual-mode converters, 21-80
Dual-mode ferrite phase shifters, 13-64, 18-14 to 18-15, 21-38 to 21-40, 21-42
Dual-mode horn antennas
 conical, 8-71 to 8-72,
  radiation pattern of, 15-95, 15-100 to 15-101
 for satellite antennas, 21-82, 21-84 to 21-89
Dual-offset reflector antennas, 15-6 to 15-15
Dual polarizations
 circular, 21-73
 microstrip arrays with, 17-109
 reflector antennas with, 15-5
 satellite, 21-73 to 21-74, 21-78 to 21-80
Dual-reflector antennas, 15-61 to 15-62
 cross-polarization with, 15-69 to 15-71
 lens antennas as, 16-6
 millimeter-wave antennas, 17-9 to 17-11
 parameters of, 15-67 to 15-68
 for satellite antennas, 21-21
 scan performance of, 15-71 to 15-73
 shaped reflectors for, 15-73 to 15-80
 surface errors with, 15-109 to 15-111
Dual substrates with microstrip dipoles, 17-129
Dual-toroid ferrite phase shifters, 21-40 to 21-41
DuFort optical technique, 19-73 to 19-76

DuFort-Uyeda lenses, 19-49 to 19-53
Dummy cables
 for conical log-spiral antennas, 9-95, 9-103
 with log-spiral antennas, 9-75, 9-80
Dyadic coefficients
 edge-diffraction, 4-27, 4-101 to 4-105
 uniform, 4-32 to 4-33
 for surface reflection and diffraction, 4-52
Dyadic transfer function, 4-30
Dynamic programming technique, 14-5

Earth
 bulge of, and line-of-sight propagation, 29-34
 modeling of, 32-76 to 32-77
 resistivity of, 23-4 to 23-8
Earth coverage satellite antennas, 21-80 to 21-81
 horns
  dielectric-loaded, 21-84
  multistepped dual-mode, 21-84 to 21-89
  stepped, 21-82 to 21-84
  shaped beam, 21-89
Earth receiving antennas, 6-25 to 6-32
Earth-substended angles for satellites, 21-3
Eccentricity
 of conic sections, 15-25 to 15-26
 and intersection curve, 15-30
Eccentric line TEM transmission lines, 28-15
ECM. *See* Equivalent current method
EDC. *See* Effective dielectric constant method
Edge-coupled transmission lines, 28-16, 28-17
Edge-of-coverage gain with satellite antennas, 21-5
Edges and edge-diffracted fields, 4-32 to 4-33, 4-38
 and boundaries, 15-12
 corner-diffracted, 4-45 to 4-47
 currents on, 4-100 to 4-101
  magnetic, 10-55 to 10-57
 and ECM, 4-97 to 4-102
 elements with, and conformal arrays, 13-52
 GTD for, 4-25 to 4-28, 15-69
 and horn antenna fields, 8-3 to 8-4, 8-50

and image-plane ranges, 32-23
and Maxwell equations, 1-9
PO for, 4-22
PTD for, 4-104 to 4-108
slope diffraction by, 4-38 to 4-40
UAT for, 4-40 to 4-43
UTD solutions for, 4-32 to 4-38, 20-10 to 20-12
  curved surfaces, 20-21 to 20-23
  noncurved surfaces, 20-37
Edge-slot array antennas, 17-26 to 17-27
Edge slots for slot arrays, 12-26 to 12-28
Edge tapers with reflector antennas, 15-36 to 15-38, 15-41 to 15-45
  and cross-polarization, 15-47
  with off-focus feeds, 15-53
  and surface errors, 15-109
  tapered-aperture, 15-16 to 15-18, 15-21, 15-23
Effective area of receiving antennas, 22-4
  and aperture efficiency, 5-27 to 5-29
Effective dielectric constants
  of fin lines, 28-56, 28-58
  for microstrip patch antennas, 24-19 to 24-21
  of transmission lines
    coplanar strip planar quasi-TEM, 28-35
    coplanar waveguide planar quasi-TEM, 28-33
Effective dielectric constant (EDC) method
  for dielectric grating antennas, 17-65, 17-68
  for phase constants, 17-53 to 17-57
Effective heights. *See also* Vector effective height
  of hf antennas, 30-30
  of L-band antennas, 30-19
  of uhf communication antennas, 30-21
Effective isotropically radiated power, 1-27
Effective length, 2-34
Effective loss tangent with cavities, 10-20
Effective phase center of log-spiral antennas, 9-91 to 9-92, 9-94
Effective radiated power
  of fm antennas, 27-4, 27-6
  of tv antennas, 27-3 to 27-5
Effective receiving area and EMP, 30-10

Effective relative dielectric constant, 10-7
Efficiency. *See also* Aperture efficiency; Beam efficiency; Radiation efficiency
  of arrays, 11-74, 11-76
    planar, 11-65
    spaced-tapered, 14-6 to 14-7
  beam-shaping, with satellite antennas, 21-5 to 21-6
  of corrugated horn antennas, 8-50, 8-62
  of dielectric grating antennas, 17-63
  and inductive loading, 6-13
  of integrated antennas, 17-135
  of loop antennas, 25-8
  of microstrip antennas, 10-6, 10-49 to 10-50
    dipole, 17-124
    and matching filters, 10-52
  of microstrip patch resonator arrays, 17-109 to 17-115
  of microwave radiometers, 22-28 to 22-30, 22-32 to 22-33
  polarization, 1-26, 2-36, 2-39, 32-42, 32-56
  and Q-factor, 6-22 to 6-23, 10-6
  of receiving antennas, 6-22 to 6-24
  of reflector antennas
    dual-reflector, 15-73
    offset parabolic, 15-42
    tapered-aperture, 15-16, 15-23
  of small transmitting antennas, 6-9 to 6-10
E-4 aircraft, antennas on, 30-29, 30-30
Ehf (extra high frequencies), lens antennas for, 16-5, 16-28
EIRP (effective isotropically radiated power), 1-27
Electrically rotating patterns
  high-gain, 25-14 to 25-16
  null, 25-11 to 25-12
Electrically scanning microwave radiometer, 22-10
  calibration of, 22-26
  as conical scanning device, 22-40
  modulating frequency of, 22-24
  as planar scanning device, 22-37
Electrical scanning, 21-5
Electrical well logging, 23-24 to 23-25
Electric dipole antennas, 3-13 to 3-14
  and EMP, 30-9
  far-field pattern of, 1-29, 4-60

Electric fields. *See also* Electric-field strength
  with arrays, 3-36
  and cavities, 10-12
  conical-wave, 4-99
  distribution of, 12-4, 12-6 to 12-10, 12-35
  and equivalent currents, 5-8 to 5-10
    edge, 4-101
  in far zone, 6-3 to 6-4
  GO for, 4-13, 4-17
  Green's dyads for, 3-46, 3-48
  GTD for, 4-23
    for convex surfaces, 4-28
    for edged bodies, 4-25
    for vertices, 4-30
  integral equations for, 3-44, 3-46 to 3-47, 3-51
  of isotropic medium, 4-9
  with plane-wave spectra, 1-32, 5-5 to 5-8
  PO, 4-19 to 4-21
  probes for, 24-37 to 24-41
    calibration of, 24-54 to 24-57
    implantable, 24-51 to 24-54
  PTD for, for edged bodies, 4-106 to 4-107
  of rectangular apertures, 5-11 to 5-12
  reflection coefficient for, 2-28
  scattering. *See* Scattering
  of short monopole, 4-65
  of small dipoles, 6-13
  of small loops, 6-18
  of spherical wave, 4-6 to 4-7
  of surface fields, 4-84
  of transmitting antenna, 6-3, 20-6
  UAT for, for edged bodies, 4-39
  in unbounded space, 2-8 to 2-9
  UTD for
    corner-diffracted, 4-44
    for edged bodies, 4-32, 4-37, 4-83
  vector for, 1-14, 7-46 to 7-47
  in waveguides, 12-4
Electric-field strength
  of hemispherical radiators, 26-31 to 26-32
  horizontal, with two towers, 26-30 to 26-34
  of isotropic radiators, 26-26
  of uniform hemispherical radiators, 26-27 to 26-29
  with UTD solutions, 20-6
Electric line dipole for UTD solutions, 20-19 to 20-20

Electric surface current, 2-26 to 2-27
Electric vector potential, 3-6 to 3-7, 3-22, 5-8 to 5-9
  of rectangular-aperture antennas, 3-22
Electrode arrays for resistivity measurements, 23-4 to 23-8
Electroformed nickel for satellite antennas, 21-28 to 21-29
Electromagnetic fields
  computer model for, 3-53
  equations for, 3-6 to 3-8
  integral vs. differential formulation of, 3-55 to 3-58
  models for
    materials for, 32-69 to 32-70
    theory of, 32-65 to 32-69
Electromagnetic pulses, 30-3
  and aircraft antennas, 30-17 to 30-31
  analyzing effects of, 30-8 to 30-16
  and mounting structures, 30-4 to 30-7
  and principal elements, 30-7 to 30-8
Electromagnetic radiation monitors, 24-37
Electromechanical measurements, 32-4
Electromechanical phase shifters, 21-38, 21-41 to 21-43
Electromechanical power dividers, 21-48 to 21-49
Electromechanical switches, 21-37 to 21-38
Electronically Scanning Airborne Intercept Radar Antenna, 19-55
Electronic scanning, 21-5, 22-36 to 22-37
Elements, array
  and EMP, 30-7 to 30-8
  for feed arrays for satellite antennas, 21-27
  interaction of, with aperiodic arrays, 14-3
  mutual impedance between, 11-5
  for periodic arrays, 13-57 to 13-60
    patterns of, 13-6
  in phased arrays, 21-12
  in planar arrays, 11-53
  spacing of, 3-36, 14-4 to 14-5
Elevated antenna ranges, 32-19 to 32-21
Elevated H Adcock arrays, 25-9 to 25-10
Elevation and noise in tvro antennas, 6-31
Elevation-over-azimuth positioners, 32-12
Elf. *See* Extremely low frequencies
Ellipses, 15-25 to 15-26
  and intersection curve, 15-27, 15-29 to 15-30
  polarization, 1-14 to 1-16

Elliptical arrays, transformations with, 11-33 to 11-36, 11-48
    uniformly excited, 11-44 to 11-45
Elliptical cylinders, 4-75, 4-77
Elliptical loops, 6-19
Elliptical patches, 10-18
Elliptical polarization, 1-15 to 1-16
    gain measurements for, 32-49 to 32-51
    with log-spiral antennas, 9-74
EM Scattering modeling code, 3-88 to 3-91
Emissivity in state, 2-40
Emissivity of surfaces, 22-7 to 22-9
EMP. *See* Electromagnetic pulses
End-correction network for input admittance, 7-7
End-fire arrays, 3-41, 11-12, 11-15
    maximum directivity of, 11-65 to 11-71
    radiation pattern for, 11-76, 11-77
Engines, aircraft, model of, 20-37
Environments. *See also* Complex environments
    measurements for, 32-4
    numerical simulation of, 20-7 to 20-53
E-plane gain correction factor for biconical horns, 21-104
E-plane horn antennas, 8-4
    aperture fields of, 8-5 to 8-7
    design procedure for, 8-19 to 8-20
    directivity of, 8-14 to 8-18
    field intensity of, 8-13 to 8-14
    gain of, 8-18 to 8-19
    half-power beamwidth of, 8-14 to 8-15
    phase center of, 8-76 to 8-77
    radiated fields of, 8-7 to 8-10
    for reflector feeds, 15-92 to 15-93, 15-96 to 15-97
    universal curves of, 8-10, 8-13 to 8-14
E-plane radiation patterns, 1-28, 5-12
    of apertures
        circular, 5-22
        rectangular, 5-12
    of dielectric grating antennas, 17-74, 17-78
    of DuFort-Uyeda lenses, 19-51
    of horn antennas
        corrugated, 8-55 to 8-59
        E-plane, 8-9 to 8-12
        H-plane, 8-22 to 8-25
        pyramidal, 8-37 to 8-40
        square, 17-20
    for log-periodic antennas
        diode arrays, 9-64 to 9-66
        metal-arm, 9-11 to 9-12
        zigzag, 9-39 to 9-42, 9-45, 9-46
    for microstrip antennas, 10-24
    and UTD solutions, 20-13 to 20-16
Equal group delay lenses, 16-45 to 16-46
Equiangular spirals, 9-72, 9-110
Equivalent current method, 4-5, 4-96 to 4-102
    and surface-diffracted rays, 4-83
Equivalent currents for surface fields, 5-8
Equivalent cylinder techniques, 27-37 to 27-38
Equivalent radii for thin-wire antennas, 7-10 to 7-11
Equivalent receiving area
    for aircraft blade antenna, 30-12
    for annular slot antenna, 30-13
    for center-fed ellipsoid antenna, 30-10 to 30-12
    for spherical dipole antenna, 30-14
ERP. *See* Effective radiated power
Errors. *See also* Aberrations; Path length constraint and errors; Phase errors; Random errors; Systematic errors
    in antenna ranges, 32-34 to 32-35
    array, 13-52 to 13-57
    boresight, 31-3 to 31-4, 31-7 to 31-9, 31-27 to 31-29
    component, for phased arrays, 18-26 to 18-28
    with far-field measurements, 32-5
    with microwave radiometers, 22-23, 22-25 to 22-28
    modeling, 3-69
    pattern, 21-65
    quantization, 13-52 to 13-57, 21-12
    with scanning ranges, 33-15 to 33-18
ESAIRA (Electronically Scanning Airborne Intercept Radar Antenna), 19-55
ESMR. *See* Electrically scanning microwave radiometer
Eulerian angles, 15-116 to 15-117, 15-120
Evaluation of antenna ranges, 32-33 to 32-39
Even distributions in rectangular apertures, 5-14
Excitations
    array element, 11-23, 13-6 to 13-7
    linear transformations on, 11-39 to 11-41
    nonuniform, 11-29 to 11-30

Excitations (*cont.*)
  of overlapped subarrays, 13-37
  in planar arrays, 11-52 to 11-53
  with TEM transmission lines, 28-11, 28-19
Excited-patch radiators, 18-8 to 18-9
Experimentation
  and modeling errors, 3-69
  and validation of computer code, 3-72 to 3-76
External fields in longitudinal slots, 12-7
External networks and EMP, 30-8
External noise, 6-24 to 6-25, 29-50
External numerical validation of computer code, 3-77 to 3-78
External quality factor with cavities, 10-20
External validation of computer codes, 3-77 to 3-78
Extra high frequencies
  lens antennas for, 16-5
    aplanatic dielectric, 16-28
Extrapolation techniques for field measurements, 33-26 to 33-28
  of gain, 32-47 to 32-48
Extraterrestrial noise, 29-50 to 29-51
Extremely low frequencies
  with geophysical applications, 23-3
  ionospheric propagation of, 29-27 to 29-28

Fade coherence time and ionospheric scintillations, 29-18
Faraday rotation
  depolarization from, 29-16
  with phase shifters, 13-64
  variable power dividers using, 21-44 to 21-46
Far-field diagnostics for slot array design, 12-36 to 12-37
Far fields and far-field radiation patterns, 1-13, 1-17 to 1-18, 3-10, 3-22 to 3-24, 6-3 to 6-4. *See also* Radiation fields and patterns
  of antennas vs. structures, 20-53 to 20-54, 20-61 to 20-70
  of aperture antennas, 3-22 to 3-24, 15-21 to 15-24
    circular, 5-20 to 5-24
    rectangular, 3-22 to 3-24, 5-11 to 5-12, 5-14 to 5-15
    sinusoidal, 3-24
  of arrays, 11-5 to 11-8
    circular, 21-91 to 21-92, 21-94
    cylindrical, feeds for, 19-99 to 19-100, 19-104
    periodic, 13-5
    phased, 21-9 to 21-10
  calculations of, 1-21 to 1-23
  with cavities, 10-15, 10-19 to 10-20
  components of, 8-73, 8-75
  coordinates of, transformations with, 15-115, 15-117 to 15-120
  of dipole antennas
    folded, 7-41
    half-wave, 6-10 to 6-11
    log-periodic, 9-19 to 9-20
    short, 6-10 to 6-11, 20-8
    sleeve, 7-27 to 7-28
  of E-plane horn antennas, 8-7 to 8-10
  and equivalent currents, 5-10
  expressions for, 3-10
  Friis transmission formula for, 2-38 to 2-39
  integral representation for, 3-8 to 3-9
  with lens antennas, 16-6, 16-13 to 16-14
    coma-free, 16-22 to 16-23
    obeying Abbe sine condition, 19-97 to 19-99
    reflector, limited-scan feed concept, 19-70 to 19-71
  measurements of. *See* Measurements of radiation characteristics
  with microstrip antennas, 10-10, 10-22, 10-24
  with multiple element feeds, 19-13 to 19-16
  with plane-wave spectra, 5-7
  polarization of, 1-23 to 1-24
  power density in, 6-7
  with radial transmission line feeds, 19-24
  with radomes, 31-3
  with reflector antennas, 15-8 to 15-10, 15-13 to 15-14
    with axial feed displacement, 15-52 to 15-55
    with lateral feed displacement, 15-56 to 15-66
    offset parabolic, 15-34 to 15-35, 15-37 to 15-39, 15-43 to 15-45
    surface errors on, 15-110 to 15-112

Index

with tapered apertures, 15-15 to 15-23
with reflector-lens limited-scan feed
 concept, 19-70 to 19-71
representation of, 1-18 to 1-21, 1-28 to
 1-29
scattered, and PO, 4-20
with single-feed CP microstrip antennas,
 10-58 to 10-59
for sleeve monopoles, 7-31
for slot array design, 12-36 to 12-37
of space-fed subarray systems, 19-69
from thin-wire antennas, 7-8, 7-15 to
 7-17
 loops, 7-44 to 7-46
for triangular-aperture antennas, 3-24
with UTD solutions, 20-4 to 20-6, 20-13
 to 20-14
Fat, human
 phantom models for, 24-55
 power absorption by, 24-8, 24-9, 24-47
F/D ratios. *See* Focus length to diameter
 ratio
FDTD modeling code, 3-88 to 3-91
Feed blockage with offset reflector, 21-19,
 21-23. *See also* Aperture blockage
Feed circuits. *See also* Beam-forming feed
 networks; Cassegrain feed systems;
 Constrained feeds; Optical feeds;
 Semiconstrained feeds; Unconstrained
 feeds
 for am antennas, 26-37 to 26-47
 for arrays
  dipole, 9-15 to 9-18, 9-21, 9-27, 9-29 to
   9-31, 9-61 to 9-63
  monopole, 9-66 to 9-67
  periodic, synthesis of, 13-61
  phased, design of, 18-3 to 18-4, 18-6,
   18-17 to 18-26
  power dividers for, 13-61
  spaced-tapered, 14-6 to 14-7
  subarrays, 13-39 to 13-41
  of tapered dielectric-rod antennas,
   17-47 to 17-48
 for conical horn antennas, 8-46
  log-spiral, 9-95 to 9-96, 9-99 to 9-104,
   9-111 to 9-112
 cylindrical array, 19-98 to 19-119
 displacements of, with offset parabolic
  antennas, 15-49 to 15-61
 for log-periodic dipole arrays, 9-15 to
  9-18, 9-21, 9-27, 9-29 to 9-31, 9-61

 for log-spiral antennas, 9-75 to 9-78,
  9-80
  dipole arrays, 9-62 to 9-63
  mono arrays, 9-66 to 9-67
 for microstrip antennas, 10-5, 10-6,
  10-28, 17-108 to 17-119
  and impedance matching, 10-51 to
   10-52
 for millimeter-wave antennas, 17-12,
  17-20, 17-47 to 17-48
 offset, 17-9, 19-54, 19-63 to 19-66
 reactance of, with microstrip antennas,
  10-31 to 10-34
 for reflector antennas, 15-5 to 15-6,
  15-13, 15-84 to 15-89
  complex, 15-94 to 15-99
  cos type patterns of, 15-99 to 15-105
  radiation patterns of, 15-89 to 15-94
  rays with, 15-11 to 15-12, 21-21 to
   21-22
 for satellite antennas, 21-5 to 21-7, 21-27
  to 21-29
 for subarrays, 13-39 to 13-41
Feed coordinates
 with reflector antennas, 15-7, 15-36,
  15-88 to 15-89
 transformations with, 15-115, 15-117 to
  15-120
Feed elements
 horn antennas as, 8-3
 location of, for feed arrays for satellite
  antennas, 21-27
Feed gaps with log-spiral antennas, 9-75 to
 9-77
Feed patterns
 for reflector antennas
  dual-reflector, 15-77, 15-80
  offset parabolic, 15-34 to 15-36, 15-47
 transformation of, with taper-control
  lenses, 16-37
Feed tapers, 15-36 to 15-37, 15-39 to
 15-40, 17-40
Feedthrough with lens antennas, 16-5
Feed transmission media for beam-forming
 networks, 21-56 to 21-63
Feed types of reflector antennas, 15-5 to
 15-6
Fence guide transmission type, 21-62
Fermat's principle and rays, 4-8 to 4-9
Ferrite absorbers with UTD solutions,
 20-17

Ferrite loop antennas, 6-18, 6-20 to 6-22
  radiation resistance of, 6-19
Ferrite phase shifters
  for arrays
    periodic, 13-62 to 13-64
    phased, 18-12, 18-14 to 18-17
  with planar scanning, 22-37
  variable, for beam-forming networks, 21-38 to 21-40, 21-42
Ferrite switches for beam-forming networks, 21-7, 21-37 to 21-38
Ferrite variable power dividers, 21-43 to 21-46
F-4 fighter aircraft, simulation of, 20-78 to 20-82
Fiber optics
  dielectric waveguides for, 28-47
  with implantable temperature probes, 24-50
Fictitious isotropic radiator, 2-35
Field curvature lens aberrations, 21-13
Field distributions
  for circular dielectric waveguides, 28-50
  measurements of, with radomes, 31-6 to 31-7
Field equations, electromagnetic, 3-6 to 3-8
Field equation pairs, 3-7
Field probes for evaluating antenna ranges, 32-33
Fields. *See also* Aperture fields; Electric fields; Incident fields; Magnetic fields; Reactive fields
  with arrays, 3-33, 3-35 to 3-36
  circular-phase, 9-109
  dual, 2-5 to 2-6, 3-9, 3-11
  and Huygen's principle, 2-18 to 2-19
  using integration to determine, 3-56 to 3-57
  and Kirchhoff approximation, 2-19 to 2-21
  in longitudinal slots, 12-7
  of perfect conductors, 3-39
  of periodic structures, 1-33 to 1-37
  plane-wave spectrum representations of, 1-31 to 1-33
  in radome-bounded regions, 31-5 to 31-6
  of rectangular-aperture antennas, 3-21 to 3-24
  with short dipole, 3-34
  surface, 4-84 to 4-89, 5-8
  tangential, 5-7 to 5-9, 7-5 to 7-6

  TE and TM, 1-30 to 1-31
  time-harmonic, 1-5 to 1-8, 3-6 to 3-8
  of transmission lines, 2-27 to 2-29
  transmitting, 2-19 to 2-21, 2-34 to 2-35
  of waveguides
    circular, 1-44 to 1-50
    circular TE/TM, 28-43
    rectangular, 1-37 to 1-44
    rectangular TE/TM, 28-39
Field strength of am antennas, 26-8
Figure of merit
  EIRP as, 1-27
  for horn antennas, 8-14
  for tvro antenna, 6-30 to 6-31
Filamentary electric current with cavities, 10-20 to 10-21
Filter type diplexers, 21-54
Fin line transmission types, 21-59, 21-61, 28-55 to 28-57
Finite sources, far fields due to, 1-32
Finite waveguide arrays and coupling, 13-44 to 13-45
Five-wire transmission lines, 28-13
Fixed beam-forming networks, 21-34, 21-37, 21-49 to 21-56
Fixed line-of-sight ranges, 32-9 to 32-11
Fixed-wire antennas, EMP responses for, 30-23 to 30-27
Flared-notch antennas, 13-59
Flaring with horn antennas
  conical, 8-46
  corrugated, 8-60, 8-64
  E-plane, 8-9
  H-plane, 8-20, 8-29 to 8-30
  and phase center, 8-76 to 8-77
Flat dielectric sheets, propagation through, 31-10 to 31-18
Flat plate array antennas, 17-26
Flat-plate geometry for simulation of environments, 20-8, 20-18
Flat-plate Luneburg lenses, 19-44, 19-46 to 19-47
Flat surface lenses, 16-9 to 16-11
Flexible coaxial transmission lines, 28-22
Flood predicting with remote sensing, 22-15
Floquet space harmonics
  and scattered fields, 1-34 to 1-35
  and UP structures, 9-32
Fluctuation of aperiodic array beams, 14-26 to 14-28

FM broadcast antennas, 27-3 to 27-8, 27-32 to 27-40
Focal length
  of conic sections, 15-24 to 15-26
  of offset parabolic reflectors, 15-31
  for reflector satellite antennas, 21-22 to 21-24
  with taper-control lenses, 16-36
Focal points. *See* Caustics
Focal region of reflector satellite antennas, 21-19
Fock radiation functions, 4-68 to 4-69, 4-73
  for mutual coupling, 4-95
  for surface fields, 4-88 to 4-89
  for surface-reflection functions, 4-52 to 4-53, 4-66
Fock-type Airy functions, 4-53, 4-69
Fock type transition functions, 4-50n
Focus length to diameter (F/D) ratio
  with lens antennas, 16-12
    and dielectric constants, 16-24
    planar surface, 16-56
    for satellites, 21-16
    taper-control, 16-37
    and zoning, 16-43
  with microwave radiometers, 22-30
  and pyramidal horn antenna dimensions, 8-45
  with reflector satellites
    and cross-polarization, 15-47
    with off-focus feeds, 15-51, 15-53, 15-55 to 15-56, 15-58, 15-62
    offset parabolic reflectors, 15-32 to 15-33, 15-37
    for satellites, 21-22 to 21-23
    and wide-angle scans, 19-34 to 19-35
Fog, scintillation due to, 29-36
Folded antennas
  dipole, 3-29, 7-36 to 7-40
  slot, and Babinet principle, 30-15 to 30-16
  unipole, 3-30
Folded Luneburg lenses, 19-49
Folded pillbox feeds, 19-17 to 19-23
Folded-T waveguide power division elements, 21-50 to 21-52
Footprints. *See also* Coverage area
  of microwave radiometers, 22-30 to 22-32
  of tvro signal, 6-26, 6-27
Formulation for computer models, 3-53 to 3-54, 3-81

Four-electrode arrays for earth resistivity, 23-4 to 23-8
Fourier transformers, 19-95
Four-port hybrid junctions, 19-5 to 19-6
Four-tower systems, 26-33
Four-wire transmission lines, 28-12
Fox phase shifter, 13-64
Fraunhofer region, 1-18
Free space, resistance of, 26-25 to 26-26
Free-space loss, 2-39, 29-5 to 29-6
  of satellite-earth path, 29-8
Free space vswr evaluation method for anechoic chambers, 32-38 to 32-39
Frequencies
  and active region
    with conical log-spiral antennas, 9-83
    of log-periodic dipole antennas, 9-17 to 9-19
  allocation of
    for satellite antennas, 21-6
    for tv, 6-28, 27-3
  and beamwidth, 9-87
  and computer time, 3-62
  dependence on, by reflector antennas, 17-21 to 17-13
  and image impedance, 9-55
  and input impedance, 10-9
  and noise temperature, 2-43
  and phase center, 9-92
  and reactance, 10-30, 10-65
  and space harmonics, 9-35
  of standard waveguides, 28-40 to 28-41
    circular dielectric, 1-49, 28-50
    rectangular, 1-42
Frequency-agile elements, 10-66 to 10-69
Frequency-domain
  compared to time-domain, 3-65 to 3-66, 3-82
  and EMP, 30-4
  method of moments in, 3-57 to 3-62, 3-64
Frequency-independent antennas, 9-3 to 9-12. *See also* Log-periodic dipole antennas; Log-periodic zigzag antennas; Log-spiral antennas; Periodic structures
  millimeter-wave, 17-27 to 17-28
Frequency-modulation broadcast antennas, 27-3 to 27-8, 27-32 to 27-40
Frequency-scanned array antennas, 17-26 to 17-27

Frequency selective surface diplexer, 21-53
Fresnel ellipsoid and line-of-sight propagation, 29-36
Fresnel fields, 1-18. *See also* Near fields
Fresnel integrals with near-field patterns, 5-25
Fresnel-Kirchhoff knife-edge diffraction, 29-42 to 29-43
Fresnel reflection coefficients, 3-27, 3-33, 3-48
Fresnel zone-plate lens antennas, 17-20
Friis transmission formulas, 6-8 to 6-9
  for far fields, 2-38 to 2-39
Front ends, integrated, 17-134
Front-to-back ratios
  for log-periodic dipole arrays, 9-60 to 9-61
  for log-spiral conical antennas, 9-89, 9-92
F-16 fighter aircraft, simulation of, 20-82, 20-85 to 20-88
Fuel tanks, aircraft, simulation of, 20-79, 20-82
Fundamental waves and periodic structures, 9-33 to 9-35, 9-55 to 9-56
Fuselages of aircrafts
  and EMP, 30-4, 30-6
  simulation of, 20-70, 20-78

Gain, 1-24 to 1-27, 3-12 to 3-13, 3-28 to 3-31. *See also* Directive gain
  aperture, 5-26 to 5-34
    circular, 5-23
    rectangular, 5-29
    uniform fields, 5-14
  of arrays, 3-36, 3-38 to 3-43, 13-10 to 13-11
    array columns, 13-54
    edge-slot, 17-27
    feed, for satellite antennas, 21-27
    multimode element, 19-12
    phased, 18-11, 19-59, 21-11 to 21-12
  axial near-field, 5-32
  and bandwidth, with tv antennas, 27-7
  with DuFort optical technique, 19-76
  and effective area, 5-27 to 5-28
  of horn antennas
    biconical, 21-105 to 21-106, 21-108 to 21-109
    conical, 8-48
    correction factors with, 21-104
    E-plane, 8-18 to 8-19
    pyramidal, 8-43 to 8-44, 15-98
  of lens antennas, 17-18
  and lens F/D ratio, 21-16
  with limited scan feeds, 19-57
  of log-periodic zigzag antennas, 9-45 to 9-47
  measurement of, 32-42 to 32-51
  of microwave antennas, 33-20 to 33-21
  of millimeter-wave antennas
    lens, 17-18
    maximum-gain surface-wave, 17-40 to 17-41
    reflector, 17-12 to 17-14
    tapered dielectric rod, 17-40 to 17-41, 17-47
  reduction factor for, near-field, 5-30 to 5-32
  of reflector antennas
    boresight, 15-81
    millimeter-wave, 17-12 to 17-14
    surface errors on, 15-106 to 15-107
  ripple of, with feed arrays, 21-27
  with satellite antennas, 21-5 to 21-7, 21-12
    lateral feed displacement with, 21-20
    tvro, 6-30 to 6-31
  of short dipole, 6-10 to 6-11
  of small loops, 6-13, 6-17
Gain-comparison gain measurements, 32-49 to 32-51
Gain-to-noise ratio of tvro antenna, 6-31
Gain-transfer gain measurements, 32-49 to 32-51
Galactic noise, 6-25, 29-50
Galerkin's method
  with methods of moments, 3-60
  as weight function, 3-83
Gallium arsenide
  in integrated antennas, 17-132, 17-137 to 17-138, 17-140
  in sensors with implantable temperature probes, 24-50
Gap capacitance and input admittance, 7-8, 7-15
Gaseous absorption
  and line-of-sight propagation, 29-34 to 29-35
  of millimeter waves, 17-5
  of satellite-earth propagation, 29-8, 29-10

Gaseous emissions, noise from, 29-50 to 29-51
GEMACS modeling code, 3-88 to 3-91
Generalized equation for standard reference am antennas, 26-20 to 26-22
Geodesic lens antennas, 17-29 to 17-32
  matrix-fed, 19-108 to 19-119
  for phased array feeds, 19-17, 19-41, 19-44
Geodesic paths, 20-19
Geometrical models, 32-69
Geometrical optics, 4-3
  and computer solutions, 3-54
  fields with, 4-8 to 4-9
    incident, 4-10 to 4-13
    reflected, 4-13 to 4-18
  for lens antennas, 16-6
  and PTD, 4-103
  with reflector antennas, 15-7, 15-13, 15-14
    dual-reflector, 15-69 to 15-73, 15-77, 15-80
  for simulation of environment, 20-10
  with UTD solutions
    for curved surfaces, 20-35 to 20-36
    for noncurved surfaces, 20-45 to 20-46
Geometrical theory of diffraction, 4-3, 4-23 to 4-24. *See also* Uniform geometrical theory of diffraction
  for apertures on curved surfaces, 13-50 to 13-51
  and computer solutions, 3-54
  for convex surfaces, 4-28 to 4-30
  and ECM, 4-97 to 4-98
  for edges, 4-25 to 4-28
  for pyramidal horn fields, 15-92
  with reflector antennas, 15-7 to 15-8, 15-11 to 15-14
    dual-reflector, 15-69 to 15-72, 15-76 to 15-77
  for vertices, 4-30 to 4-31
Geophysical applications, antennas for, 23-3, 23-23 to 23-25
  electrode arrays, 23-4 to 23-8
  grounded wire, 23-8 to 23-17
  loop, 23-17 to 23-23
Geostationary Operational Environmental Satellite, 22-34, 22-43
Geosynchronous satellite antennas, 6-26
  microwave radiometers for, 22-43 to 22-46
G for microstrip antennas, 10-34 to 10-41

Gimbaling
  angles of, and radome field measurements, 31-6
  with satellite antennas, 21-5
Glass plates for microstrip arrays, 17-108
GOES (Geostationary Operational Environmental Satellite), 22-34, 22-43
GO methods. *See* Geometrical optics
Goniometers, rotating, 25-15 to 25-16
Good conductors, modeling of, 32-70 to 32-71
Good dielectrics, modeling of, 32-70 to 32-75
Goubau antenna, 6-24
Gradient synthesis method, 21-65
Graphite epoxy material for satellite antenna feeds, 21-28 to 21-29, 22-46
Grating antennas, 17-34 to 17-35, 17-48 to 17-82
Grating lobes
  with arrays
    aperiodic, 14-3, 19-57
    blindness in, 13-47
    Dolph-Chebyshev, 11-18 to 11-19
    for feeding satellite antennas, 21-27
    periodic, 11-32, 13-10
    phased, 18-9 to 18-10, 21-10 to 21-12
    scanned in one plane, 13-13 to 13-14
    scanned in two planes, 13-15 to 13-16
    space-fed, 19-69
    and subarrays, 13-32, 13-35, 19-69
  with multiple element feeds, 19-15
  with radial transmission line feeds, 19-31
  and space harmonics, 1-35
Grating period
  and leakage constants, 17-62
  and radiation angle in leaky modes, 17-56 to 17-58
Great-circuit cuts, 32-6
Green's dyads, 3-46 to 3-48
Green's function
  for field propagation, 3-56
  in unbounded space, 2-6 to 2-11
  with UTD solutions, 20-4 to 20-5
Gregorian feed systems
  offset-fed, 19-63 to 19-66
  with reflector satellite antennas, 21-21 to 21-22
Gregorian reflector systems
  cross polarization in, 15-69 to 15-70
  gain loss in, 15-73
Gridded reflector satellite antennas, 21-21

Groove depths
  of dielectric grating antennas, 17-77 to 17-78
  and leakage constants, 17-60 to 17-61
Groove guide transmission type, 17-35, 21-59, 21-62
  for leaky-wave antennas, 17-82 to 17-92
Ground-based hf antennas, modeling of, 32-76 to 32-77
Grounded wire antennas, 23-8 to 23-17
Ground-plane antenna ranges, 32-19
Ground planes
  and dipole arrays, 13-42
  imperfect, 3-27, 3-32 to 3-33, 3-47 to 3-50
  infinite, in slot array design, 12-34 to 12-35
  with microstrip antennas, 10-5, 10-6, 10-23 to 10-24
    circular polarization, 10-59
  perfect, 3-18 to 3-20, 3-47 to 3-48
  in UTD solutions, 20-7
Ground-reflection antenna ranges, 32-21 to 32-23
  gain measurement on, 32-48 to 32-49
Ground systems for am antennas, 26-37, 26-38
Ground-wave attenuation factor, 29-45
Group delay and satellite-earth propagation, 29-15 to 29-16
GTD. *See* Geometrical theory of diffraction
Guided waves from radomes, 31-5
Guidelines with modeling codes, 3-87, 3-95
Gyros for satellite antennas, 22-46
Gysel hybrid ring power division element, 21-50, 21-51

HAAT. *See* Height above average terrain
Half-power beamwidths, 1-20, 1-29
  of aperture antennas, 8-67
    circular, 5-22 to 5-24
    rectangular, 5-13, 5-16 to 5-17, 5-20, 15-91
  of arrays
    aperiodic, 14-15, 14-18, 14-23, 14-27 to 14-32
    Dolph-Chebyshev, 11-19
    linear, 11-15
    scanned in one plane, 13-12 to 13-15
  of classical antennas, 3-28 to 3-31
  and directivity, 32-41
  of horn antennas
    corrugated, 8-55, 8-59, 15-102 to 15-104
    E-plane, 8-14 to 8-15
    H-plane, 8-29 to 8-30
  of log-periodic zigzag antennas, 9-39 to 9-46
  of log-spiral conical antennas, 9-86, 9-87, 9-89, 9-97 to 9-98, 9-104
  and microwave radiometer beam efficiency, 22-27
  of millimeter-wave antennas
    dielectric grating, 17-79 to 17-80
    maximum-gain surface-wave, 17-40 to 17-42
    tapered dielectric-rod, 17-39, 17-47
  of reflector antennas
    lateral feed displacement, 21-20
    offset parabolic, 15-37, 15-41
    tapered-aperture, 15-16 to 15-17, 15-23
  of waveguide feeds, open-ended
    circular, 15-105
    rectangular, 15-104
Half-wave antennas, 26-8
  dipole, 3-17, 6-14 to 6-15
    radiation pattern of, 6-11
    receiving current of, 7-18 to 7-19
Half-wave radomes, 31-17
Hallen antenna, 6-24
Hansen-Woodyard end-fire linear array, 11-15
HAPDAR radar, 19-56
Harmonic distortion from pin diodes, 21-38
Harvesting of crops, remote sensing for, 22-15
H/D (height-to-diameter ratio), 15-32 to 15-33, 15-37, 15-40
Health care. *See* Medical applications
Heat sinking with integrated antennas, 17-133
Heating patterns, phantoms for checking, 24-54 to 24-57. *See also* Hyperthermia
Height. *See also* Vector effective height
  effective, 30-19, 30-21, 30-30
  for grating antennas
    dielectric, 17-64
    metal, 17-68 to 17-69
  and leakage constants, 17-58 to 17-59
  of log-spiral conical antennas, 9-101
  offset, of parabolic reflectors, 15-31 to 15-32

of substrates, 17-107, 17-109, 17-111, 17-124, 17-126
of test antennas in elevated antenna ranges, 32-20
Height above average terrain
and fm antenna ERP, 27-4, 27-6
and tv antenna ERP, 27-3 to 27-5
Height-to-diameter ratio, 15-32 to 15-33, 15-37, 15-40
Height-to-radius ratio
and impedance of dipoles, 9-61 to 9-62
of log-periodic antennas, and directivity, 9-27, 9-30
Helical antennas, 3-31
circularly polarized tv, 27-15 to 27-19
Helical-coil hyperthermia applicators, 24-48 to 24-49
Helical geodesic surface-ray paths, 4-75 to 4-76
Helix phased arrays, 21-12, 21-68, 21-70 to 21-71
Hemispherical lenses, 19-81 to 19-82
Hemispherical radiation pattern of am antennas, 26-5
Hemispherical radiators, 26-9, 26-27 to 26-29, 26-32
Hemispherical reflector antennas
high-resolution, 19-85, 19-87, 19-89 to 19-91
with wide-angle systems, 19-80 to 19-81
HEMP. See Electromagnetic pulses
Hertzian source, characteristics of, 3-13 to 3-14
Hexagonal arrays, 11-30 to 11-31
H-fields. See Magnetic fields
H-guide transmission type, 21-59, 21-62
modification of, for leaky-wave antennas, 17-93 to 17-94
High-attenuation rf cable, 28-25
High-delay rf cable, 28-25
Higher-order modes
junction coupling and scattering, 12-35 to 12-36
for leaky-wave antennas, 17-87 to 17-92, 17-98 to 17-103
with microstrip antennas, 10-25
High frequencies, ionosphere propagation of, 29-21 to 29-25
High-frequency antennas
fixed-wire, 30-23 to 30-27, 30-29 to 30-30
modeling of, 32-64, 32-76 to 32-77

High-frequency limit, 9-3
of log-spiral antennas, 9-75
High frequency techniques, 4-3 to 4-6
equivalent current method, 4-96 to 4-102
geometrical optics fields, 4-6 to 4-8
geometrical theory of diffraction, 4-23 to 4-96
physical optics field, 4-8 to 4-23
physical theory of diffraction, 4-102 to 4-115
High-gain antennas
direction finding, 25-14 to 25-16
dual-reflector, 15-73
lens, 16-5
limited scan, 19-57
microstrip dipole, 17-122 to 17-126
millimeter-wave antennas, 17-9 to 17-27
multimode element array, 19-12
rotating, 25-10 to 25-11
High-resolution antennas, 19-85, 19-87, 19-89 to 19-91
High-temperature rf cable, 28-24
HIHAT (high-resolution hemispherical reflector antennas), 19-85, 19-87, 19-89, to 19-91
Holey plate experiment, 14-27 to 14-35
Hollow-tube waveguides, 28-35 to 28-36
Holographic antennas, 17-106, 17-129 to 17-131
Homology with millimeter-wave antennas, 17-14
Horizon, radio, 29-34
Horizontal field strength with two towers, 26-30 to 26-34
Horizontal optimization, 11-82 to 11-83
Horizontal polarization
with leaky-wave antennas, 17-97
with loop fm antennas, 27-33
for microwave radiometers, 22-44 to 22-46
and surface emissivity, 22-8
with tv antennas, 27-23
Horizontal profiling for earth resistivity, 23-4
Horizontal scanners, 33-19
Horizontal stabilizers, simulation of, 20-82 to 20-84
Horizontal transmitting loop antennas, 23-17

Horn antennas, 3-31, 8-5. *See also* Aperture-matched horn antennas; Conical horn antennas; Corrugated horn antennas; Dielectric loading, with horn antennas; E-plane horn antennas; H-plane horn antennas; Pyramidal horn antennas
   as array elements, 8-4
   as calibration standard, 8-3
   millimeter-wave, 17-23 to 17-24
   multimode, 8-69 to 8-73, 15-95
   phase center, 8-73 to 8-84
   with satellite antennas, 21-7, 21-80, 21-82 to 21-84
Horn feeds, 15-5
Horn launchers, 17-46
Horn waveguides, 2-27
H-plane folded and septum T waveguide power division elements, 21-51 to 21-52
H-plane gain correction factor for biconical horns, 21-104
H-plane horn antennas, 8-4
   aperture fields of, 8-20
   design of, 8-23 to 8-34
   directivity of, 8-29 to 8-33
   half-power beamwidth of, 8-29 to 8-30
   phase center of, 8-76 to 8-77
   radiated fields of, 8-20 to 8-26, 8-28
   for reflector feeds, 15-92 to 15-93, 15-96 to 15-97
   universal curves of, 8-23, 8-27 to 8-29
H-plane radiation patterns, 1-28, 5-12
   of apertures
      circular, 5-22
      rectanguar, 5-12
   with cavity-backed slot arrays, 9-68 to 9-71
   of dielectric grating antennas, 17-73 to 17-74, 17-78 to 17-79
   of DuFort-Uyeda lenses, 19-52
   of horn antennas
      corrugated, 8-53 to 8-59
      E-plane, 8-9, 8-10
      H-plane, 8-23 to 8-26, 8-28
      pyramidal, 8-37 to 8-40
   for log-periodic antennas
      diode arrays, 9-63 to 9-66
      zigzag, 9-39 to 9-42, 9-45, 9-46
   with microstrip antennas, 10-25
   of monofilar zigzag antenna, 9-36
   and UTD solutions, 20-13 to 20-14

Hughes approaches with cylindrical array feeds
   matrix-fed Meyer geodesic lens, 19-108 to 19-119
   phased lens, 19-107 to 19-108
Human-made noise, 29-51 to 29-52
Humidity
   and brightness temperature, 22-10
   remote sensing of, 22-6 to 22-8
   and scintillations, 29-20
   sounders for, 22-14, 22-22
Huygens-Fresnel principle, 1-32, to 1-33
Huygen's principle and source, 2-18 to 2-20
Hybrid antennas with reflector antennas, 17-11
Hybrid beam-forming networks, 21-34
Hybrid-coupled phase shifters, 13-62 to 13-63, 18-13
Hybrid-mode horn feeds, 15-5, 15-97, 15-99
Hybrid-mode waveguides, 28-47, 28-50 to 28-58
Hybrid-ring coupler power division element, 21-50, 21-52
Hydrology, remote sensing for, 22-15 to 22-16
Hyperbolas, 15-25 to 15-26
Hyperbolic lenses, 16-10, 16-53 to 16-54
Hyperbolic reflectors, feeds with, 19-61 to 19-62
Hyperthermia for medical therapy, 24-5 to 24-7, 24-9
   applicators for, 24-43 to 24-49
   heating patterns with, 24-50 to 24-57
   microstrip antennas for, 24-18 to 24-25
   waveguide antennas for, 24-9 to 24-18
Hypodermic monopole radiators, 24-26 to 24-27

Ice, 22-9
   remote sensing of, 22-14 to 21-15
   and sidefire helical tv antennas, 27-26
Illuminating fields and imperfect ground planes, 3-48
Illumination control of apertures, 13-30 to 13-32
Image-guide-fed slot array antenna, 17-50
Image guides, 21-59, 21-61, 28-50 to 28-53

Image impedances
    for log-periodic mono arrays, 9-66
    with periodic loads, 9-50, 9-55
Image-plane antenna ranges, 32-11, 32-19, 32-22 to 32-24
Images
    and imperfect ground planes, 3-27, 3-32, 3-48
    and perfect ground planes, 3-20, 3-47
    theory of, 2-11 to 2-13
Imaging arrays, 17-137 to 17-141
Impedance. *See also* Characteristic impedance; Input impedance; Mean impedance; Wave impedance
    base, with am antennas, 26-8 to 26-17
    of bow-tie dipoles, 17-138
    of cavities, 10-20 to 10-21
    of complementary planar antennas, 2-15 to 2-16
    of dipoles, 9-56
        and height-to-radius ratio, 9-61 to 9-62
        skewed, 27-18
    driving point
        with directional antenna feeder systems, 26-43 to 26-45
        with microstrip antennas, 10-20, 10-34, 10-53
    and EMP, 30-4
    feeder, with log-periodic diode arrays, 9-65 to 9-66
    of ideal sources, 2-29 to 2-30
    of log-periodic planar antennas, 9-10
    of log-spiral conical antennas, 9-107
    for medical applications, 24-49 to 24-50
    of microstrip antennas, 10-26 to 10-44
        resonant, 10-28, 10-30 to 10-31
    of scale models, 32-64, 32-68, 32-76 to 32-77
    of short monopole probe, 24-31 to 24-32
    surface, 3-54, 32-8
    of transmission lines, 21-63, 28-5
        coplanar, 28-35
        triplate stripline TEM, 28-22, 28-28 to 28-29
        two-wire, 28-11, 28-19
    and UTD solutions, 20-5
    waveguide, 2-28
Impedance function of cavities, 10-11
Impedance-loaded antennas
    dipole, 7-19 to 7-20
    monopole, 7-20 to 7-21

Impedance matching
    with directional antenna feeder systems, 26-40 to 26-41
    and gain measurements, 32-42
    and lens antennas, 16-5
        transformers for, 21-19
    of microstrip antennas, 10-50 to 10-52
    with receiving antennas, 6-6 to 6-7
    of split-tee power dividers, 13-61
    of waveguide arrays, 13-57
Impedance matrices, 3-61
    and rotational symmetry, 3-67 to 3-68
Imperfect ground planes, 3-27, 3-32 to 3-33
    integral equations with, 3-47 to 3-50
Implantable antennas for cancer treatment, 24-25 to 24-29
    E-field probes, 24-51 to 24-54
    temperature probes, 24-50 to 24-51
Incident fields
    GO, 4-10 to 4-13
    reciprocity of, 2-32 to 2-33
    and UTD solutions, 20-8 to 20-9
Incident powers and gain, 1-24 to 1-25
Incident shadow boundary, 4-10 to 4-11, 4-17
Incident wavefronts, 4-15, 4-16
Inclined edge-slot radiators, 18-8
Inclined-slot array antennas, 17-26
Inclined slots, 12-4, 12-24
    narrow wall arrays, 12-25 to 12-28
Incremental conductance technique, 12-26, 12-34
Index of refraction. *See* Refractive index
Indoor antenna ranges, 32-26 to 32-30
Inductive loading of short dipole, 6-13
Inductive susceptance of microstrip antennas, 10-26
Infinite array solutions, 13-48 to 13-49
Infinite baluns, 9-103, 9-105
Infinite ground plane assumptions in slot array design, 12-34 to 12-35
Inhomogeneous lenses, 16-51 to 16-54
In-line power dividers for array feeds, 13-61
Input features in modeling codes, 3-84 to 3-85
Input impedance, 2-29
    of array elements, 11-5
    of biconical antennas, 3-26
    of coaxial sleeve antennas, 7-32 to 7-33, 7-36

Input impedance (*cont.*)
 of dipole antennas
  folded, 7-36, 7-39 to 7-41
  linear, 7-14 to 7-17
  microstrip, 17-125
  short, 6-11 to 6-12
  sleeve, 7-26 to 7-29
 of L-band antennas, 30-18
 of loaded microstrip elements, 10-54 to 10-55
 of log-periodic antennas
  dipole arrays, 9-20 to 9-21, 9-24, 9-27 to 9-30, 9-58, 9-62
  zigzag, 9-39, 9-42 to 9-44
 of log-spiral antennas, 9-78 to 9-79
  conical, 9-95 to 9-97, 9-99
 of loop antennas
  ferrite, 6-21 to 6-22
  small, 6-17 to 6-18
 of microstrip antennas, 10-9, 10-26 to 10-28, 10-31, 10-65
 of models, 32-85 to 32-86
 of open-ended coaxial cable, 24-35
 of self-complementary antennas, 9-5, 9-7
 of sleeve monopoles, 7-30 to 7-31, 7-34
 of thin-wire antennas, 7-7 to 7-8, 7-44, 7-47 to 7-48
 of transmission lines, 28-7
  two-wire, 7-40
 of uhf communication antennas, 30-21
 validation of computer code for, 3-74 to 3-75, 3-77
 of vhf antennas
  communication, 30-24
  localizer, 30-26
  marker beacon, 30-28
Input susceptance of linear thin-wire antenna, 7-15 to 7-17
Insertion delay phase and dielectric sheets, 31-14
Insertion loss
 and beam-forming network switches, 21-37
 of phase shifters, 18-15 to 18-17, 18-21, 21-41
 and power division elements, 21-51
  diode varible, 21-47
 of transmission lines, 21-63
 of waveguides, 21-7
Insertion-loss gain measurement method, 32-44 to 32-46

Instantaneous direction-finding patterns, 25-12 to 25-14
Instantaneous field of view, 22-30 to 22-32
Instrumentation for antenna ranges, 32-30 to 32-33
Insular waveguide transmission types, 21-59, 21-61
 for microstrip arrays, 17-117 to 17-118
Integral equations, 3-36, 3-38 to 3-39, 3-44 to 3-52
 codes for, 3-80 to 3-96
 compared to differential, 3-55 to 3-58, 3-82
 computation with, 3-68 to 3-72
 for far fields, 3-8 to 3-9
 numerical implementation of, 3-52 to 3-68
 for unbounded space, 2-9 to 2-11
 validation of, 3-72 to 3-80
Integrated antennas, 17-105 to 17-106, 17-131 to 17-141
Integrated optics, 28-47
Interaction terms and computational effort, 3-64
Interfaces, between media, 1-8 to 1-9
 reflections at, 3-32, 24-7, 31-12 to 31-13
Interference
 in antenna ranges, 32-19
 with satellite antennas, 21-5
Interferometers, 25-21 to 25-23
Intermodulation distortion from pin diodes, 21-38
Internal networks and EMP, 30-8
Internal noise, 6-24, 29-50
Internal validation of computer code, 3-78 to 3-80
Interpolation of measured patterns, 20-6
Interrupt ability of modeling codes, 3-86
Intersection curve of reflector surface, 15-27 to 15-30
Inverted strip guide transmission type, 21-62
 dielectric, 17-49 to 17-50
In-vivo measurement, antennas for, 24-30 to 24-35
Ionosondes, 29-24
Ionosphere
 propagation via, 29-21 to 29-30
  satellite-earth, 29-14 to 29-15
 scintillation caused by, 29-16 to 29-21

Irrigation control, remote sensing for, 22-15
ISB (incident shadow boundary), 4-10 to 4-11, 4-17
Isolation
  with ferrite switches, 21-38
  for satellite antennas, 21-5 to 21-7
Isosceles beams, 19-11 to 19-12
Isosceles triangular lattices, 1-35 to 1-36
Isotropic radiators
  characteristics of, 3-28
  electric-field strength of, 26-26
  radiation intensity of, 1-27
  receiving cross section of, 2-35
Isotropic response for E-field probes, 24-39 to 24-40

Jamming signals, arrays for, 11-10
Johnson compact antenna range, 32-29 to 32-30
Junction coupling in slot array design, 12-36
Junction effect with sleeve antennas, 7-23 to 7-26

KC-135 aircraft, patterns for, 20-65 to 20-67
Keller cone, 4-33
Keller's diffraction coefficients, 15-13
Kirchhoff approximation, 2-19 to 2-21
  and physical optics approximation, 2-27
Knife-edge diffraction, 29-42 to 29-44
Kolmogroff turbulence, 29-20
Ku-band systems, 29-8

Land
  emissivity of, 22-10
  modeling of, 32-74 to 32-77
  remote sensing of, 22-15 to 22-16
Large arrays, design of, 12-28 to 12-34
LASMN3 modeling code, 3-88 to 3-91
Latching Faraday rotator variable power dividers, 21-44 to 21-46
Latching phase shifters, 13-63 to 13-64
Lateral feed displacement, 15-51 to 15-61, 21-20
Lateral width of dielectric grating antennas, 17-64
Lattices
  field representation of, 1-35
  for phased-array radiators, 18-9 to 18-10
  spacing of, for feed arrays, 21-27

Launching coefficients
  for conducting cylinder, 21-97
  surface-ray, 4-70
Law of edge diffraction, 4-33
L-band blade antennas, 30-17 to 30-19
Leakage constants
  for dielectric grating antennas, 17-65 to 17-69, 17-74 to 17-79, 18-82
  with leaky mode antennas, 17-53 to 17-63, 17-85, 17-87 to 17-90, 17-95 to 17-97
Leakage radiation in medical applications, 24-5, 24-50
Leaky modes with periodic dielectric antennas, 17-51 to 17-62
Leaky-wave antennas, 17-34 to 17-35
  uniform-waveguide, 17-82 to 17-103
Least squares moment method, 3-60
Left-hand circular polarization, 1-15, 1-20
  of arrays, 11-7
  dual-mode converters for, 21-78, 21-80
  far-field pattern for, 1-29
  for microstrip antennas, 10-59, 10-69
  with offset parabolic reflectors, 15-36, 15-48 to 15-50
  power dividers for, 21-78, 21-80
Left-hand elliptical polarization, 1-16, 9-74
Lenses and lens antennas, 16-5 to 16-6. See also Constrained lenses; Optical feeds; Semiconstrained feeds
  aberrations and tolerance criteria of, 16-12 to 16-19
  with analytic surfaces, 16-9 to 16-12
  in compact ranges, 32-29
  design principles of, 16-7 to 16-9
  with horn antennas
    conical, 8-46
    diagonal, 8-70
    H-plane, 8-23
  inhomogeneous, 16-51 to 16-54
  millimeter-wave, 17-14 to 17-22
  for satellite antennas, 21-7, 21-8, 21-12 to 21-19, 21-71 to 21-77
  surface mismatches of, 16-54 to 16-57
  taper-control, 16-33 to 16-38
  wide-angle, 16-19 to 16-33
  zoning of, 16-38 to 16-41
LEO (low earth orbiting) satellites, 22-35 to 22-43
Level curves, 14-12 to 14-14
Lightning and tv antennas, 27-8, 27-9

Limited-scan unconstrained feeds, 19-12, 19-52, 19-56 to 19-76
Lindenblad tv antenna, 27-12
Linear arrays, 11-8 to 11-23
  broadside, maximum directivity of, 11-64 to 11-68
  cophasal end-fire, 11-69, 11-70
  of longitudinal slots, 12-13 to 12-16
  millimeter-wave, 17-33
    uniform leaky-wave, 17-35
Linear dipole antennas, 3-13 to 3-14, 7-6, 7-11 to 7-23
Linear errors with lenses, 16-14
Linear polarization, 1-14 to 1-16, 1-20, 1-28 to 1-29
  conversion of, from circular, 13-60 to 13-61
  measurement of, 32-6, 32-8, 32-53, 32-59
  for microwave radiometers, 22-44 to 22-46
  with phased arrays, 18-6
  with satellite antennas, 21-7
    gridded reflector, 21-21
Linear transformations in arrays, 11-23 to 11-48
Linear wire antenna, polarization of, 1-24
Line-of-sight propagation, 29-30, 29-32 to 29-36
  with millimeter-wave antennas, 17-5
Line-of-sight ranges, 32-9 to 32-11
Line source illumination, 4-12 to 4-13
Line source patterns, 13-13
Link calculations for satellite tv, 6-27, 6-29 to 6-31
Liquid artificial dielectric, 17-22, 17-72
Lit zone, 4-75
  Fock parameter for, 4-73
  terms for, 4-71 to 4-72
Load current of receiving antennas, 6-3, 6-4
Loaded-line diode phase shifters, 13-62 to 13-63, 18-13
Loaded-lines, 9-46 to 9-62
Loaded loop antennas, 7-47 to 7-48
Loaded microstrip elements, 10-52 to 10-56
Loaded thin-wire antennas, 7-9
Lobes, 1-28. *See also* Grating lobes; Side lobes
  with log-spiral antennas, 9-75, 9-77
  shoulder, 32-16
Localizer antennas, 30-23, 30-25 to 30-26

Logarithmic spiral antennas, 17-28
Logarithmic spiral curve, 9-72
Log-periodic arrays, 9-3 to 9-4, 9-8 to 9-11
  periodic structure theory designs for, 9-62 to 9-71
  wire diameter of, 32-76
Log-periodic dipole antennas
  active regions in, 9-17 to 9-21, 9-24
  bandwidth of, 9-17, 9-21, 9-24
  circular, 9-3, 9-4
  design of, 9-20 to 9-32
  directivity of, 9-21 to 9-24, 9-27, 9-61, 9-66
  feeder circuits for, 9-16 to 9-18
  front-to-back ratios of, 9-60 to 9-61
  impedance of, 9-21, 9-24, 9-27 to 9-30, 9-58
  parameters for, 9-12, 9-14, 9-15, 9-24
  radiation fields for, and Q, 9-19 to 9-20, 9-62 to 9-66
  swr of, 9-58 to 9-59
  transposition of conductors in, 9-14 to 9-16
Log-periodic mono arrays, 9-64 to 9-67
Log-periodic zigzag antennas
  active region of, 9-35 to 9-36
  design procedures for, 9-37 to 9-46
  directivity of, 9-39 to 9-40
  half-power beamwidths for, 9-39 to 9-46
  radiation fields of, 9-39 to 9-40
Log-spiral antennas. *See also* Conical log-spiral antennas; Logarithmic spiral antennas
  feeds for, 9-75 to 9-78, 9-80
  geometry of, 9-8
  input impedance of, 9-78 to 9-79
  polarization of, 9-74, 9-77
  radiated fields of, 9-75
  truncation of, 9-72 to 9-73
Long-slot array antennas, 17-26
Longitude amplitude taper in antenna ranges, 32-15
Longitudinal broadwall slots, 12-4
Longitudinal currents in waveguides, 12-3
Longitudinal section electric mode, 8-73
Longitudinal slot arrays
  aperture distribution of, 12-24
  design equations for, 12-10 to 12-13
  E-field distribution in, 12-6 to 12-10, 12-35
  mutual coupling with, 12-30 to 12-31
  nonresonantly spaced, 12-16 to 12-20
  planar, 12-20 to 12-24

resonantly spaced, 12-13 to 12-16
shunt-slot antennas, 17-33
Look angles for tvro, 6-29
Loop antennas
  coaxial current, 24-45 to 24-46
  ferrite, 6-18 to 6-22
  fm, 27-33 to 27-34
  for geophysical applications, 23-17 to 23-23
  as magnetic-field probes, 24-40 to 24-43
  rotating systems, 25-4 to 25-9
  small, 3-16, 3-19, 6-13, 6-17 to 6-19, 7-46 to 7-47
  spaced, 25-9 to 25-10
  thin-wire, 7-40 to 7-48
  vector effective area of, 6-5 to 6-6
Loop mutual impedance, 26-12 to 26-13
Looper radiators for medical applications, 24-20 to 24-22
Lorentz reciprocity theorem, 2-16 to 2-18
  and planar scanning, 33-7
Losses. *See also* Conduction losses; Path losses
  with conical horn antennas, 8-46
  free space, 2-39, 29-5 to 29-6, 29-8
  with microwave radiometers, 22-32 to 22-33
  system, power for, with am antennas, 26-36 to 26-37
  tangents for, 1-6
Lossy dielectrics, modeling of, 32-71 to 32-75
Low capacitance rf cable, 28-25
Low earth orbiting satellites, 22-35 to 22-43
Low frequencies, surface propagation by, 29-44
Low-frequency limit, 9-3
  of log-periodic planar antennas, 9-8 to 9-9
  of log-spiral antennas, 9-75
Low-frequency techniques, 3-5
  for classical antennas, 3-13 to 3-36
  for EMP analysis, 30-9 to 30-14
  integral equations for. *See* Integral equations
  theory with, 3-6 to 3-13
Low-frequency trailing-wire antennas, 30-31
Low-noise amplifiers for satellite tv, 6-26
Low-noise detectors, 17-136
Low-profile arrays. *See* Conformal arrays

Low-Q magnetic-field probes, 24-43, 24-44
Lowest usable frequency of ionospheric propagation, 29-25
LSE (longitudinal section electric) mode, 8-73
L-section power dividing circuits, 26-39 to 26-40
Ludwig's definition for polarization, 1-23 to 1-24
Lumped impedance for thin-wire antennas, 7-9
Luneburg lenses, 16-23
  for feeding arrays, 19-18, 19-44, 19-46 to 19-49
  as inhomogeneous lenses, 16-51 to 16-52
  vs. two-layer lenses, 17-18n
  with wide-angle multiple-beam, 19-80

Magic-T power division elements, 21-50 to 21-52
Magnetic cores. *See* Ferrite loop antennas
Magnetic current distributions
  and Kirchhoff approximation, 2-20 to 2-21
  for microstrip antennas, 10-22 to 10-25
  for dual-band elements of, 10-66
  edge of, in, 10-55 to 10-57
Magnetic dipoles
  and EMP, 30-9
  radiation pattern of, 4-59
Magnetic fields
  with arrays, 3-36
  conical-wave, 4-99
  and edge conditions, 1-9, 4-27 to 4-28, 4-101
  and equivalent currents, 4-101, 5-8 to 5-10
  in far zone, 6-3 to 6-4
  of GO reflected ray, 4-14
  Green's dyads for, 3-46, 3-48
  GTD for, 4-23
  for edge conditions, 4-27 to 4-28
  integral equation for, 3-46 to 3-47, 3-51
  with plane-wave spectra, 1-32, 5-7 to 5-8
  PO, 4-19 to 4-20
  probes for, 24-37 to 24-38, 24-40 to 24-44
  calibration of, 24-54 to 24-57
  reflection coefficient of, 2-28
  of small dipoles, 6-13
  of small loops, 6-18

Magnetic fields (*cont.*)
  surface, 4-84
    with cavities, 10-12 to 10-13
    for physical optics approximation, 2-26 to 2-27
  tangential, 1-32, 5-7
  with UTD, 4-83, 4-84, 4-95
  for vertices, 4-31
Magnetic loss tangent, 1-6
Magnetic vector potential, 3-6 to 3-7
  with arrays, 3-33, 3-35, 11-6
Magnetometers, 23-23 to 23-24
Magnetrode for hyperthermia, 24-45
Magnitude pattern for linear arrays, 11-9, 11-11
Main beam, 1-20
  with aperiodic arrays, 14-24, 14-26 to 14-34
  in array visible region, 11-31 to 11-32
  of corrugated horns, 8-55
  directivity of, 1-29
  with rectangular apertures, 5-12 to 5-14, 5-19
  and space harmonics, 1-35
Main line guide for planar arrays, 12-21 to 12-23
Main Scattering Program modeling code, 3-88 to 3-91
Maintenance of computer models, 3-68
Mangin Mirror with wide-angle multiple-beam systems, 19-85, 19-89
Mapping transformation, 11-25
Marine observation satellite (MOS-1), 22-47
Mariner-2 satellite, 22-17, 22-18
Marker beacon antennas, 30-23, 30-27 to 30-28
Masts, shipboard, simulation of, 20-94 to 20-96
Matched feeds and cross-polarization, 15-99
Matched four-port hybrid junctions, 19-5 to 19-6
Matched loads calibration method, 22-25 to 22-27
Matching. *See* Impedance matching
Matching filters, 10-52
Matching transformers for lenses, 21-19
Mathematical models, 32-65
Matrices. *See also* Admittance matrices; Blass matrices; Butler matrices
  impedance, 3-61, 3-67 to 3-68
  scattering, 12-31, 33-7
  transformation, 15-117, 15-119
Matrix-fed cylindrical arrays, 19-101 to 19-106
  with conventional lens approach, 19-107 to 19-108
  with Meyer geodesic lens, 19-108 to 19-119
Maximum effective height
  of short dipoles, 6-12, 6-14 to 6-15
  of small loops, 6-18
Maximum-gain antennas, 17-40 to 17-41
Maximum receiving cross section
  of short dipoles, 6-12 to 6-15
  of small loops, 6-18
Maximum usable frequency of ionospheric propagation, 29-22 to 29-23
Maxson tilt traveling-wave arrays, 19-80
Maxwell equations, 1-5 to 1-7
  for electromagnetic fields, 3-6
  using MOM to solve, 3-55
  in unbounded space, 1-9
Maxwell fish-eye lenses, 16-51 to 16-53
Meanderline polarizers, 21-29
Mean effective permeability of ferrite loops, 6-20 to 6-21
Mean impedance
  of log-periodic antennas
    dipole, 9-27
    zigzag, 9-46
  with self-complementary antennas, 9-10 to 9-11, 9-13
  of spiral-log antennas, 9-78
Mean resistance of log-periodic antennas
  dipole, 9-28
  zigzag, 9-40
Mean spacing parameter for log-periodic dipole antennas, 9-28
Measurement of radiation characteristics, 32-3 to 32-5, 32-39 to 32-63, 33-3 to 33-5
  compact ranges for, 33-22 to 33-24
  current distributions for, 33-5 to 33-6
  cylindrical scanning for, 33-10 to 33-12
  defocusing techniques for, 33-24 to 33-26
  errors in, 33-15 to 33-17
  extrapolation techniques for, 33-26 to 33-28
  interpolation of, 20-6

modeling for, 32-74 to 32-86
planar scanning for, 33-7 to 33-10
plane-wave synthesis for, 33-21 to 33-22
and radiation cuts, 32-6 to 32-8
ranges for, 32-8 to 32-39, 33-15 to 33-21
for slot array design, 12-26 to 12-37
spherical scanning for, 33-12 to 33-15
Mechanical scanning
 conical, 22-39 to 22-40
 with millimeter-wave antennas, 17-9, 17-10, 17-24 to 17-26
 with satellite antennas, 21-5
Mechanical switches with satellite antennas, 21-7
Medical applications, 24-5 to 24-9
 characterization of antennas used in, 24-49 to 24-57
 hyperthermia applicators for, 24-43 to 24-49
 implantable antennas for, 24-25 to 24-29
 in-vivo antennas for, 24-30 to 24-35
 microstrip antennas for, 24-18 to 24-25
 monitoring rf radiation in, 24-36 to 24-43
 waveguide- and radiation-type antennas for, 24-9 to 24-18
Medium frequencies, propagation of
 ionospheric, 29-25 to 29-27
 surface, 29-44
Medium-gain millimeter-wave antennas, 17-9 to 17-27
Medium propagation loss, 29-6
Menzel's antennas, 17-102 to 17-104
Metal grating antennas, 17-49, 17-51, 17-68 to 17-69
 bandwidth with, 17-74
Metal waveguides, 17-117 to 17-118
Meteorology and remote sensing, 22-13 to 22-15
Meteors, ionization trails from, 29-30
Meteor satellite, 22-17
Meters for receiving antennas, 2-30 to 2-32
Method of moments
 for complex environment simulations, 20-4
 for coupling, 20-68
 frequency-domain, 3-57 to 3-62, 3-64
 with Maxwell's equations, 3-55
 radomes, 31-27 to 31-29
 with reflector antennas, 15-7
 time-domain, 3-62 to 3-64

Meyer lenses for feeds, 19-19, 19-32 to 19-37
 matrix-fed, 19-108 to 19-119
MFIE modeling code, 3-88 to 3-91
Microbolometers, 17-133, 17-137 to 17-138, 17-141
Microstrip antennas, 10-5 to 10-6
 applications with, 10-56 to 10-70
 arrays of, for medical applications, 24-25
 circular polarization with, 10-57 to 10-63
 coupling of, to waveguides, 17-199 to 17-121
 efficiency of, 10-49 to 10-50
 elements for
  dual-band, 10-63 to 10-66
  frequency-agile, 10-66 to 10-69
  loaded, 10-52 to 10-56
  polarization-agile, 10-69 to 10-70
 feeds for, 10-28, 10-51
 impedance of, 10-28 to 10-44
 leaky modes with, 17-35
 loop radiators, 24-20 to 24-22
 matching of, 10-50 to 10-52
 for medical applications, 24-18 to 24-20
 models of
  circuit, 10-26 to 10-28
  physical, 10-7 to 10-21
 radiation patterns of, 10-21 to 10-26
 resonant frequency of, 10-44 to 10-49
 slot, for medical applications, 24-22 to 24-24
Microstrip array antennas, 17-25
Microstrip baluns, 18-8
Microstrip dipole antennas, 17-106, 17-118 to 17-119, 17-122 to 17-126
Microstrip excited-patch radiators, 18-8 to 18-9
Microstrip lines, 13-58, 21-59, 21-60
 for leaky-wave antennas, 17-82 to 17-83, 17-98 to 17-103
 for planar transmission lines, 28-30 to 28-33
Microstrip patch resonator arrays, 17-106 to 17-118
Microstrip radiator types, 13-60
 dipole, for phased arrays, 18-7
Microstrip resonator millimeter-wave antennas, 17-103 to 17-106
 with thick substrates, 17-122 to 17-129
 with thin substrates, 17-107 to 17-122

Microstrip techniques for integrated antennas, 17-131, 17-134 to 17-137
Microstrip waveguides for microstrip arrays, 17-117 to 17-118
Microwave-absorbing material for indoor ranges, 32-25
Microwave antennas
  constrained lenses for, 16-42
  measurements for, 33-20
Microwave diode switches, 21-38
Microwave fading, 29-40 to 29-41
Microwave radiation
  antennas to monitor, 24-36 to 24-43
  transfer of, 22-4 to 22-7
Microwave radiometers and radiometry
  antenna requirements, 22-16 to 22-46
  fundamentals of, 22-3 to 22-9, 22-22 to 22-23
  future of, 22-47 to 22-52
  for remote sensing, 22-9 to 22-16
  spacecraft constraints with, 22-46 to 22-47
Microwave sounder units, 22-14
  modulation frequency of, 22-24 to 22-25
  onboard reference targets for, 22-26
  for planar scan, 22-37
  as step scan type, 22-43
  on TIROS-N, 22-37
Microwave transmissions, 29-32
Military aircraft, simulation of, 20-78 to 20-88
Military applications, antennas for, 17-6 to 17-7, 17-15
Millimeter-wave antennas, 17-5 to 17-8
  fan-shaped beam, 17-28 to 17-32
  high-gain and medium-gain, 17-9 to 17-27
  holographic, 17-129 to 17-131
  integrated, 17-131 to 17-141
  microstrip
    monolithic, 17-134 to 17-137
    with thick substrates, 17-122 to 17-129
    with thin substrates, 17-107 to 17-122
  near-millimeter-wave imaging array, 17-137 to 17-141
  omnidirectional, 17-32 to 17-34
  periodic dielectric, 17-48 to 17-82
  printed-circuit, 17-103 to 17-131
  spiral, 17-27 to 17-28
  tapered-dielectric rod, 17-36 to 17-48, 17-133 to 17-134
  uniform-waveguide, leaky-wave, 17-82 to 17-103
Milimeter-wave Sounders, 22-47, 22-50 to 22-51
Mine communications, 23-3
  grounded wire antennas for, 23-8 to 23-17
  loop antennas for, 23-17 to 23-23
Minimax synthesis method, 21-65
Mismatches
  with lenses, 16-54 to 16-55
  and signal-to-noise ratio, 6-25
  transmission line and antenna, 1-24
Missile racks, aircraft, simulation of, 20-79 to 20-80, 20-82
Mixer diodes for integrated antennas, 17-133 to 17-134, 17-138 to 17-139, 17-141
MMIC (monolithic microwave integrated circuit technology), 17-12
Modal field distributions of loaded elements, 10-53
Modal transverse field distributions
  in circular waveguides, 1-47 to 1-48
  in rectangular waveguides, 1-40 to 1-41
Mode converters, odd/even, 21-54 to 21-56
Mode distributions of patches, 10-16 to 10-19
Modeling. *See also* Scaling
  for antenna ranges, 32-63 to 32-86
  with computers
    codes for, 3-80 to 3-96
    computation with, 3-68 to 3-72
    validation of, 3-72 to 3-80
Modes, patch, 10-7 to 10-21, 10-24 to 10-26, 10-35 to 10-37
Mode voltage in common waveguides, 12-13, 12-16
Modulating radiometers, 22-24
MOM. *See* Method of moments
Moment Method Code for antenna currents, 20-91
Monitors
  for directional antenna feeder systems, 26-41 to 26-43
  of rf radiation, 24-36 to 24-43
Monofilar zigzag wire antennas, 9-34, 9-36
Monolithic microstrip antennas, 17-134 to 17-137
Monolithic microwave integrated circuit (MMIC) technology, 17-12

Monopole antennas
  on aircraft, 20-62 to 20-68
    Boeing 737, 20-70, 20-73 to 20-75
    F-16 fighter, 20-87 to 20-88
  bandwidth of, 6-24
  hypodermic, 24-26 to 24-27
  images with, 3-20
  impedance-loaded, 7-20 to 7-21
  for in-vivo measurements, 24-30 to 24-32
  millimeter-wave, 17-32
  radiation of, on convex surface, 4-64 to 4-66, 4-71, 4-83 to 4-85
  short, 4-94, 6-12
  sleeve, 7-29 to 7-31
Monopulse pyramidal horn antennas, 8-72 to 8-75
MOS (marine observation satellite), 22-47
Mounting structures, 30-4 to 30-7
Movable line-of-sight ranges, 32-9 to 32-11
MSU. *See* Microwave sounder units
Multiarm self-complementary antennas, 9-6 to 9-8
Multibeam applications, lens antennas for, 16-6, 17-21
  wide-angle dielectric, 16-19
Multibeam satellite antennas, 15-5, 15-80 to 15-87, 21-57 to 21-58, 21-62, 21-64 to 21-67
  zoning in, 16-38
Multidetectors with integrated antennas, 17-137
Multifocal bootlace lenses, 16-46 to 16-48
Multifrequency Imaging Microwave Radiometer (MIMR), 22-47, 22-48 to 22-25
Multimode circular arrays, 25-17 to 25-20
Multimode element array beam feed technique, 19-12 to 19-15
Multimode feeds, 9-110, 9-112
  radial transmission line, 19-23 to 19-32
Multimode generators, Butler matrix as, 19-10
Multimode horn antennas, 8-69 to 8-73
  for reflector antennas, 15-95
Multipath propagation, 29-30, 29-37 to 29-41
Multipath reflections with scanning ranges, 33-16, 33-18
Multiple-access phased arrays, 21-68 to 21-71
Multiple-amplitude component polarization measurement method, 32-61 to 32-62

Multiple-antenna installations, 27-36 to 27-40
Multiple-beam antennas, 15-80
  reflector antennas with, 15-5
Multiple-beam constrained lenses, 19-79 to 19-85
Multiple-beam-forming networks, 19-23
Multiple-beam matrix feeds, 19-8 to 19-12
Multiple-feed circular polarization, 10-61 to 10-63
Multiple quarter-wavelength impedance matching transformers, 21-19
Multiple reflections
  with compact ranges, 33-27
  with thin-wire transmitting antennas, 7-13
Multiple-signal direction finding, 25-24 to 25-25
Multiport impedance parameters, 10-34, 10-44
Multiprobe launcher polarizers, 21-30
Multistepped dual-mode horns, 21-82, 21-84 to 21-89
Muscle, human
  phantom models for, 24-55
  power absorption by, 24-8, 24-9, 24-24, 24-47
Mutual admittance between slots, 4-91
Mutual base impedance of am antennas, 26-12 to 26-17
Mutual conductance with microstrip antennas, 10-9
Mutual coupling, 4-49
  and array transformations, 11-23
  of arrays
    aperiodic, 14-20 to 14-27
    circular, on cylinder, 21-100
    conformal, 13-49 to 13-52
    inclined narrow wall slot, 12-25
    large, 12-28 to 12-33
    periodic, 13-7, 13-41 to 13-46
    phased, 21-8 to 21-10, 21-12
    planar, of edge slots, 12-27 to 12-28
    scanned in one plane, 13-12
    slots in, 12-5
    of tapered dielectric-rod antennas, 17-47
  of convex surfaces, UTD for, 4-84 to 4-96
  with integrated antennas, 17-137
  with microstrip antennas, 10-7

Mutual coupling (cont.)
  of millimeter-wave antennas
    dielectric grating, 17-79
    dipoles, 17-128 to 17-129
    tapered dielectric-rod, 17-47
  nulls from, 13-45 to 13-46
  and open-periodic structures, 9-56
  and phase shift, 9-58
  with reflector antennas, 15-6
    offset parabolic, 15-83 to 15-84
  of slots
    longitudinal, 12-15, 12-18
    radiating, 12-35 to 12-36
Mutual impedance, 6-8 to 6-9
  with am antennas, 26-12 to 26-13
  with aperiodic arrays, 14-3, 14-20 to 14-21
  of array elements, 11-5
  and dipole spacing, 9-58
  of microstrip antennas, 10-34
  of monopoles, 4-94
  and radiation resistance, 9-56, 14-3

Narrowband conical horn antennas, 8-61 to 8-62
Narrow wall coupler waveguide power division elements, 21-51 to 21-52
Narrow wall slots, inclined, arrays of, 12-25 to 12-28
National Oceanic and Atmospheric Administration (NOAA)—series satellites, 22-51
Natural modes for dipole transient response, 7-21 to 7-22
Navigation systems, ionospheric propagation for, 29-27
Near-degenerate modes, 10-28, 10-29, 10-60
Near fields and near-field radiation patterns, 1-16 to 1-18
  of aperture antennas, 5-24 to 5-26, 15-21 to 15-24
    axial gain of, 5-29 to 5-33
  with Cassegrain feed systems, 19-62 to 19-63
  of dipole arrays, 13-44
  of dipoles, 20-69
  experimental validation of computer code for, 3-76
  gain reduction factor for, 5-30 to 5-32
  measurement of. *See* Measurement of radiation characteristics
  in medical applications, 24-5 to 24-8
  for monopole on aircraft, 20-62 to 20-65
  in periodic structure theory, 9-33
  with pyramidal horns, 15-92
  with reflector antennas, 15-8, 15-21 to 15-23
  transformation of, to far-field, 20-61 to 20-62
  with UTD solutions, 20-4, 20-6, 20-13 to 20-14
  in zigzag wire antennas, 9-33
Near-millimeter-wave imaging arrays, 17-137 to 17-141
NEC (Numerical Electromagnetic Code) modeling code, 3-87 to 3-95
Needle radiators, 24-26 to 24-27
Nickel for satellite antenna feeds, 21-28 to 21-29
Nimbus-5 satellite, 22-17
  planar scanning on, 22-37
Nimbus-6 satellite, 22-17 to 22-18
Nimbus-7 satellite, 22-17
  SMMR on, 22-22
NNBW (null-to-null beamwidths), 22-28
Nodal curves with frequency-agile elements, 10-68 to 10-69
Nodal planes with circular polarization, 10-62
Noise and noise power
  with arrays, 11-59
  emitters of, 2-40 to 2-42
  and propagation, 29-47 to 29-52
  and receiving antennas, 6-24 to 6-25
  at terminal of receiver, 2-42 to 2-43
  at tvro antenna, 6-26, 6-30
Noise temperature, 2-39 to 2-43
  effective, 22-27
  and figure of merit, 1-27, 6-31
  and microwave radiometers, 22-22 to 22-23
Noncentral chi-square distribution for array design, 14-10 to 14-12
Nonconverged solutions and computer solutions, 3-54
Nondirectional antennas for T T & C, 21-89
Nondirective couplers, 19-4 to 19-5
Nonoverlapping switch beam-forming networks, 21-30 to 21-31
Nonperiodic configuration with multibeam antennas, 21-66
Nonradiative dielectric guides, 17-93 to 17-98
Nonreciprocal ferrite phase shifters, 13-63, 18-14 to 18-15, 21-38 to 21-39

Index

Nonresonantly spaced longitudinal slots, 12-16 to 12-20
Nontrue time-delay feeds, 19-54 to 19-56
Nonuniform component of current, 4-103
Nonuniform excitation of array elements, 11-29 to 11-30, 11-47 to 11-48
Normal congruence of rays, 4-6
Normalized aperture distributions
  circular, 5-21 to 5-22
  rectangular, 5-19 to 5-20
Normalized intensity of far shields, 1-20 to 1-21
Normalized modal cutoff frequencies
  for circular waveguides, 1-46, 1-49
  for rectangular waveguides, 1-42
Normalized pattern functions, 11-11 to 11-13
Normalized resonant modes of microstrip antennas, 10-13, 10-15
  frequency of, 10-46
Norton's theorem for receiving antenna loads, 6-3, 6-4
Nose section of aircraft, simulation of, 20-78
N-port analogy, 3-64, 3-85
NRD guides, 17-35, 17-82 to 17-83, 17-92 to 17-98
Nuclear detonations. *See* Electromagnetic pulses
Null-to-null beamwidths, 22-28
Nulls
  with apertures
    circular, 5-22 to 5-23
    rectangular, 5-13 to 5-15, 5-19
  in Bayliss line source pattern synthesis, 13-28
  electrically rotating, 25-11 to 25-12
  filling in of, with tv antennas, 27-7
  with linear arrays, 11-15, 11-40
  with log-spiral conical antennas, 9-102
  with loop antennas, 25-5 to 25-6
  with microstrip antennas, 10-25 to 10-26
  with multiple element feeds, 19-13
  from mutual coupling, 13-45 to 13-46
  with offset reflector antennas, 15-37, 15-41
  prescribed, arrays with, 11-9 to 11-10
  with two-tower antennas, 26-22 to 26-24
Numerical Electromagnetic Code, 3-87 to 3-95
Numerical implementation for electromagnetic field problems, 3-53 to 3-68
Numerical modeling errors, 3-69 to 3-71
Numerical simulations for antennas, 20-5 to 20-7
Numerical solutions
  for aircraft antennas, 20-70 to 20-90
  for ship antennas, 20-90 to 20-97
Numerical treatment in modeling codes, 3-81, 3-83 to 3-84, 3-90
Numerical validation of computer code, 3-76 to 3-80

Obstacles, scattering by, 2-21 to 2-27
Ocean
  emissivity of, 22-10 to 22-11
  modeling of, 32-79
  reflection coefficient over, 29-38 to 29-40
  remote sensing of, 22-14 to 22-15
  and remote sensing of humidity, 22-6 to 22-7
  simulation of, 20-91
Odd distribution in rectangular apertures, 5-15
Odd/even converters, 21-7, 21-54 to 21-56
Odd-mode amplitude control, 19-14 to 19-15
Off-focus feeds with reflectors, 15-49 to 15-61
Offset distance with reflector antennas, 21-23
Offset feeds
  for Gregorian feed systems, 19-63 to 19-66
  for reflectarrays, 19-54
  for reflector antennas, 17-9
Offset height of offset reflectors, 15-31 to 15-32
Offset long-slot array antennas, 17-26
Offset parabolic reflectors
  edge and feed tapers with, 15-36 to 15-37
  feed patterns for, 15-34 to 15-36
  geometrical parameters for, 15-31 to 15-34
  off-focus feeds for, 15-49 to 15-61
  on-focus feeds for, 15-37 to 15-49
Offset phased array feeds reflector, 19-61 to 19-62
Offset reflector antennas, 15-6, 15-23. *See also* Cassegrain offset antennas

Offset reflector antennas (*cont.*)
   dual-reflector, 15-6 to 15-15, 15-73, 17-11
   satellite, 21-19 to 21-26
Ohmic losses with remote-sensing microwave radiometers, 22-32
Omega navigation systems, 29-27
Omnidirectional antennas
   dielectric grating, 17-80 to 17-82
   millimeter-wave, 17-32 to 17-34
   for satellite T T & C, 21-90
   tv, 27-5
On-axis gain of horn antennas, 8-3 to 8-4
One-parameter model for reflector antennas, 15-18 to 15-21
One plane, arrays scanned in, 13-12 to 13-15
On-focus feeds with reflectors, 15-37 to 15-49
Open-circuit voltage, 2-32
   with UTD solutions, 20-7
   and vector effective height, 6-5
Open-ended coaxial cable, 24-32 to 24-35
Open-ended waveguides
   arrays of, 13-44
   circular
      for reflector feeds, 15-90 to 15-93
      with reflector satellite antennas, 21-74
      half-power beamwidths of, 15-105
   dielectric-loaded, for medical applications, 24-9
   radiators of, for phased arrays, 18-6 to 18-7
   rectangular
      electric-field distribution of, 5-14 to 5-15
      half-power beamwidths of, 15-104
      for reflector feeds, 15-89 to 15-92
   TEM, for medical applications, 24-9 to 24-13
Open-periodic structures, 9-55 to 9-56
Open-wire transmission lines, 28-11
   two-wire, impedance of, 28-12, 28-13
Optical devices as Fourier transformers, 19-95 to 19-98
Optical feeds. *See also* Unconstrained feeds
   for aperiodic arrays, 19-57 to 19-59
   corporate, for beam-forming feed networks, 18-22
   transform, 19-91 to 19-98
Optics fields. *See* Geometrical optics; Physical optics

Optimization array problems, 11-81 to 11-86
Optimum pattern distributions, 5-18 to 5-20
Optimum working frequency and ionospheric propagation, 29-25
Orbits of satellite antennas, 6-26
Orthogonal beams, pattern synthesis with, 13-22 to 13-23
Orthogonal polarization, 21-7
Orthomode junctions, 21-7
Orthomode transducers, 22-45
Out-of-band characteristics and EMP, 30-8
Outdoor antenna ranges, 32-19 to 32-24, 33-28
Outer boundaries of radiating near-field regions, 1-18
Output features in modeling codes, 3-86
Overfeeding of power with am antennas, 26-36 to 26-37
Overlapped subarrays, 13-35 to 13-39
   space-fed system, 19-68 to 19-76
Overlapping switch beam-forming networks, 21-31 to 21-32
Overreach propagation, 29-41
Oversized waveguide transmission type, 21-60
Owf (optimum working frequency), 29-25
Oxygen
   absorption by, 29-35
   remote sensing of, 22-12 to 22-13
   temperature sounders for, 22-16, 22-21

Pancake coils, 24-47 to 24-48
Panel fm antennas, 27-35 to 27-36
Parabolas, 15-25 to 15-26
Parabolic patches, 10-18
Parabolic pillbox feeds, 19-19 to 19-20
Parabolic reflector antennas, 3-31. *See also* Offset parabolic reflectors
   far-field formulas for, 15-15 to 15-23
   near-field radiation pattern of, 4-109
Paraboloidal lenses, 21-18
Paraboloidal surfaces and intersection curve, 15-29 to 15-30
Paraboloid reflectors, 15-76
   with compact antenna ranges, 32-28 to 32-29
   phased array feeds with, 19-58 to 19-61
Parallel feed networks, 18-19, 18-21 to 18-22, 18-26, 19-5 to 19-6
Parallel plate optics. *See* Semiconstrained feeds

Parallel polarization, 31-11 to 31-15
Parallel-resonant power dividing circuits, 26-39 to 26-40
Parasitic arrays, 3-42
Parasitic reflectors, 9-64 to 9-65
Partial gain, 1-26
Partial time-delay systems, 19-52, 19-68
Partial zoning, lenses with, 16-28
Passive components for periodic arrays, 13-60 to 13-62
Patches and patch antennas. *See also* Microstrip antennas
  elements for, 10-5 to 10-7, 13-60
  medical applications of, 24-18 to 24-20
  parameters for, 10-16 to 10-17
  resonant frequency of, 10-47 to 10-49
Patch radiators for phased arrays, 18-6
Path length constraint and errors with lens antennas, 16-7 to 16-10, 16-21, 21-18
  bootlace, 16-47 to 16-48
  constrained, 16-41 to 16-51, 19-82 to 19-83
  geodesic, 19-115, 19-119
  microwave, 19-43
  and phase errors, 16-13
  with pillbox feeds, 19-22
  spherical cap, 16-25
  taper-control, 16-34 to 16-35
Path losses
  from ionospheric propagation, 29-25
  and line-of-sight propagation, 29-34 to 29-35
  with offset parabolic reflectors, 15-36 to 15-37
  of tvro downlink, 6-31
Pattern cuts and radiation patterns, 32-6 to 32-8
Pattern error with multibeam antennas, 21-65
Pattern footprint of tvro signal, 6-26, 6-27
Pattern functions. *See* Array pattern functions
Patterns. *See also* Far fields; Fields; Near fields; Radiation fields and patterns
  of arrays
    aperiodic, 14-30 to 14-32
    multiplication of, 3-36, 11-8
    periodic, 13-12 to 13-29
    synthesis of, 11-76 to 11-86
  distortion of. *See* Distortion
  with microstrip antennas, 10-21 to 10-26
  periodic, 13-20 to 13-29
Peak gains, 1-26 to 1-27

Pekeris functions, 4-52 to 4-54
Pencil-beam reflector antennas, 15-5 to 15-6
  AF method with, 15-13
  with direction-finding antennas, 25-20
  far-field formulas for, 15-15 to 15-23
  for satellites, 21-5
Perfect conductors, 3-36, 3-38 to 3-39, 3-44, 3-46
  time-domain analyses with, 3-51
Perfect ground planes, 3-18 to 3-20
  Green's functions for, 3-47 to 3-48
Periodic arrays, 13-5 to 13-11. *See also* Phased arrays
  linear transformations with, 11-31 to 11-33
  organization of, 13-23 to 13-29
  patterns of, 13-12 to 13-23
  practical, 13-30 to 13-64
Periodic configuration with multibeam antennas, 21-66
Periodic dielectric antennas, 17-34 to 17-35, 17-48 to 17-82
Periodic structures, 1-33 to 1-37
  theory of, 9-32 to 9-37
    log-periodic designs based on, 9-62 to 9-71
    and periodically loaded lines, 9-37 to 9-46, 9-46 to 9-62
Period of surface corrugations, 17-64
Permeability, 1-6
  of core in ferrite loop, 6-20
Permittivity
  of corrugation regions, 17-53 to 17-54
  and dielectric modeling, 32-72 to 32-73
  of human tissue, 24-12, 24-30 to 24-35, 24-47
  and leakage constants, 17-61 to 17-63
  of microstrip patch antennas, 17-109
  of open-ended coaxial cable, 24-33
  of substrates, 17-114, 17-124, 17-127
  of troposphere, 29-31
Perpendicular polarization, 31-11 to 31-12, 31-14 to 31-20
Personnel dosimeters, rf, 24-43
Phantoms for heating patterns of antennas, 24-54 to 24-57
Phase-amplitude polarization measurement methods, 32-57 to 32-60
Phase angle of ocean surface, 29-38
Phase center
  with horn antennas, 8-73 to 8-75
    conical, 8-61, 8-76, 8-78 to 8-80

Phase center (*cont.*)
  E-plane, 8-76 to 8-77
  H-plane, 8-76 to 8-77
  of log-periodic dipole antennas, 9-24
  of log-spiral conical antennas, 9-83, 9-91 to 9-92, 9-94
  technique to measure, 8-80 to 8-84
  testing for, 32-9
Phase constants
  for dielectric grating antennas, 17-68, 17-71
  with leaky mode antennas, 17-51, 17-53 to 17-62
  for leaky-wave antennas, 17-84 to 17-85, 17-89 to 17-92, 17-95 to 17-96, 17-100
  of TE/TM waveguides, 28-37
  for transmission lines, 28-7
Phase constraint with dielectric lenses, 16-21
Phase control, array, 13-62 to 13-64
Phased arrays. *See also* Beam-forming feed networks; Periodic arrays
  bandwidth of, 13-19 to 13-20
  conical scanning by, 22-39
  design of, 18-3 to 18-5
    and component errors, 18-26 to 18-28
    feed network selection in, 18-17 to 18-26
    phase shifter selection in, 18-12 to 18-17
    radiator selection in, 18-6 to 18-12
  feeds for, 19-58 to 19-62
  horn antennas in, 8-3
  with integrated antennas, 17-134 to 17-137
  for medical applications, 24-16 to 24-18
  for satellite antennas, 21-7 to 21-12, 21-68 to 21-71
Phase delay
  with equal group delay lenses, 16-45
  with series feed networks, 19-3
Phased lens approach, 19-107 to 19-108
Phase differences
  and direction-finding antennas, 25-3
  with interferometers, 25-21 to 25-22
Phase errors
  in antenna ranges, 32-16 to 32-18
  of array feeds, 13-61
  with arrays
    aperiodic, 14-32
    microstrip patch resonator, 17-109
    periodic, 13-55
    phased, 18-26 to 18-28, 21-12
  with beam-forming feed networks, 18-25
  with horn antennas
    conical, 8-46
    H-plane, 8-23
    pyramidal, 8-39 to 8-40
  with lenses, 16-12 to 16-20, 16-33, 21-13 to 21-14
    bootlace, 16-47 to 16-48
    constrained, 19-82
    equal group delay, 16-46
    geodesic, 19-115
    spherical, 16-24 to 16-25, 16-29
    surface tolerance, 21-18
    zone constrained, 16-43 to 16-45
  with offset parabolic antennas, 15-49, 15-55
  with pillbox feeds, 19-21
  from reflector surface errors, 15-107 to 15-108
  with satellite antennas, 21-5
Phase fronts with reflector antennas, 15-75 to 15-76, 15-80
Phase pattern functions, 11-7
Phase progression with longitudinal slots, 12-17 to 12-18
Phase quantization and array errors, 13-52 to 13-57, 21-12
Phase shift
  and mutual coupling, 9-58
  with periodic loads, 9-49 to 9-54
Phase-shifted aperture distributions, 5-17 to 5-18
Phase shifters
  with arrays
    periodic, 13-20, 13-60 to 13-64
    phased, 18-3 to 18-4, 18-6, 18-12 to 18-17
  with beam-forming networks
    scanned, 21-7, 21-38 to 21-43
    switched, 21-32
  with Butler matrix, 19-9
  with feed systems
    broadband array, 13-40 to 13-41
    directional antenna, 26-40 to 26-43, 26-45 to 26-47
    for gain ripple, 21-27
    parallel, 19-6
    series, 19-3 to 19-4
  with planar scanning, 22-37
  quantization errors with, 21-12

Index

with subarrays
  contiguous, 13-32 to 13-35
  space-fed, 19-68
in true-time-delay systems, 19-7, 19-80
with Wheeler Lab approach, 19-105, 19-107
Phase squint with subarrays, 13-35
Phase steering, 13-7 to 13-8, 13-10
Phase taper in antenna ranges, 32-15 to 32-19
Phase term for aperture fields, 8-5
Phase velocity of TEM transmission lines, 28-10, 28-22
Phasors, Maxwell equations for, 1-5 to 1-7
Physically rotating antenna systems, 25-4 to 25-11
Physical models
  errors with, 3-69 to 3-71
  for microstrip antennas, 10-7 to 10-21
Physical optics and physical optics method, 4-5, 4-18 to 4-23
  and computer solutions, 3-54
  and PTD, 4-103
  with reflector antennas, 15-7 to 15-10, 15-14
    dual-reflector, 15-69, 15-72, 15-75 to 15-77
  for scattering, 2-25 to 2-27
Physical theory of diffraction, 4-5, 4-102 to 4-103
  for edged bodies, 4-104 to 4-108
Pillbox antennas, 17-28 to 17-30
Pillbox feeds, 18-22, 19-17 to 19-23
Pinched-guide polarizers, 21-30
Pin diodes
  for beam-forming network switches, 21-38
  for dielectric grating antennas, 17-71 to 17-72
  with integrated antennas, 17-135
  with microstrip antennas, 10-69
  in phase shifters, 13-62, 18-13
    variable, 21-40
  with variable power dividers, 21-46 to 21-47
Pin polarizers, 21-30
Pitch angle of log-periodic antennas, 9-39
Planar antennas
  impedance of, 2-15 to 2-16
  log-periodic antennas, 9-10 to 9-12
    radiated fields of, 9-8 to 9-9
    truncation with, 9-8

log-spiral antennas, 9-72
  with polar orbiting satellites, 22-35 to 22-38
  power radiated by, 5-26 to 5-27
Planar apertures
  feed elements with, 15-88
  and plane-wave spectra, 5-5
  radiation patterns of, 5-10 to 5-26
  for reflector antennas, 15-14
Planar arrays, 11-48 to 11-58
  directivity of, 11-63 to 11-65, 13-11
  millimeter-wave antennas, 17-26
  mutual coupling in, 12-32 to 12-33
  optimization of, 11-63 to 11-65
  periodic, transformations with, 11-26 to 11-29
  rectangular, patterns of, 13-15 to 13-17
  slot, 12-4 to 12-5
    edge, mutual coupling of, 12-27 to 12-28
    longitudinal, 12-4, 12-20 to 12-24
  triangular, grating lobes of, 13-15 to 13-17
  of vertical dipoles, geometry of, 11-85
  waveguide, 17-26
Planar curves
  conic sections as, 15-23 to 15-24
  for reflector surfaces, 15-27
Planar lenses, 21-18
Planar quartz substrates, 17-137
Planar scanning for field measurements,

Planar transmission lines, 28-30 to 28-35
Plane symmetry and computer time and storage, 3-65, 3-67
Plane-wave illumination
  for distance parameters, 4-56
  and GO incident fields, 4-11 to 4-12
Plane wave propagation
  and dielectric sheets, 31-20 to 31-25
  flat, 31-10 to 31-18
Plane waves
  polarization of, 1-13 to 1-16
  propagation direction of, 1-16
  reciprocity of, 2-33 to 2-34
  from spherical wavefronts, 16-6, 16-9 to 16-10
  synthesis of, for field measurements, 33-21 to 33-22
Plane-wave spectra, 5-5 to 5-8
  representation of, 1-31 to 1-33
Plano-convex lenses, 16-20 to 16-21

Plate-scattered fields, 20-23
PO. *See* Physical optics
Pockington's equation for dipole arrays, 13-43
Poincare sphere, 32-54 to 32-56
Point-current source, 3-13 to 3-14
Point matching
  with methods of moments, 3-60
  weight function with, 3-83
Point-source illumination
  for distance parameter, with UTD, 4-56
  and GO incident field, 4-11
Polar cap absorptions and ionospheric propagation, 29-25
Polar coordinates with lens antennas, 16-8 to 16-9
Polarization, 1-14 to 1-16, 1-19 to 1-21, 1-28 to 1-29, 3-28 to 3-31. *See also* Circular polarization; Cross polarization; Horizontal polarization; Linear polarization; Vertical polarization
  for am antennas, 26-3
  of arrays, 11-7
    phased, 18-6, 21-12
  and diagonal horns, 8-70
  for earth coverage satellite antennas, 21-82
  efficiency of, 1-26, 2-36, 32-56
    and gain measurements, 32-42
    and mismatch, 2-39
  and Faraday rotation, 29-16
  and Fresnel reflection coefficients, 3-27
  and gain, 1-25 to 1-26
  of leaky-wave antennas, 17-90, 17-94, 17-97
  of log-periodic antennas, 9-8
  with log-spiral antennas, 9-74, 9-77
    conical, 9-86, 9-90 to 9 91, 9-99
  measurements of, 32-51 to 32-64
  of microstrip antennas, 10-57 to 10-63, 10-69 to 10-70
  for microwave radiometers, 22-43 to 22-45
  parallel, 31-11 to 31-15
  perpendicular, 31-11 to 31-12, 31-14 to 31-20
  positioners for, 32-12, 32-14
  and radome interface reflections, 31-11 to 31-13
  and rain, 29-14
  reference, 1-23
  of reflector antennas, 15-5 to 15-6, 15-89, 21-19 to 21-20, 21-73
  for satellite antennas, 21-6 to 21-7, 21-80, 21-82
  and surface emissivity, 22-7 to 22-8
Polarization-agile elements, 10-69 to 10-70
Polarization ellipses, 1-14 to 1-16
Polarization-matching factor, 2-36, 6-7 to 6-8
Polarization pattern, 32-6
  polarization measurement method, 32-62 to 32-63
Polarization-transfer methods, 32-56 to 32-57
Polarizers
  for satellite antennas, 21-7, 21-29, 21-30, 21-74, 21-79 to 21-80
  waveguide, 13-60 to 13-61
Polar orbiting satellites, 22-35 to 22-43
Poles of cavities, 10-11
Polystyrene antennas, 17-44 to 17-46
Polytetrafluoroethylene bulb for implantable antennas, 24-27, 24-29
Porous pots with earth resistivity measurements, 23-8
Positioning systems for antenna ranges, 32-8 to 32-14, 32-30 to 32-32
Potter horns
  earth coverage satellite antennas, 21-82
  for reflector antennas, 15-95, 15-100 to 15-101
Power
  accepted by antennas, 1-24, 1-25, 22-4
  complex, 1-7
  conservation of, numerical validation of, 3-79
  incident to antennas, 1-24, 1-25
  input to antenna, 6-6
  overfeeding of, with am antennas, 26-36 to 26-37
  received by antenna, 2-38, 32-42, 33-22
  reflected, 2-29
  reflection coefficient for, 1-24
  sources of, for millimeter-wave antennas, 17-11
  time-averaged, 1-7 to 1-8, 3-11
    for modal fields, 1-46
    in waveguides, 1-39
  transfer ratio for, 6-8 to 6-9
  transformation of, with taper-control lenses, 16-37
  with transmission lines, 28-8, 28-10

Index

Power amplifiers, integration of, 17-136
Power conservation law, 16-33 to 16-34
Power density
  with arrays, 11-59
  with bistatic radar, 2-37 to 2-38
  near-field, 5-29 to 5-33
  of plane waves, 1-14
  of receiving antenna, 6-7
Power detectors, integration of, 17-136
Power dividers
  for array feeds, 13-61
  for beam-forming networks
    scanned, 21-7, 21-43 to 21-49
    switched, 21-32 to 21-33
  for directional antenna feeder systems, 26-39 to 26-40
  fixed beam-forming networks, 21-49 to 21-53
  for gain ripple with feed arrays, 21-27
  LHCP and RHCP, 21-80
  quantiziation errors with, 21-12
Power flow integration method for am antennas, 26-22 to 26-33
Power gain with multibeam antennas, 21-64
Power handling capability
  measurements for, 32-4
  of transmission lines, 21-63
    circular coaxial TEM, 28-21
    microstrip planar quasi-TEM, 28-32 to 28-33
    rectangular TE/TM, 28-39, 28-42
    triplate stripline TEM, 28-28 to 28-29
    two-wire TEM, 28-19
  of waveguides
    circular TE/TM, 28-44
Power law coefficients for rain attenuation, 29-12
Power lines, noise from, 29-51
Power radiated. *See* Radiated power
Power rating
  of rf cables, 28-27
  of standard waveguides, 28-40 to 28-41
  of tv antennas, 27-7 to 27-8
Power ratio with bistatic radar, 2-36 to 2-39
Power transmittance, 2-29
  of A-sandwich panels, 31-18 to 31-19
  and dielectric sheets, 31-14 to 31-16
  efficiency of, 2-43
  with radomes, 31-3, 31-7
Poynting theorem, 1-7 to 1-8
Poynting vector for radiated power, 3-11

P-percent level curves, 14-12 to 14-14
Precipitation distributions, remote sensing of, 22-14
Prescribed nulls, 11-9 to 11-10
Pressurized air with transmission lines, 28-10
Principal polarization. *See* Reference polarization
Principle of stationary phase, 4-6 to 4-7
Printed circuits
  dipoles, 13-58
    bandwidth of, 17-122 to 17-127
    and integrated antennas, 17-130 to 17-131
    millimeter-wave, 17-25, 17-103 to 17-131
  waveguides, 21-57
    ridged, 28-57 to 28-58
    TEM, 21-57
  zigzag, 24-39, 24-40
Probabilistic approach to aperiodic arrays, 14-5, 14-8 to 14-35
Probability mean, 11-81
Probes
  for alignment, 32-33
  E-field, 24-37 to 24-41
    calibration of, 24-54 to 24-57
    implantable, 24-51 to 24-54
    in-vivo, 24-30 to 24-35, 24-49 to 24-57
  magnetic field, 24-37 to 24-38, 24-40 to 24-44
    calibration of, 24-54 to 24-57
  and polarization, 32-6 to 32-8
  for rf radiation detection, 24-37 to 24-43
  temperature, 24-40 to 24-51
Prolate spheroid
  geometry of, 4-82
  rectangular slot in, 4-79 to 4-82
Propagation, 29-5 to 29-6
  in computer models, 3-55 to 3-56
  ionospheric, 29-21 to 29-30
  with millimeter-wave antennas, 17-5
  noise, 29-47 to 29-52
  and polarization, 1-13 to 1-16
  satellite-earth, 29-7 to 29-21
  tropospheric and surface, 29-40 to 29-47
Propagation constants
  with conical log-spiral antennas, 9-80 to 9-84
  for dielectric grating antennas, 17-65
  of microstrip patch resonator arrays, 17-110
  for plane waves, 29-6

Propagation constants (*cont.*)
   for transmission lines, 28-7
   and UP structures, 9-32
Protruding dielectric waveguide arrays, 13-48 to 13-49
PTD. *See* Physical theory of diffraction
Ptfe bulb for implantable antennas, 24-27, 24-29
Pulse rf cable, 28-24 to 28-25
Purcell-type array antennas, 17-27
Push-pull power dividing circuits, 26-39
Pyramidal horn antennas, 8-4
   aperture and radiated fields of, 8-34 to 8-40
   corrugated, 8-51 to 8-59
   design procedure for, 8-43 to 8-45
   directivity of, 8-37 to 8-42
   for feeds for reflectors, 15-92 to 15-93, 15-96 to 15-98
   gain of, 8-43 to 8-44
   as surface-wave launcher, 17-46
Pyramidal log-periodic antenna, 9-4

Q. *See* Quality factor
Quadratic phase errors
   in antenna ranges, 32-16 to 32-18
   and conical horn antennas, 8-46
   with lenses, 16-3, 16-14, 16-16 to 16-18, 16-33
Quadratic ray pencil, 4-8
Quadrature hybrids for circular polarization, 10-61
Quadrature power dividing circuits, 26-39 to 26-40
Quadrifocal bootlace lenses, 16-47 to 16-48
Quality factor
   of arrays, 11-61 to 11-62
      dipoles in, 9-61 to 9-63
      directivity in, 11-67 to 11-68, 11-76
      planar, 11-65
      and SNR, 11-70 to 11-72
   of cavities, 10-11 to 10-13, 10-20, 10-21
   of dipoles
      in arrays, 9-61 to 9-63
      and phase shift, 9-58
   of ferrite loop, 6-22
   with log-periodic structures, 9-55 to 9-56
      dipole arrays, radiation fields in, 9-62 to 9-63
   of microstrip antennas, 10-6, 10-9, 10-49, 10-53

   of receiving antennas, 6-22 to 6-23
   and stopband width, 9-49
Quantization errors with phased arrays, 13-52 to 13-57, 21-12
Quarter-wave matching layers for lenses, 21-19
Quarter-wave vertical antennas, 26-9
Quartic errors with lenses, 16-16, 16-18
Quartz substrate with integrated antennas, 17-137

Radar cross section, 2-23 to 2-24
   measurements of, 32-64
Radar equation, 2-36 to 2-39
Radar requirements and phased array design, 18-4 to 18-5
Radial slots in cone, 4-78 to 4-80
Radial transmission line feeds, 19-23 to 19-32
Radiated power, 1-25, 3-11 to 3-12, 5-26 to 5-27, 6-6 to 6-7
   with arrays, 11-60
   of biconical antennas, 3-25
   and cavities, 10-20
   and directive gain, 5-26
   of fm antennas, 27-4, 27-6
   into free space, 2-29
   of space harmonics, 17-52
   of tv antennas, 27-3 to 27-5
Radiating edges, 10-24
Radiating elements
   for feed arrays, 21-27 to 21-28
   for phased array design, 18-3 to 18-12
Radiating near-field region, 1-18
Radiation angle and grating period in leaky modes, 17-56 to 17-58
Radiation conditions and Maxwell equations, 1-9
Radiation efficiency, 1-25, 3-12
   of microstrip dipole antennas, 17-119, 17-124 to 17-127
   of microstrip patch resonator arrays, 17-108 to 17-109
   of receiving antennas, 6-22 to 6-24
Radiation fields and patterns, 1-11 to 1-13, 1-20 to 1-21. *See also* Apertures and aperture distributions, radiation from; Far fields, Near fields; Patterns; Reactive fields
   of am antennas, 26-5 to 26-9, 26-17, 26-22 to 26-33
   standard reference, 26-3 to 26-8

Index

with arrays, 3-36, 3-37
   aperiodic, 14-15, 14-17, 14-19 to 14-21
   circular, on cylinder, 21-100
   conformal, 13-52
   Dolph-Chebyshev, 11-19 to 11-20
   end-fire, 11-76, 11-77
   periodic, 13-5, 13-8 to 13-10, 13-12 to 13-17, 13-45 to 13-48
   subarrays, 13-30 to 13-32, 13-35 to 13-39
of biconical antennas, 3-26, 21-107, 21-109
blackbody, 22-3 to 22-4
with circular loop antennas, 23-21
of convex surfaces, 4-63 to 4-84
of dipole antennas
   short, 20-8 to 20-9
   skewed, 27-14 to 27-18
with dual fields, 2-5 to 2-6, 3-9, 3-11
with DuFort optical technique, 19-76, 19-78
of feed systems
   radial transmission line, 19-31 to 19-32
   reflector-lens limited-scan concept, 19-72 to 19-73
   for reflectors, 15-11 to 15-12, 15-86 to 15-94
   semiconstrained, 19-49, 19-51 to 19-53
   simple, 15-89 to 15-94
with grounded wire antennas, 23-9 to 23-16
with ground planes
   imperfect, 3-27, 3-33
   perfect, 3-22
of horn antennas, 8-3
   aperture-matched, 8-58, 8-66 to 8-67
   biconical, 21-107, 21-109
   corrugated, 8-53 to 8-59, 8-64, 8-65
   dielectric-loaded, 21-84 to 21-86
   dual-mode, 15-95, 15-100 to 15-101
   E-plane, 8-7 to 8-10
   E-plane-flared biconical, 21-107
   H-plane, 8-20 to 8-26, 8-28
   H-plane-flared biconical, 21-107
   pyramidal, 8-34 to 8-39, 15-92 to 15-93
and Huygen's principle, 2-18 to 2-19
and image theory, 2-11 to 2-13
of implantable antennas, 24-27 to 24-29
of lenses
   constrained, 19-82, 19-86 to 19-87
   effect of aperture amplitude distributions on, 21-14
   microwave, 19-45
   modified Meyers, 19-115 to 19-118
   pillbox, 17-28
   for satellites, 21-73, 21-75 to 21-77
   spherical thin, 16-30
   Teflon sphere, 17-188
of log-periodic antennas
   dipole, 9-17, 9-18, 9-62 to 9-63
   dipole arrays, 9-64 to 9-66
   planar, 9-8 to 9-9
   zigzag antennas, 9-39 to 9-40
of log-spiral antennas, 9-75
   conical, 9-85 to 9-86, 9-89 to 9-92, 9-99 to 9-109
of loop antennas, 25-6 to 25-7
measurements for. *See* Measurement of radiation characteristics
of microstrip antennas, 10-10
of millimeter-wave antennas
   dielectric grating, 17-69 to 17-70, 17-72 to 17-81
   dipoles, 17-118 to 17-119, 17-127 to 17-128
   holographic, 17-132
   leaky-wave, 17-86
   spiral, 17-28 to 17-29
   tapered dielectric-rod, 17-41 to 17-43
with models, 32-64
of monofilar zigzag wire antennas, 9-34, 9-36
of monopole on aircraft, 20-64, 20-74 to 20-75, 20-86 to 20-88
of multibeam antennas, 21-64
and polarization, 1-20 to 1-21, 1-25 to 1-26
with radomes, 31-25 to 31-27
reciprocity of, 2-32 to 2-33
with reflector antennas, 15-8 to 15-10, 21-20 to 21-21
   random surface distortion on, 21-23
   slotted cylinder, 21-110 to 21-111
with satellite antennas, 21-5
   lens, 21-73, 21-75 to 21-77
   multibeam, 21-64
   slotted cylinder reflector, 21-110 to 21-111
of side-mount tv antennas, 27-20, 27-22
for slot antennas, 27-27
   axial, 21-95, 21-100
   circumferential, 21-92 to 21-93, 21-96
of small loops, 6-13, 6-17

Radiation fields and patterns (*cont.*)
  of subarrays, 13-30 to 13-32, 13-35 to 13-39
  for TEM waveguides, 24-11
  in tissue, 24-51 to 24-54
  of transmission lines, 21-63
    radial feeds for, 19-31 to 19-32
    TEM, 28-11
  with traveling-wave antennas, 3-16, 3-18
    slot, 27-27
  and UTD solutions, 20-5
    for convex surfaces, 4-63 to 4-84
    with curved surface, 20-18, 20-20, 20-25 to 20-26, 20-31 to 20-36
    for noncurved surfaces, 20-51
  of vee-dipole antennas, 27-11
  of waveguide-slot antennas, 27-30 to 27-31
Radiation resistance, 3-12
  of ferrite loop, 6-21 to 6-22
  of microstrip patch antennas, 17-107
  and mutual impedance, 9-56
  of short dipole, 6-11 to 6-12, 6-14 to 6-15
  of small loops, 6-19, 25-7 to 25-8
  of vertical antenna, 6-16
Radiation sphere, 1-19, 1-21, 32-6 to 32-7
Radiation-type antennas for medical applications, 24-9 to 24-18
Radiative transfer, microwave, 22-4 to 22-7
Radio-astronomy
  feeds for, 15-94
  and microwave remote sensing, 22-9
  millimeter-wave antennas for, 17-14, 17-16
Radio-frequency cables, list of, 28-23 to 28-25
Radio-frequency power absorption, 24-30 to 24-35
Radio-frequency radiation, monitoring of, 24-36 to 24-43
  personnel dosimeters for, 24-43
Radio horizon, 29-34
Radius vector with conical log-spiral antennas, 9-79, 9-98, 9-100
Radome Antenna and RF Circuitry, 19-54 to 19-55
Radomes, 31-3 to 31-5
  on aircraft, numerical solutions for, 20-74
  and boresight error, 31-27
  design of, 31-18 to 31-20
  and flat dielectric sheets, 31-10 to 31-18, 31-20 to 31-25
  materials for, 31-29 to 31-30
  modeling of, 32-82 to 32-83
  and moment method, 31-27 to 31-29
  with omnidirectional dielectric grating antennas, 17-81 to 17-82
  patterns with, 31-25 to 31-27
  physical effects of, 31-5 to 31-10
Rain
  line-of-sight propagation path loss from, 29-35
  and millimeter-wave antennas, 17-5
  and noise temperature, 29-49, 29-50
  remote sensing of, 22-12, 22-14
  and satellite-earth propagation
    attenuation of, 29-10 to 29-13
    depolarization of, 29-14
Random errors
  with arrays
    periodic, 13-53 to 13-57
    phased, 18-26 to 18-28
  with microwave radiometers, 22-23
  with reflector surfaces, 15-105 to 15-114, 21-23 to 21-24
Random numbers with aperiodic array design, 14-18 to 14-19
Range equation, 2-38
Ranges, antenna
  design criteria for, 32-14 to 32-19
  errors with, 33-15 to 33-18
  evaluation of, 32-33 to 32-39
  for field measurements, 33-15 to 33-24
  indoor, 32-25 to 32-30
  instrumentation, 32-30 to 32-33
  outdoor, 32-19 to 32-25
  positioners and coordinate systems for, 32-8 to 32-14
RAR (Reflect Array Radar), 19-55
RARF (Radome Antenna and RF Circuitry), 19-54 to 19-55
Rayleigh approximation and computer solutions, 3-54
Rayleigh criterion for smoothness of antenna ranges, 22-30 to 22-32, 32-22
Rayleigh-Jeans approximation with blackbody radiation, 22-3
Rays, 4-6 to 4-8
  caustics of
    distance of, 4-8
    reflected and transmitted, 4-15
  construction of, with reflector antennas, 15-11 to 15-12

paths of, with UTD solutions, 20-48 to 20-49, 20-51
RCS (radar cross-section), 2-23 to 2-24
  measurements of, 32-64
Reactive fields, 1-16 to 1-18
  of aperture and plane-wave spectra, 5-5
  coupling of, in antenna ranges, 32-14
Reactive loads with dual-band elements, 10-63 to 10-66
Realized gain, 1-25
Real poles of cavities, 10-11
Real space of arrays scanned in one plane, 13-14
Received power, 2-38, 32-42, 33-22
Receiving antennas
  bandwidth and efficiency of, 6-22 to 6-24
  effective area of, 22-4
  equivalent circuit of, 6-3
  ferrite loop, 6-18 to 6-22
  Friis transmission formula for, 6-8 to 6-9
  grounded-wave, 23-16 to 23-17
  impedance-matching factor of, 6-6 to 6-7
  linear dipole, 7-18 to 7-19
  loops, 23-22 to 23-24
  meters for, 2-30 to 2-32
  mutual impedance between, 6-9
  and noise, 2-42 to 2-43, 6-24 to 6-25
  polarization-matching factor of, 6-7 to 6-8
  power accepted by, 1-24, 1-25, 22-4
  receiving cross section of, 6-6
  reciprocity of, 2-32 to 2-36
  satellite earth stations, 6-25 to 6-32
  small. *See* Small receiving antennas
  thin-wire, 7-8 to 7-9
  thin-wire loop, 7-46 to 7-47
  vector effective height of, 6-3 to 6-6
Receiving cross section, 2-35 to 2-36
  of receiving antennas, 6-6
  of short dipoles, 6-12 to 6-15
  of small loops, 6-18
Receiving polarization, 32-53
Receiving systems for antenna ranges, 32-30 to 32-32
Receptacles and plugs with modeling, 32-83
Reciprocal bases in transformations, 11-25
Reciprocal ferrite phase shifters
  dual-mode, 18-14 to 18-15, 21-38 to 21-40
  for periodic arrays, 13-64
Reciprocity, 2-32 to 2-36
  numerical validation of, 3-79
  and reciprocity theorem, 2-16 to 2-18, 33-7
  with scattering, 2-24 to 2-25
Recording systems for antenna ranges, 32-30 to 32-32
Rectangular anechoic chambers, 32-25 to 32-26
Rectangular apertures
  antennas with, 3-21 to 3-24
  compound distributions with, 5-16 to 5-17
  directive gain of, 5-27
  displaced, phase-shifted distributions with, 5-17 to 5-18
  effective area of, 5-27 to 5-28
  efficiency of, 5-28 to 5-29
  gain of, 5-29
  near-field gain reduction factors with, 5-30 to 5-31
  near-field pattern with, 5-25 to 5-26
  optimum pattern distributions with, 5-18 to 5-20
  radiation fields of, 5-11 to 5-20, 5-27
  simple distributions with, 5-13 to 5-15
  and uniform aperture distribution, 5-11 to 5-13
Rectangular coaxial TEM transmission lines, 28-18
Rectangular conducting plane, RCS of, 2-23 to 2-24
Rectangular coordinates with lens antennas, 16-7 to 16-8
Rectangular lattices
  field representation for, 1-35
  for phased-array radiators, 18-9
Rectangular loops
  for mine communications, 23-20
  radiation resistance of, 6-19
Rectangular patch antennas, 10-41 to 10-45, 24-19
  characteristics of, 10-16, 10-21 to 10-26
  medical applications of, 24-19 to 24-20
  principle-plane patterns of, 10-22
  resonant frequency of, 10-46 to 10-47
Rectangular planar arrays, patterns of, 13-15 to 13-17
Rectangular slots
  in cones, mutual coupling of, 4-91 to 4-92
  in prolate spheroid, radiation pattern of, 4-79 to 4-82

Rectangular tapered dielectric-rod antennas, 17-37 to 17-38
Rectangular waveguides, 1-37 to 1-44, 21-60
  currents in, 12-3 to 12-4
  dielectric, 28-51 to 28-54
  impedance of, 2-28
  with infinite array solutions, 13-48 to 13-49
  mutual coupling with, 13-44 to 13-45
  open-ended
    feeds for, half-power beamwidths of, 15-104
    radiators of, for phased arrays, 18-6
    for reflectors, 15-89 to 15-92, 15-104
  radiation from, 1-28
  TE/TM, 28-38 to 28-42
  voltage and current in, 2-28 to 2-29
Rectangular XY-scanners, 33-19
Reduction factor for line-of-sight propagation, 29-36
Redundant computer operations and symmetry, 3-65
Reference polarization, 1-23
  of microwave antennas, 33-20
  with probe antennas, 32-8
Reflect Array Radar, 19-55
Reflectarrays, 19-54 to 19-55
Reflected fields. *See* Reflections and reflected fields
Reflected power, 2-29
Reflection boundaries and UTD, 4-34
Reflection coefficients
  approximation of, for imperfect grounds, 3-48
  with coaxial sleeve antennas, 7-36
  for dielectric grating antennas, 17-71, 17-81
  and imperfect ground planes, 3-27
  with linear dipole antennas, 7-23
  of ocean surface, 29-38 to 29-40
  with tapered dielectric-rod antennas, 17-44 to 17-46
  with transmission lines, 2-28, 28-7, 28-8
  with unloaded transmitting antennas, 7-13
  with UTD noncurved surface solutions, 20-43, 20-44
Reflections and reflected fields
  from aircraft wings, 20-78
  in antenna ranges, 32-34 to 32-37
  errors from, in scanning ranges, 33-16

and Fresnel ellipsoid, 29-36
GO, 4-13 to 4-18
by human tissue, 24-7
and imperfect ground planes, 3-27, 3-33
lobes from, 19-5
and multipath propagation, 29-37
and plane boundaries, 31-10 to 31-12
with radomes, 31-5, 31-6, 31-10 to 31-17
reflected-diffracted, 20-29, 20-33, 20-37, 20-43, 20-45, 20-55
reflected-reflected, 20-29, 20-33, 20-45
in scattering problems, 2-13 to 2-14
shadow boundary, 4-13
space-fed arrays with, 21-34, 21-36
space feed types, 18-17 to 18-19
and spatial variations in antenna ranges, 32-19
with UTD solutions, 20-8 to 20-9, 20-15
  for curved surfaces, 20-21, 20-23 to 20-25, 20-29, 20-31
  for noncurved surfaces, 20-37, 20-42, 20-44, 20-49, 20-53 to 20-54, 20-59 to 20-60
Reflectivity of surfaces, 22-8 to 22-9
Reflector antennas, 15-5
  for antenna ranges, 32-28 to 32-30
  basic formulations for, 15-6 to 15-15
  bifocal lens, 16-30 to 16-33
  contour beam, 15-80 to 15-84
  and coordinate transformations, 15-115 to 15-120
  diameter of, 21-24
  dual. *See* Dual-reflector antennas
  far-field formulas for, 15-15 to 15-23
  with feed, rotating antenna systems, 25-4
  feeds for, 15-84 to 15-105
  generated, 15-23 to 15-31
  with implantable antennas, 24-27 to 24-28
  millimeter-wave, 17-9 to 17-14
  with off-focus feeds, 15-49 to 15-61
  offset parabolic, 15-31 to 15-61
  with on-focus feeds, 15-37 to 15-49
  phased arrays, 21-12
  random surface errors on, 15-105 to 15-114
  satellite, 21-7, 21-8, 21-19 to 21-26, 21-73 to 21-74, 21-77 to 21-81
  surfaces for, 15-26 to 15-29
  types of, 15-5 to 15-6, 21-20 to 21-22

Reflector coordinates
  with reflector antennas, 15-7, 15-88 to 15-89
  transformations with, 15-115, 15-117 to 15-119
Reflector-lens limited-scan feed concept, 19-67 to 19-73
Refraction
  and line-of-sight propagation, 29-32 to 29-33
  and plane boundaries, 31-10 to 31-12
Refractive index, 29-7
  of atmosphere, 29-30 to 29-32
  and ionosphere, 29-14 to 29-15, 29-21 to 29-22
  lenses with varying, 16-51 to 16-54
  of short monopoles, 24-31
  of troposphere, 29-46
Regularization synthesis method, 21-65
Relabeling of bases in transformations, 11-25
Relative dielectric constant, 1-7
Relative permeability, 1-7
Remote sensing
  antenna requirements for, 22-16 to 22-46
  microwave radiometry for, 22-9 to 22-16
  sleeve dipole for, 7-26
Reradiated fields and imperfect ground planes, 3-48
Reradiative coupling in antenna ranges, 32-14
Resistance. *See also* Radiation resistance
  of free space, 26-25 to 26-26
  measurements of, electrode arrays for, 23-4 to 23-8
  surface, 1-43
Resonance
  in inclined narrow wall slot arrays, 12-26
  with planar arrays, 12-21 to 12-22
  in slot arrays, 12-14 to 12-15
  and thin-wire antennas, 7-10
  validation of computer code for, 3-73 to 3-76
Resonant arrays, 17-114, 17-118 to 17-119
Resonant frequency
  of am towers, 26-9 to 26-12
  for microstrip antennas, 10-9, 10-44 to 10-49
  of loaded elements, 10-53 to 10-54
Resonant impedance, 10-28, 10-30 to 10-31

Resonant loads with dual-band elements, 10-63 to 10-66
Resonant modes of cavities, 10-13, 10-15
Resonantly spaced longitudinal slots, 12-13 to 12-16
Resonator millimeter-wave antennas, 17-35 to 17-36
Rexolite with lenses, 16-22, 16-36 to 16-37
Rf. *See* Radio-frequency
RG-type cables, 28-22 to 28-25
Rhombic antennas, 3-31
  dielectric plate, 17-23, 17-25
Richmond formulation with radomes, 31-27 to 31-29
Ridged-waveguide antennas
  for medical applications, 24-12 to 24-16
  phased arrays for satellite antennas, 21-12
  TE/TM, 28-45 to 28-49
Ridge-loaded waveguide arrays, 13-48 to 13-49
Ridge waveguide feed transmissions, 21-56 to 21-58
Right-hand circular polarization, 1-15, 1-20
  of arrays, 11-7
  dual-mode converters for, 21-78, 21-80
  far-field pattern for, 1-29
  for microstrip antennas, 10-59, 10-69
  with offset parabolic reflectors, 15-36, 15-48 to 15-50
  power dividers for, 21-78, 21-80
Right-hand elliptical polarization, 1-16, 9-74
Rinehart-Luneburg lenses, 19-41 to 19-49
Ring focus
  with bifocal lenses, 16-32
  with DuFort-Uyeda lenses, 19-49
Ring-loaded slots, 8-64
Ripples
  with conformal arrays, 13-50 to 13-52
  with feed arrays, 21-27
Rock conductivity, dielectric logging for, 23-24 to 23-25
Rod antennas, effective receiving area of, 30-10
Roll-over-azimuth positioners, 32-12
Rotary-field phase shifters, 13-64
Rotary joints for beam-forming networks, 21-53, 21-55
Rotating antenna patterns, 25-4 to 25-11

Rotating reflector
   with conical scanning, 22-40
   with direction-finding antennas, 25-10
   with feed rotating antenna systems, 25-4 to 25-5
Rotating-source polarization measurement method, 32-63
Rotational symmetry
   and computer time and storage, 3-67 to 3-68
   dielectric grating antennas with, 17-80 to 17-81
Rotman and Turner line source lenses, 19-37 to 19-41
ROTSY modeling code, 3-88 to 3-91
Rounded-edge triplate striplines, 28-17
RSB (reflection shadow boundary), 4-13
R-3 singularity, integration involving, 2-9 to 2-10
R-2R lenses, 16-48 to 16-49
Run-time features in modeling codes, 3-86
Ruze lenses, 19-39

Saltwater, modeling of, 32-79, 32-80. *See also* Ocean
Sam-D Radar, 19-55
Sample directivity in aperiodic array design, 14-15, 14-17
Sample radiation patterns in aperiodic array design, 14-15, 14-17, 14-19 to 14-21
Sampling systems, 26-41 to 26-44
Sampling theorem, 15-10
SAR, 22-30
Satellite antennas and systems, 21-3
   communication. *See* Communication satellite antennas
   conformal array for, 13-49
   contour beam antennas for, 15-5, 15-80 to 15-87
   earth coverage, 21-80 to 21-89
   earth receiving antennas for, 6-25 to 6-32
   lenses for, 16-22
      millimeter-wave, 17-15, 17-19 to 17-20
      spherical-thin, 16-28
      taper-control, 16-36
      zoning of, 16-38
   limited scan antennas for, 19-56
   low earth orbiting 22-35 to 22-43
   for microwave remote sensing, 22-9 to 22-10
   millimeter-wave antennas for, 17-7, 17-11 to 17-12, 17-14
      lens, 17-15, 17-19 to 17-20
   modeling of, 32-86
   offset parabolic antennas for, 15-58
   polar orbiting, 22-34 to 22-42
   propagation for, 29-7 to 29-21
   reflector antennas for, 15-5
   testing of, 33-4
   tracking, telemetry, and command, 21-89 to 21-111
   weather, 22-14
S-band phased array satellite antennas, 21-68 to 21-71
Scalar conical horn antennas, 8-61
Scalar wave equation, 2-6 to 2-7
Scaling. *See also* Modeling
   with log-periodic dipole antennas, 9-12, 9-14
      and directivity, 9-20, 9-22 to 9-24
      and element length, 9-31 to 9-32
      and swr, 9-27, 9-28
   with log-periodic planar antennas, 9-9 to 9-10
   for millimeter-wave antennas, 17-7 to 17-9
   with modeling, 32-74 to 32-86
   with offset parabolic antennas, 15-81
   and principal-plane beamwidths, 9-9 to 9-10
SCAMS (scanning microwave spectrometer), 22-26, 22-37, 22-43
Scan angles vs. directivity, with arrays, 11-78
Scan characteristics, array, 14-24 to 14-25
Scanned beams
   beam-forming feed networks for, 18-21, 18-24, 21-27 to 21-34, 21-37 to 21-49
   with offset parabolic antennas, 15-51, 15-53
   with phased array satellite antennas, 21-71
Scanning. *See* Beam scanning
Scanning microwave spectrometers, 22-26, 22-37, 22-43
Scanning multichannel microwave radiometer, 22-22
   calibration of, 22-25 to 22-26
   compared to SSM/I, 22-48
   as conical scanning device, 22-40 to 22-42
   as continuous scan type, 22-43

modulating frequency of, 22-24
momentum compensation devices in, 22-46 to 22-47
onboard reference targets for, 22-26
as step scan type, 22-43
Scanning ranges. *See* Ranges, antenna
Scanning thermographic cameras, 24-51
Scan performance
  of dual-reflector antennas, 15-71 to 15-73, 15-78 to 15-79
  of zoned lenses, 16-28
Scattering and scattered propagation fields, 2-21 to 2-27, 4-49
  and Babinet principle, 2-13 to 2-15
  cross section of, 2-22 to 2-25
  and dielectric thickness, 10-6
  with dual fields, 2-5 to 2-6
  field representation of, 1-33 to 1-37
  ionospheric, 29-28 to 29-30
  in longitudinal slots, 12-7, 12-10 to 12-11
  losses from, with remote-sensing microwave radiometers, 22-32, 22-33
  plate-scattered fields, 20-23
  and PO, 4-20, 4-22
  by radomes, 20-74, 31-5
  with reflector antennas, 15-7, 15-11
  with shipboard antennas, 20-90 to 20-97
  in slot array design, 12-35 to 12-36
  tropospheric, 29-46 to 29-47
  for UTD solutions
    centers of, 20-4 to 20-6
    for convex surfaces, 4-49 to 4-63
    for curved surfaces, 20-18
    for noncurved surfaaces, 20-47, 20-49, 20-51 to 20-52, 20-57 to 20-60
Scattering matrix for planar arrays, 12-21
Scattering matrix planar scanning, 33-7
Schelkunoff's induction theorem, 31-25
Schiffman phase shifters, 13-62
Schlumberger array, 23-5 to 23-7
Schmidt corrector with wide-angle lenses, 16-23 to 16-28
Schottky diodes for integrated antennas, 17-133 to 17-134, 17-138 to 17-139, 17-141
Schumaun resonance and ionospheric propagation, 29-27
Scintillation
  from fog and turbulence, 29-36
  index for, 29-16 to 29-17
  and satellite-earth propagation, 29-16 to 29-21

SCS (scattering cross section), 2-24 to 2-25
Sea. *See* Ocean
Seasat satellite, 22-17
  SMMR on, 22-22
Second-order effects in slot array design, 12-34 to 12-36
Second-order scattering, 20-49
Security, transmission, with millimeter-wave antennas, 17-6 to 17-7
Seidel lens aberrations, 16-14, 21-13
Self-admittance with microstrip antennas, 10-7
Self-baluns, 9-16, 9-17
Self base impedance of am antennas, 26-8 to 26-12
Self-complementary antennas, 9-5 to 9-7
  log-spiral, 9-98 to 9-99
Self-conductance with microstrip antennas, 10-7
Self-resistance of am antennas, 26-8
Semicircular arrays, 11-78 to 11-80
Semicircular-rod directional couplers, 21-50
Semiconstrained feeds, 19-15 to 19-19
  DuFort-Uyeda lens, 19-49
  Meyer lens, 19-32 to 19-37
  pillbox, 19-19 to 19-23
  radial transmission line, 19-23 to 19-32
  Rinehart-Luneburg lens, 19-41 to 19-49
  Rotman and Turner line source microwave lens, 19-37 to 19-41
Sense of rotation with polarization ellipse, 1-16
Sensing systems, millimeter-wave antennas for, 17-5. *See also* Remote sensing
Sensitivity and reception, 6-10
  with planar arrays, 11-65
SEO-I and -II satellites, 22-17
Septum polarizers, 21-30
  with reflector satellite antennas, 21-74, 21-79 to 21-80
  tapered, 13-60 to 13-61
Serial shift registers, 18-3 to 18-4
Series couplers, 19-4 to 19-5
Series-fed arrays
  for microstrip dipole antennas, 17-119
  microstrip patch resonator, 17-113 to 17-116
Series feed networks, 18-9, 18-21, 18-26, 19-3 to 19-5

Series impedance of transmission lines, 28-5
Series resonant magnetic field probes, 24-43, 24-44
Series-resonant power dividing circuits, 26-39 to 26-40
Series slot radiators with phased arrays, 18-8
Shadow boundaries
 and GO, 4-10, 4-17
 and PO, 4-22
 and PTD, 4-103 to 4-104
 and UTD, 4-32, 4-34, 4-36, 4-74, 20-8, 20-12
Shadow regions, 4-5, 4-75
 and diffracted rays, 4-3, 4-108
 Fock parameter for, 4-69
 and GO incident field, 4-10
 terms for, 4-67
 and UTD, 4-52, 4-63, 4-65
  with curved surfaces, 20-20 to 20-21
Shaped-beam antennas
 in antenna ranges, 32-19
 horns, 21-5, 21-82, 21-87 to 21-89
Shaped lenses, 21-18
Shaped reflectors, 15-73, 15-75 to 15-80
Shaped tapered dielectric-rod antennas, 17-37 to 17-38
Shapes for simulation of environments, 20-8
Shaping techniques for lens antennas, 17-19
Sheleg method, 19-103 to 19-105
Shielded-wire transmission lines, 28-11
 two-wire, impedance of, 28-15
Shields
 with loops, 25-9
 with scanning ranges, 33-16
Shift registers, 18-3 to 18-4
Ships and shipboard antennas
 models for, 32-74, 32-77 to 32-79
 numerical solutions for, 20-90 to 20-97
 simulation of, 20-7 to 20-8
Short antennas
 dipole, 3-13 to 3-15, 6-10 to 6-13, 6-15
  array of, 3-38 to 3-39
  compared to small loop antenna, 3-16
  fields with, 3-34
  UTD solutions for, 20-8 to 20-15
 monopole
  impedance of, 6-12

 for in-vivo measurements, 24-30 to 24-32
 mutual impedance of, 4-94
Short-circuit current, 2-32
 and vector effective height, 6-5
Shorted patch microstrip elements, 13-60
Shoulder lobes and phase taper, 32-16
Shunt admittance of transmission lines, 9-47 to 9-49, 28-5
Shunt couplers, 19-4
Shunt-slot array antennas, 17-26
Shunt slot radiators, 18-8
Side firing log-periodic diode arrays, 9-63 to 9-65
Side lobes, 1-20
 and antenna ranges
  compact, 33-23
  measurement of, 33-17
  phase errors in, 32-19
 and aperture field distibutions, 21-14, 21-16
 with apertures
  circular, 5-22 to 5-23
  rectangular, 5-12 to 5-14, 5-16, 5-18 to 5-20
 with arrays, 11-15
  aperiodic, 14-4, 14-5, 14-8, 14-14 to 14-19, 14-23 to 14-32
  binomial, 11-11
  conformal, 13-52
  Dolph-Chebyshev, 11-13 to 11-14, 11-18, 11-21, 11-49 to 11-50
  edge-slot, 17-27
  feed, for satellite antennas, 21-27
  periodic, 9-44, 9-45, 13-10 to 13-11, 13-18
  phased, 18-3, 18-26 to 18-28, 19-60, 19-61
  from random errors, 13-53 to 13-55
  scanned in one plane, 13-12
 with beam-forming networks, 18-25, 21-57
 and beamwidth, 9-44, 9-45, 13-18
 with DuFort optical technique, 19-76
 with feed systems
  arrays for satellite antennas, 21-27
  constrained, 18-20
  Gregorian, 19-65
  parallel, 19-5
  transmission line, radial, 19-26, 19-31, 19-33

with horn antennas
  corrugated, 8-55
  multimode, 8-70
and lateral feed displacement, 21-20
with lens antennas, 16-5
  coma aberrations, 16-22 to 16-23, 16-41, 21-14
  distortion in, 16-17
  DuFort-Uyeda, 19-53
  millimeter-wave, 17-15, 17-19, 17-20
  obeying Abbe sine condition, 19-98
  power distribution of, 21-18
  quadratic errors in, 16-14
  taper-control, 16-33
  TEM, 21-16
and log-periodic zigzag antennas, 9-44, 9-45
measurement of, 33-17
and microwave radiometer beam efficiency, 22-28
with millimeter-wave antennas
  dielectric grating, 17-69, 17-75 to 17-77
  geodesic, 17-32
  lens, 17-15, 17-19, 17-20
  metal grating, 17-69
  microstrip patch resonator arrays, 17-115 to 17-116
  tapered dielectric-rod, 17-38 to 17-43, 17-47
with offset parabolic antennas, 15-33, 15-37, 15-41 to 15-42
with orthogonal beam synthesis, 13-22
with parabolic pillbox feeds, 19-20, 19-22
from radomes, 31-6
and reflections, 32-34 to 32-37
with reflector antennas
  millimeter-wave, 17-9
  satellite, 21-23
  surface errors on, 15-105 to 15-106, 15-109 to 15-110, 15-112 to 15-114, 21-26
  tapered-aperture, 15-16 to 15-17, 15-21 to 15-23
and reradiative coupling, 32-14
with satellite antennas, 21-5, 21-7
with Schiffman phase shifters, 13-62
and series feed networks, 19-5
slots for, 8-63
with space-fed arrays, 19-55
and subarrays, 13-30, 13-32, 13-39
with synthesized patterns, 13-23 to 13-29
with waveguides, open-ended
  circular, 15-94 to 15-95
  rectangular, 15-91 to 15-92
Side-mount antennas
  circularly polarized tv, 27-20 to 27-23
  for multiple-antenna installations, 27-37
Sidefire helical tv antennas, 27-24 to 27-26
Sidewall inclined-slot array antennas, 17-26
Signal-power to noise-power ratio for tvro link, 6-30
Signal-to-noise ratio and mismatches, 6-25
  with arrays, 11-58 to 11-63, 11-70 to 11-76
  with satellite antennas, 6-31
Significant height and ocean reflection coefficient, 29-38 to 29-40
Simple distributions
  with circular apertures, 5-23
  with rectangular apertures, 5-13 to 5-15
Simple lenses, 16-9 to 16-12
Simulation
  of antennas, 20-5 to 20-7
  of environment, 20-7 to 20-53
Simultaneous multi-beam systems, 19-79 to 19-80
Single-dielectric microstrip lines, 28-16
Single-feed circular polarization, 10-57 to 10-61
Single-wire transmission lines, 28-13 to 28-15
Sinusoidal-aperture antennas, 3-24
Sinusoidal current distribution, 3-14, 3-17
Skewed dipole antennas
  fm, 27-34 to 27-35
  tv, 27-11 to 27-18
Skin-effect resistance, 1-43
Skylab satellite, 22-17
Slant antenna ranges, 32-24 to 32-25
Sleeve antennas
  coaxial sleeve, 7-29 to 7-37
  dipole, 7-26 to 7-29
  with implantable antennas, 24-26 to 24-27
  junction effect with, 7-23 to 7-26
  monopole, 7-29 to 7-31
Slope of lenses, 16-9
Slope diffraction, 4-38 to 4-40, 4-108

Slot antennas, 3-31
  on aircraft wing, 20-85 to 20-86, 20-90
  annular, 30-13
  Babinet principle for, 30-14 to 30-16
  balanced, 9-72 to 9-73
  effective receiving area of, 30-10
  impedance of, 2-15 to 2-16, 9-78 to 9-79
  microstrip, for medical applications, 24-22 to 24-24
  radiation pattern of, 4-61
  stripline, 13-58 to 13-59
  tv, 27-26 to 27-31
Slot arrays
  cavity-backed, 9-68 to 9-72
  on curved surfaces, 13-51 to 13-52
  millimeter-wave, 17-26
  waveguide-fed, design of, 12-3 to 12-6
    aperture distribution in, 12-24
    and center-inclined broad wall slots, 12-24 to 12-25
    and E-field distribution, 12-6 to 12-10
    and equations for slots, 12-10 to 12-13
    far-field and near-field diagnostics for, 12-36 to 12-37
    and inclined narrow wall slots, 12-25 to 12-28
    for large arrays, 12-28 to 12-34
    and nonresonantly spaced slots, 12-16 to 12-20
    for planar arrays, 12-20 to 12-24
    and resonantly spaced slots, 12-13 to 12-16
    second-order effects in, 12-34 to 12-36
Slot line transmission types, 21-59, 21-61
Slot line waveguides, 28-53 to 28-57
Slot radiators
  for aperture antennas, 4-64
  with phased arrays, 18-8
Slots, 3-31
  axis, 4-75, 4-77, 21-92, 21-95
  in circular cylinder, 4-77 to 4-79
  mounted in plate-cylinder, 4-61
  for side lobes, 8-63
  in sphere, 4-77 to 4-79
  voltages in, 12-10 to 12-12
Slotted cylinder reflector antennas, 21-110 to 21-111
Slotted waveguide array antennas, 17-26, 17-33
Small dipole antennas, 3-28
  and EMP, 30-9
Small loop antennas, 3-16, 3-19, 6-13, 6-17 to 6-19, 7-46 to 7-47
Small receiving antennas, 6-9
  bandwidth of, 6-22 to 6-24
  short dipole, 6-10 to 6-16
  small loop, 3-16, 3-19, 6-13, 6-17 to 6-18, 7-46 to 7-47
SMMR. *See* Scanning multichannel microwave radiometer
Snell's law of refraction, 32-70
  and lenses, 21-18
    constrained, 16-51
    dielectric, 16-6, 16-8 to 16-9
    taper-control, 16-34
  with rays, 4-8 to 4-9
Snow and snowpack
  depolarization by, 29-14
  probing of, 23-25
  remote sensing of, 22-7, 22-15 to 22-16
SNR. *See* Signal-to-noise ratio
Soil
  emissivity of, 22-9 to 22-10
  probing of, 23-25
  remote sensing of moisture in, 22-13, 22-15 to 22-16
Solar radiation and ionospheric propagation, 29-14, 29-25
  noise from, 29-50
Solid bodies
  integral equations for, 3-46
  time-domain solutions with, 3-63
Solid-state components
  in integrated antennas, 17-131
  for millimeter-wave antennas, 17-11
Solid thin-wire antennas, 7-10 to 7-11
Sommerfield treatment, 3-49 to 3-50
Sounders. *See also* Microwave sounder units; Temperature sounders
  humidity, 22-14, 22-22
Source field patterns with UTD solutions, 20-15
  for curved surfaces, 20-25, 20-29, 20-31
  for noncurved surfaces, 20-37 to 20-42, 20-48
Sources
  radiation from, 1-11 to 1-13
  for transmitting antennas, 2-29 to 2-30
Space applications, lens antennas for, 16-5 to 16-6
Space-combination with millimeter-wave antennas, 17-11

Space configuration with am antennas, 26-17 to 26-19
Spacecraft constraints for microwave radiometers, 22-46 to 22-47
Spaced loops for direction-finding antennas, 25-9 to 25-11
Spaced-tapered arrays, 14-6 to 14-8
Space factors, cylinder, 21-98 to 21-100
Space feeds
 for beam-forming feed networks, 18-17 to 18-19
  for arrays in, 21-34, 21-35 to 21-36
  for reflectarrays, 19-54 to 19-56
  for subarray systems, 13-26, 19-68 to 19-69, 19-70 to 19-76
Space harmonics
 for dielectric grating antennas, 17-72
 with periodic dielectric antennas, 17-51 to 17-53
 and periodic structure theory, 9-33 to 9-35, 9-55 to 9-56
 and scattered fields, 1-34 to 1-35
Space shuttle and remote sensing, 22-10
Space waves for leaky-wave antennas, 17-100 to 17-101
Spacing with arrays, 3-36
 aperiodic, 14-4 to 14-5
 with log-periodic dipole antennas, 9-28, 9-29, 9-31
S-parameter matrix, 10-34 to 10-37
Spatial coupling with UTD solutions, 20-6 to 20-7
Spatial diplexers, 21-54
Spatial resolution of microwave radiometers, 22-30 to 22-32
Spatial variations in antenna ranges, 32-19
Special sensor microwave/imager, 22-25
 momentum compensation devices in, 22-46 to 22-47
 polarization with, 22-48
 scanning by, 22-46 to 22-47
Special sensor microwave/temperature sounder, 22-14
Spectrum functions and aperture fields, 5-6
Specular reflections in indoor ranges, 32-27
Speed of phase propagation, 29-7
Sphere
 radiation, 1-19, 1-21, 32-6 to 32-7
 slot in, radiation pattern of, 4-78

Spherical aberrations with lenses, 16-15, 16-20
 spherical thin, 16-29
Spherical coordinates, 3-24
 positioning system for, 32-8 to 32-13
 transformations of, 15-115 to 15-120
Spherical dipole antennas, 30-14
Spherical lenses, 21-18. *See also* Luneburg lenses
 aberrations with, 21-13 to 21-14
 cap, 16-23 to 16-27
 Maxwell fish-eye, 16-52 to 16-53
 symmetrical, 17-15
 thin, 16-28 to 16-30
Spherical near-field test, 33-20
Spherical radiation pattern of am antennas, 26-4, 26-8 to 26-9
Spherical reflector systems
 with satellite antennas, 21-20 to 21-21
 with wide-angle multiple-beam systems, 19-80
Spherical scanning for field measurements, 33-12 to 33-15
Spherical surfaces
 and intersection curve, 15-29
 for lenses, 16-11 to 16-12
 for offset parabolic antennas, 15-56
Spherical-wave illumination
 for distance parameter, with UTD, 4-56
 and GO incident field, 4-11 to 4-12
Spherical waves
 and biconical antennas, 3-24 to 3-25
 transformation of, to plane waves, 16-6, 16-9 to 16-10
Spillover
 with DuFort optical technique, 19-76, 19-78
 and lens antennas, 16-5
 taper-control, 16-37
 with offset parabolic antennas, 15-37, 15-42, 15-47, 15-56
Spinning-diode circular polarization patterns, 10-60 to 10-62
Spinning geosynchronous satellites, 22-43
Spin-scan technique, efficiency of, 22-34
Spin-stabilized satellites, 21-90
Spiral angle with log-spiral antennas, 9-108
 and directivity, 9-89
Spiral antennas, 17-27 to 17-28. *See also* Log-spiral antennas
 balanced, 9-78

Spiral-phase fields, 9-109
Spiral-rate constant, 9-72, 9-102
Split-tee power dividers
  for array feeds, 13-61
  stripline, 21-50, 21-51
Spread-F irregularities
  and scintillations, 29-18 to 29-19
  and vhf ionospheric propagation, 29-29
Square apertures, 5-32 to 5-33
Square coaxial transmission lines, 28-18
Square log-periodic antennas, 9-3, 9-4
Square waveguides
  for circular polarization, 16-43
  normalized modal cutoff frequencies for, 1-42
SSB. *See* Surface shadow boundary
S-65-147 vhf antenna, 30-25 to 30-26
S-65-8262-2 uhf antenna, 30-22
SSM/I. *See* Special sensor microwave/imager
SSM/T (special sensor microwave/temperature sounder), 22-14
Stabilizers, aircraft, simulation of, 20-73, 20-78, 20-82 to 20-83
Staggered-slot array antennas, 17-26
Standard am antenna patterns, 26-35 to 26-36
Standard-gain horn, 8-43
Standard radio atmosphere, 29-32
Standard reference am antennas, 26-3 to 26-22
Standard waveguides, 28-40 to 28-41
Standing-wave fed arrays, 12-4, 12-13 to 12-16
Stationary phase point, 4-7
Steering
  with periodic arrays, 13-7 to 13-10
  with subarrays, 13-32 to 13-35
Step functions
  for convex surfaces, 4-28
  in scattering problem, 4-51
  for wedges, 4-25
Step scans compared to continuous scans, 22-42
Stepped-horn antennas, 21-82 to 21-84
Stepped-septum polarizer, 13-61
Stopbands
  with dielectric grating antennas, 17-71, 17-81
  with leaky-mode antennas, 17-52, 17-68
  for log-periodic dipole antennas, 9-61 to 9-62, 9-64 to 9-65
  with periodic loads, 9-49 to 9-50, 9-55, 9-58
Stored energy with cavities, 10-20
Straight lines and intersection curve, 15-27
Straight wire antennas, 3-49 to 3-50
Stratton-Chu formula, 2-21
Stray efficiency of microwave radiometers, 22-29
Strip antennas, impedance of, 2-15 to 2-16
Strip dielectric guide transmission type, 21-62
Strip edge-diffracted fields, 20-21 to 20-23
Stripline techniques, 17-106
  for arrays, 13-61
    blindness in, 13-47
  asymmetric, 17-84 to 17-87, 17-97 to 17-98
  for beam-forming networks, 20-50 to 20-51, 21-37
  for hybrid phase shifters, 13-62
  for printed dipoles, 13-58
  for slot antennas, 13-58 to 13-59
  for transmission types, 21-58, 21-60
    feeds for, 19-29 to 19-30
    for satellite antennas, 21-7
  TEM, 28-11, 28-16
  for waveguides, 21-57 to 21-58
Structural integrity, measurements for, 32-4
Structural stopbands, 9-49 to 9-50, 9-55
  with log-periodic arrays, 9-64 to 9-65
Structure bandwidth of log-periodic antennas, 9-17
Structures
  effect of, on radiation patterns, 20-53 to 20-54, 20-61 to 20-70
  and EMP, 30-4 to 30-7
  for simulation of environment, 20-8
Stub-loaded transmission lines
  attentuation curves for, 9-52, 9-54
  dispersion curves for, 9-52, 9-54, 9-57
Subarrays
  aperture illumination control with, 13-30 to 13-32
  overlapped, 13-35 to 13-39, 19-68 to 19-76
  time-delayed, 13-32 to 13-35
  with unconstrained feeds, 19-52
Subdomain procedures, 3-60
Submarine communications, 23-16

Subreflectors
   of Cassegrain offset antenna, 15-67, 15-69
   with reflector antennas, 15-7
Subsectional moment methods, 3-60
Substrates
   for integrated antennas, 17-131
   for microstrip antennas, 10-5, 17-107, 17-109 to 17-111, 17-114, 17-122
     dipole, 17-124 to 17-129
Substructures for simulation of environments, 20-8
Sum beams with beam-forming feed networks, 18-25
Sum-hybrid mode with microstrip antennas, 10-60 to 10-61
Sun
   and ionospheric propagation, 29-14, 29-25
   noise from, 29-51
Superdirective arrays, 11-61, 11-68 to 11-70
   aperiodic, 14-3
Supergain antennas, 5-27
   arrays, 11-61
Superstrates with microstrip dipoles, 17-127 to 17-128
Superturnstile tv antennas, 27-23 to 27-24
Surface corrugations
   for dielectric grating antennas, 17-64
   with periodic dielectric antennas, 17-48
Surface current
   decays of, with corrugated horn antennas, 8-52 to 8-54
   measurement of, 33-5 to 33-6
   and PO, 2-26 to 2-27
   and PTD, 4-103
Surface-diffracted waves with UTD, 4-57
Surface emissivity and brightness temperature, 22-7 to 22-9, 22-15
Surface errors of reflector antennas, 15-105 to 15-114
   and gain, 17-12
Surface fields
   equivalent currents for, 5-8
   with UTD, 4-84 to 4-89
Surface impedances
   and computer solutions, 3-54
   and modeling, 32-68
Surface integration method, 31-24 to 31-25
Surface mismatches with dielectric lenses, 16-54 to 16-56, 21-19
Surface Patch modeling code, 3-88 to 3-91
Surface perturbations for dielectric grating antennas, 17-70 to 17-71
Surface propagation, 29-30 to 29-47
Surface-ray field
   in GTD, 4-30
   in UTD, 4-58, 4-84 to 4-86
Surface-ray launching coefficients, 4-70
Surface representation of lens antennas, 21-13
Surface resistance, waveguide, 1-43
Surface roughness and conduction loss, 17-111, 22-31
Surface shadow boundary
   and GO fields, 4-13, 4-17
   and UTD, 4-50 to 4-51, 4-56
     transition regions, 4-58 to 4-59
Surface tolerances
   with lenses, 21-18 to 21-19
   with reflector satellite antennas, 21-23, 21-24, 21-26
Surface-wave antennas, 17-34
   periodic dielectric, 17-48 to 17-82
   tapered-rod, 17-35 to 17-48
Surface-wave launchers, 17-46
Surface waves
   with integrated antennas, 17-137
   for leaky-wave antennas, 17-100
   propagation with, 29-44 to 29-46
   from substrates, 17-122, 17-129
Susceptance
   with microstrip antennas, 10-7, 10-9
   with thin-wire antennas, 7-14 to 7-17
Suspended substrate transmission types, 21-59, 21-60
Switch beam-forming networks, 21-30 to 21-32
Switched-line phase shifters, 13-62 to 13-63, 18-13
Switches for beam-forming networks, 21-7, 21-37 to 21-38
Switching networks with Wheeler Lab approach, 19-105 to 19-106
Switching speeds of phase shifters, 18-16
Swr
   and feed cables, for log-spiral antennas, 9-78, 9-80
   and image impedance, 9-55

Swr (*cont.*)
  and log-periodic antennas, 9-27, 9-28
    dipole array, 9-58 to 9-59
    dipole scale factor, 9-27, 9-28
  and Q, in dipole arrays, 9-61
Symmetrical directional couplers, 21-50, 21-51
Symmetric parabolic reflectors. *See* Offset parabolic reflectors
Symmetry
  and computer time and storage, 3-65, 3-67
  with Dolph-Chebyshev arrays, 11-52
Synchronous satellites, 21-3
Synthesis methods
  for array patterns, 11-42 to 11-43, 11-76 to 11-86
  with multibeam antennas, 21-65
  for plane waves, 33-21 to 33-22
Synthetic aperture radar (SAR), 22-30
Systematic errors
  with microwave radiometers, 22-23, 22-25 to 22-28
  and phased array design, 18-26 to 18-28
System losses, power for, 26-36 to 26-37

TACOL structure, 19-56
Tangential fields, 5-7 to 5-9
  with thin-wire antennas, 7-5 to 7-6
Taper-control lenses, 16-33 to 16-38
Tapered anechoic chambers, 32-26 to 32-28
Tapered aperture distributions
  with parallel feed networks, 19-6
  with radial transmission line feeds, 19-31, 19-33
  for reflector antennas, 15-15 to 15-23
  with series feed networks, 19-5
Tapered-arm log-periodic zigzag antennas, 9-37 to 9-39
Tapered dielectric-rod antennas, 17-34 to 17-48
  for integrated antennas, 17-133 to 17-134
Tapered septum polarizer, 13-60 to 13-61
Taperline baluns, 9-95
Taylor line source pattern synthesis, 13-25 to 13-27
TDRSS (Tracking and Data Relay Satellite System), 21-67 to 21-71
TE field. *See* Transverse electric fields
TE/TM. *See* Transverse-electric/transverse-magnetic waveguides

TEC (total electron content), 29-15 to 29-16
Tee-bars for multiple-antenna installations, 27-37
Tee power dividers, 13-61, 26-40 to 26-41, 27-13, 27-15, 27-17
Teflon sphere lens antennas, 17-17 to 17-18
Television broadcast antennas, 27-3 to 27-8
  batwing, 27-23 to 27-24
  circularly polarized, 27-8 to 27-9
    helical, 27-15 to 27-19
    side-mount, 27-20 to 27-23
  horizontally polarized, 27-23
  multiple, 27-37 to 27-40
  sidefire helical, 27-24 to 27-26
  skewed dipole, 27-11 to 27-15
  slot
    coaxial, 27-28 to 27-29
    traveling-wave, 27-26 to 27-27
    waveguide, 27-29 to 27-31
  uhf, 27-27 to 27-28
  vee dipole array, 27-9 to 27-11
  zigzag, 27-31 to 27-32
TEM. *See* Transverse-electromagnetic fields
Temperature inversion layers and scintillations, 29-20
Temperature probes, implantable, 24-50 to 24-51
Temperature sensitivity of microwave radiometers, 22-21 to 22-22, 22-23
Temperature sounders
  and AMSU-A, 22-51
  for oxygen band, 22-16, 22-20
  for remote sensing, 22-13 to 22-14
Theoretical am antenna pattern, 26-34 to 26-36
Theory of uniform-periodic structures. *See* Periodic structure theory
Therapy, medical. *See* Hyperthermia
Thermal considerations with satellite antennas, 22-46
Thermal distortion, 21-28 to 21-29, 21-57
Thermal noise, 2-39 to 2-43, 29-50
Thermal temperature and sea surface emissivity, 22-14
Thermal therapy. *See* Hyperthermia
Thermistor detectors with implantable temperature probes, 24-50

Index

Thermocoupler detectors for rf radiation, 24-37
Thermographic cameras, 24-51
Thevenin's theorem for receiving antenna load, 6-3, 6-4
Thick dipole, 3-28
Thickness of lenses, 16-10 to 16-13. *See also* Zoning of lenses
   constrained, 16-43 to 16-44
   spherical, 16-27 to 16-29
   and surface tolerance, 21-18
   taper-control, 16-36
Thin dipole, 3-28
Thin-wall radome design, 31-17
Thin-wire antennas, 7-5 to 7-11
   and computer solutions, 3-54
   integral equations for, 3-44 to 3-46
   log-periodic zigzag, 9-45 to 9-47
   loop, 7-42 to 7-48
   receiving, 7-18 to 7-19
   time-domain analyses of, 3-51 to 3-52
   transmitting, 7-12 to 7-18
Thin-Wire Time Domain modeling code, 3-87 to 3-95
Three-antenna measurement methods
   for gain, 32-44, 32-46
   for polarization, 32-60 to 32-61
Three-axis stabilized satellites, 22-43 to 22-44
Three-wire transmission lines, 28-12
Through-the-earth communications. *See* Geophysical applications
Thunderstorms, noise from, 29-50
TID (traveling ionospheric disturbances), 29-25
Tilt angle and polarization, 32-51 to 32-53
   with polarization ellipse, 1-16
   with polarization pattern method, 32-62
   with reflector antennas, 15-48, 15-69 to 15-70, 15-72 to 15-73
Time-averaged power, 1-7 to 1-8, 3-11
   for modal fields, 1-46
   in rectangular waveguides, 1-39
Time delay
   with constrained analog of dielectric lenses, 16-51
   with constrained lenses, 16-44 to 16-46
   feed systems with
      matrix feed for beam-forming feed networks, 18-23, 18-25
      partial, 19-52, 19-68
      true, 19-7 to 19-8, 19-54, 19-76 to 19-91
   lenses for, 16-45 to 16-46, 16-54
   offset beams with, 13-39 to 13-41
   steering with
      with periodic arrays, 13-7 to 13-8
      with subarrays, 13-32 to 13-35
   with subarrays
      and contiguous subarrays, 13-32 to 13-35
      wideband characteristics of, 13-32
Time-domain
   compared to frequency-domain, 3-65 to 3-66, 3-82
   and EMP, 30-4
   integral equations in, 3-50 to 3-52
   method of moments in, 3-62 to 3-64
Time-harmonic excitation
   for grounded wire antennas, 23-8 to 23-12
   with loop antennas, 23-17 to 23-20
Time-harmonic fields, 1-5 to 1-8, 3-6 to 3-8
Tin-hat Rinehart lenses, 19-44, 19-46 to 19-47
TIROS-N satellite, 22-18
   planar scanning on, 22-36 to 22-37
   sounder units on, 22-14
Tissue, human
   permittivity of, 24-12, 24-30 to 24-35, 24-47
   phantoms for, 24-54 to 24-57
   power absorption by, 24-6 to 24-8
   radiation patterns in, 24-51 to 24-54
TM field. *See* Transverse magnetic fields
TMI microwave radiometer, 22-48 to 22-49
T-network as power dividers and phase shifters, 13-61, 26-40 to 26-41, 27-13, 27-15, 27-17
Toeplitz matrix, 3-68
Tolerance criteria for lenses, 16-16 to 16-20
Top loading of short dipole, 6-13
Topology with beam-forming networks, 21-57
Top-wall hybrid junction power division elements, 21-51 to 21-52
Toroidal beams for T T & C, 21-89
Toroidal ferrite phase shifters, 18-14, 21-40 to 21-42
Torsion factor, 4-87
Total electron content and satellite-earth propagation, 29-15 to 29-16

Towers for am antennas. *See also* Two-tower antenna patterns
   height of, and electric-field strength, 26-7
   mutual base impedance between, 26-12 to 26-17
   self-base impedance of, 26-8 to 26-12
   vertical radiation pattern of, 26-3 to 26-9
Tracking, telemetry, and command for satellite antennas, 21-89 to 21-111
Tracking and Data Relay Satellite System, 21-67 to 21-71
Trailing-wire antennas, EMP responses for, 30-27, 30-31
Transfer, microwave radiative, 22-4 to 22-7
Transformations
   of coordinates, 15-115 to 15-120
   distortion from, 11-57
   linear, 11-23 to 11-48
   of near-field data to far-field, 20-61 to 20-62
   for scale models, 32-67, 32-69
Transform feeds, 19-91 to 19-98
Transient conditions, 3-51
Transient excitation
   with grounded-wire antennas, 23-12 to 23-16
   with loop antennas, 23-20 to 23-21
Transient response
   of dipole antennas, 7-21 to 7-23
   of thin-wire antennas, 7-10
Transionospheric satellite-earth propagation, 29-14 to 29-15
Transition function with UTD solutions, 20-11
Transition regions, 4-5
   and PTD, 4-104
   and UTD, 4-32, 4-58 to 4-59
Translational symmetry and computer time and storage, 3-67 to 3-68
Transmission lines, 2-27 to 2-28. *See also* Constrained feeds
   analytical validation of computer code for, 3-76
   for beam-forming networks, 21-56 to 21-63
   with constrained lenses, 16-41
   current in, with folded dipoles, 7-38 to 7-39
   equations for, 28-5 to 28-10
   impedance of, with folded dipoles, 7-41
   and input admittance computations, 7-7
   with log-spiral conical antennas, 9-95
   for medical applications, 24-24 to 24-25
   for microstrip antennas, 10-7 to 10-10
   and mismatches, 1-24
   periodically loaded, 9-46 to 9-53, 9-57, 9-59 to 9-60
   planar quasi-TEM, 28-30 to 28-35
   radiator, for feeds, 19-23 to 19-32
   resonator for, with dual-band elements, 10-63, 10-67
   TEM, 28-10 to 28-29
Transmission space-fed arrays, 21-34, 21-35
Transmission-type space feeds, 18-17 to 18-19
Transmitting antennas
   linear dipole, 7-12 to 7-18
   reciprocity of, with receiving antennas, 2-32 to 2-36
   sources for, 2-29 to 2-30
   thin-wire, 7-8
   loop, 7-42 to 7-44
Transmitting fields
   and effective length, 2-34 to 2-35
   and Kirchhoff approximation, 2-19 to 2-21
Transmitting systems for antenna ranges, 32-30 to 32-32
Transposition of conductors with log-periodic antennas, 9-14 to 9-16, 9-59 to 9-60
Transverse amplitude taper in antenna ranges, 32-14
Transverse current in waveguides, 12-3
Transverse electric fields
   and circular waveguides, 1-44 to 1-50
   in rectangular waveguides, 1-37 to 1-41, 13-44
   representation of, 1-30 to 1-32
Transverse electric/transverse magnetic waveguides, 21-56 to 21-57, 28-35 to 28-47
   for medical applications, 24-9 to 24-12
Transverse electromagnetic fields
   and chambers for calibration of field probes, 24-54 to 24-57
   lenses for, 21-16
   transmission lines for, 28-10
      balanced, 28-12, 28-13
      circular coaxial, 28-20 to 28-22

Index

triplate stripline, 28-22 to 28-29
two-wire, 28-11, 28-19 to 28-20
and waveguide antennas
   for beam-forming networks, 21-56 to 21-58
   for medical applications, 24-9 to 24-13
Transverse magnetic fields
   and circular waveguides, 1-44 to 1-50
   in rectangular waveguides, 1-37 to 1-41
   representation of, 1-30 to 1-32
Transverse modes for leaky-wave antennas, 17-86, 17-88
Transverse slots, mutual coupling with, 12-30 to 12-31
Trapezoidal-tooth log-periodic antennas, 9-37 to 9-38
Trapped image-guide antenna, 17-49, 21-62
Trapped inverted microstrip transmission type, 21-59, 21-61
Trapped-miner problem, 23-23 to 23-24
Trapped surface waves, 29-45
Traveling ionospheric disturbances, 29-25
Traveling-wave antennas, 3-16, 3-18
   arrays
      for microstrip antennas, 17-118 to 17-119
      microstrip patch resonator, 17-103 to 17-106, 17-114 to 17-115
   television
      sidefire helical, 27-24 to 27-26
      slot, 27-26 to 27-27
Traveling-wave-fed arrays, 12-5
   linear array of edge slots, 12-26
   longitudinal slots, 12-16 to 12-20
   planar, 12-23 to 12-24
Triangular-aperture antennas, 3-24
Triangular lattices
   field representation of, 1-35
   for phased-array radiators, 18-9 to 18-10
Triangular patches, 10-18
Triangular planar arrays and grating lobes, 13-16 to 13-17
Triangular-tooth log-periodic antennas, 9-37 to 9-38
Trifocal bootlace lenses, 16-47
Triplate striplines, 28-22, 28-26 to 28-29
   impedance of, 28-17
Tropical Rainfall Measurement Mission (TRMM), 22-48 to 22-49
Troposphere and tropospheric propagation, 29-30 to 29-47
   permittivity and conductivity of, 29-31

scatter propagation by, 29-46 to 29-47
scintillations caused by, 29-20
True-time-delay feed systems, 19-7 to 19-8, 19-54, 19-76 to 19-91
Truncation
   of aperiodic arrays, 14-20
   with log-periodic antennas, 9-4 to 9-5, 9-8
   of log-spiral antennas, 9-72 to 9-73
      conical, 9-79, 9-84, 9-89 to 9-90, 9-95 to 9-97
   with microstrip antennas, 10-23 to 10-24
   and mutual coupling, 12-29, 12-34
   with optical devices, 19-98
   and scanning range errors, 33-17
T T & C (tracking, telemetry, and command), 21-89 to 21-111
Tumors. See Hyperthermia
Turbulence, scintillation due to, 29-36
Turnstile antennas, 3-29
   phased arrays for satellite antennas, 21-12
   with radial transmission line feeds, 19-24, 19-26 to 19-27
TV. See Television broadcast antennas
TVRO (television receiving only) antennas, 6-25 to 6-32
Twin-boom construction for log-periodic dipole antennas, 9-15 to 9-16
Twist reflector for millimeter-wave antennas, 17-11
Two-antenna gain measurement method, 32-43 to 32-44, 32-46
Two-dimensional arrays, 13-8, 13-9
   Dolph-Chebyshev planar, 11-49 to 11-52
Two-dimensional lenses
   aberrations in, 16-19
   waveguide, 16-42 to 16-43
Two-dimensional multiple-beam matrices, 19-11 to 19-12
Two-layer lenses, 17-15 to 17-18
Two-layer pillbox, 19-17 to 19-23
Two-parameter model for reflector antennas, 15-15 to 15-18
Two planes, arrays scanned in, 13-15 to 13-17
Two-point calibration method, 22-25 to 22-26
Two-terminal stopbands, 9-49 to 9-50
   with log-periodic diode arrays, 9-65
   phase shift in, 9-58
Two-tower antenna patterns, 26-22 to 26-24

Two-tower antenna patterns (*cont.*)
  horizontal electric-field strength with, 26-30 to 26-34
  mutual impedance of, 26-12 to 26-13
Two-wire transmission lines, 28-12
  field configuration of, 28-19
  impedance of, 7-39, 28-11, 28-14, 28-15, 28-20
  validation of computer code for, 3-76
TWTD (Thin-Wire Time Domain) modeling code, 3-87 to 3-95

UAT (uniform asymptotic theory), 4-5
  for edges, 4-32, 4-40 to 4-43
Uhf antennas, 27-3, 27-27 to 27-28
  bandwidth and gain of, 27-7
  EMP responses for, 30-17, 30-20 to 30-21
  and ionospheric propagation, 29-28 to 29-30
  slot, 27-29 to 27-31
Unconstrained feeds, 18-23 to 18-24, 19-49 to 19-54, 21-34
  limited scan, 19-56 to 19-76
  wide field of view, 19-54 to 19-56, 19-76 to 19-91
Underground communications, 23-3
  grounded-wire antennas for, 23-8 to 23-17
  loop antennas for, 23-17 to 23-23
Uniform aperture distributions, 5-11 to 5-13
  far-zone characteristics of, 3-23
  near-field reduction factors for, 5-30 to 5-31
  phase distortion with, 15-109
  radiation pattern with, 21-17
Uniform broadside linear arrays, 11-15
Uniform circular arrays, 11-46 to 11-47
Uniform dielectric layers, 17-65
Uniform diffracted fields, 20-44
Uniform dyadic edge-diffraction coefficient, 4-32 to 4-33
Uniform end-fire linear array, 11-15
Uniform geometrical theory of diffraction, 4-5
  for complex environment simulations, 20-4
  with convex surfaces
    mutual coupling, 4-84 to 4-96
    radiation, 4-63 to 4-84
    scattering, 4-50 to 4-63
  for edges, 4-32 to 4-38
  for numerical solutions. *See* Numerical solutions
  for vertices, 4-43 to 4-48
Uniform hemispherical radiators
  characteristics of, 26-9
  electric-field strength of, 26-27 to 26-29
Uniform linear arrays, 11-11 to 11-13
Uniformly perturbed millimeter waveguides, 17-35
Uniform reflected fields, 20-43
Uniform spherical radiators, 26-8
Uniform structures. *See* Periodic structures
Uniform-waveguide leaky-wave antennas, 17-35, 17-82 to 17-103
Unilateral fin lines, 28-55
Universal curves
  of E-plane horn antennas, 8-10, 8-13 to 8-14
  of H-plane horn antennas, 8-23, 8-27 to 8-29
Unknowns, and computer storage and time, 3-57
Unloaded linear dipole antennas
  receiving, 7-18 to 7-19
  transmitting, 7-12 to 7-18
Updating of computer models, 3-68
Uplink transmitters for satellite tv, 6-25 to 6-26
Upper frequency limit, 9-73
  of log-spiral antennas, 9-75
Use assistance with computer models, 3-68
UTD. *See* Uniform geometrical theory of diffraction

Validation of computer code, 3-55, 3-72 to 3-80
Varactor diodes
  with microstrip antennas, 10-67, 10-70
  in phase shifters, 18-13, 21-40
Variable amplitude networks, 21-33 to 21-34
Variable offset long-slot arrays, 17-26
Variable phase networks, 21-33 to 21-34
Variable phase shifters
  for beam-forming networks
    scanned, 21-38 to 21-43
    switched, 21-32
  for gain ripple with feed arrays, 21-27
  with Wheeler Lab approach, 19-105, 19-107

Variable power dividers
  for beam-forming networks
    scanned 21-43 to 21-49
    switched, 21-32 to 21-33
  for gain ripple with feed arrays, 21-27
  quantization errors with, 21-12
Vector effective area, 6-5 to 6-6
Vector effective height, 6-3 to 6-6
  and power ratio, 6-8 to 6-9
  of short dipole, 6-12
  with UTD solutions, 20-6
Vector far-field patterns, 7-8
Vector wave equation, 2-7 to 2-8
Vee dipole tv antennas, 27-9 to 27-11
Vegetation and surface emissivity, 22-10, 22-15
Vehicular antennas, 32-9, 32-19
Vehicular-mounted interferometers, 25-21 to 25-23
Vertical antennas, radiation resistance of, 6-16
Vertical current element, 26-9
Vertical full-wave loops, 3-31
Vertical optimization, 11-82 to 11-83
Vertical polarization
  for am antennas, 26-3
  with leaky-wave antennas, 17-94
  for microwave radiometers, 22-44 to 22-46
  and surface emissivity, 22-8
  with surface-wave propagation, 29-44 to 29-45
Vertical radiation characteristics for am antennas, 26-3 to 26-8
Vertical sounding, 23-4, 23-18
Vertical stabilizers, simulation of, 20-73, 20-82, 20-84
Vertical transmitting loop antennas, 23-17
Vertices
  GTD for, 4-30 to 4-31
  UTD for, 4-43 to 4-48
Very high frequencies
  communication antennas for, and EMP, 30-17, 30-20 to 30-24
  and ionospheric propagation, 29-21, 29-28 to 29-30
  localizer antennas for, and EMP, 30-23, 30-25 to 30-26
  marker beacon antennas for, and EMP, 30-23, 30-27 to 30-28
Very low frequencies
  ionospheric propagation of, 29-27 to 29-28
  trailing-wire antennas for, and EMP, 30-27, 30-31
Viscometric thermometers, 24-50
Visible regions
  of linear arrays, 11-9
  and log-periodic arrays, 9-56
  with planar periodic arrays, 11-27
  single main beam in, 11-31 to 11-33
Vokurka compact antenna range, 32-30
Voltage-controlled variable power dividers, 21-46 to 21-47, 21-49
Voltage reflection coefficient, 2-28
  of surfaces, 22-8
  for transmission lines, 28-7
Voltages
  in biconical transmission line, 3-24
  diagrams of, with standard reference am antennas, 26-19 to 26-20
  mode, 12-13, 12-16
  in rectangular waveguide, 2-28 to 2-29
  slot, 12-10 to 12-12
  in transmission lines, 28-6 to 28-8
  with UTD solutions, 20-7
Voltage source, ideal, 2-29 to 2-30
Voltmeters for receiving antenna, 2-30 to 2-32
Vswr
  of corrugated horns, 15-97
  and couplers, 19-4 to 19-5
  and lens surface mismatch, 16-54 to 16-55
  with log-periodic antennas
    front-to-back ratios in, 9-60 to 9-61
    zigzag, 9-42 to 9-44
  and log-spiral conical antennas impedances, 9-97, 9-107
  with microstrip antennas, 17-120
  with phase shifters, 21-40 to 21-41
  for pyramidal horns, 8-68
  and stopbands, 17-52, 17-81
  with transmission lines, 28-8
  of tv broadcast antennas, 27-3

Wall currents in waveguides, 12-3, 12-4
Water
  modeling of, 32-74 to 32-75
  remote sensing of, 22-15
Water bolus for loop radiators in medical applications, 24-20, 24-22
Water vapor
  absorption by, 29-35
  noise from, 29-51
  remote sensing of, 22-12 to 22-14, 22-22

Watson-Watt direction-finding system, 25-12 to 25-14
Wavefronts
  aberrations in, and radomes, 31-7 to 31-9
  incident, 4-15, 4-16
  and lenses, 16-6, 16-9 to 16-10
  and rays, 4-6 to 4-8
Waveguide-fed slot arrays. *See* Slot arrays, waveguide-fed
Waveguide feeds
  for beam-forming networks, 21-56 to 21-58
  for reflector antennas, 15-5
Waveguide hybrid ring waveguide power division elements, 21-50 to 21-52
Waveguide lenses, 16-42 to 16-45, 21-16
  for satellite antennas, 21-71 to 21-77
Waveguide loss. *See also* Leaky-wave antennas; Surface wave antennas
  circular, 1-49
  rectangular, 1-39
  with reflector antennas, 17-11
Waveguide nonreciprocal ferrite phase shifters, 21-38 to 21-39
Waveguide phased arrays, 21-12
Waveguides, 2-27 to 2-29
  arrays of, 13-44 to 13-45, 13-57
  blindness in, 13-45, 13-47 to 13-49
  circular. *See* Circular waveguides
  coplanar, 28-30 to 28-35
  coupling of, to microstrip antennas, 17-119 to 17-121
  hybrid-mode, 28-47 to 28-58
  impedance of, 2-28 to 2-29
  for medical applications, 24-9 to 24-18
  for microstrip patch resonator arrays, 17-117 to 17-118
  open-ended, for reflector feeds, 15-89 to 15-93, 15-104 to 15-105
  polarizers for, 13-60 to 13-61
  power divider elements for, 21-50 to 21-52
  rectangular. *See* Rectangular waveguides
  ridged, 28-45 to 28-49
  standard, 28-40 to 28-41
  TE/TM, 28-35 to 28-47
  uniformly perturbed millimeter, 17-35
Waveguide slot radiators, 18-6, 18-10
Waveguide slot tv antennas, 27-29 to 27-31
Waveguide transmission lines, 21-37
Wave impedances, 1-6, 1-13 to 1-14, 1-18 to 1-19
  of hollow-tube waveguides, 28-36, 28-37
  and modeling, 32-68
Wavelength, 1-6 to 1-7. *See also* Cutoff wavelength
Wave number, 1-6 to 1-7
Weapon-locating radar, limited scan, 19-56
Weather forecasting, remote sensing for, 22-13 to 22-15
Wedge-diffraction and UTD solutions, 20-11, 20-17 to 20-18
  coefficient of, 20-46
Wedges, conducting, 1-10
Weight functions in modeling codes, 3-83 to 3-84
Well logging, 23-24 to 23-25
Wenner array for earth resistivity, 23-5, 23-7
Wheeler Lab approach to cylindrical array feeds, 19-105 to 19-107
Wide-angle antennas, 17-29 to 17-30
Wide-angle lenses
  dielectric, 16-19 to 16-33
  multiple-beam, 19-37, 19-80 to 19-85
Wide-angle scanning
  and F/D ratios, 19-34 to 19-35
  lens antennas for, 17-15
Wide antennas, leaky mode with, 17-53 to 17-62
Wideband conical horn antennas, 8-61 to 8-62
Wideband Satellite Experiment, 29-19
Wideband scanning with broadband array feeds, 13-40 to 13-41
Wide field of view unconstrained feeds, 19-52, 19-54 to 19-56, 19-76 to 19-91
Width
  of dielectric grating antennas, 17-64
  main beam, with rectangular apertures, 5-13
  of metal grating antennas, 17-68 to 17-69
  of microstrip antennas, 10-8
Wilkinson power dividers for array feeds, 13-61
Wind speed, remote sensing of, 22-14
Wind sway, effect of, 27-38
Wings, aircraft, simulation of, 20-37, 20-70 to 20-71, 20-78, 20-80
Winter anomaly period for ionospheric propagation, 29-25

Index

Wire-antennas, 7-5 to 7-11
  folded dipole, 7-37 to 7-40
  linear dipole, 7-11 to 7-23
  loop, 7-40 to 7-48
  sleeve, 7-23 to 7-37
Wire diameter and radiation patterns, 32-76, 32-79
Wire objects, computer storage and time for, analysis of, 3-57
Wire-outline log-periodic antennas, 9-13
Wire problems
  subdomain procedures for, 3-60
  time-domain, 3-63
Wullenweber arrays, 25-14 to 25-16

X-band horn antennas, 8-41 to 8-42, 8-44 to 8-45, 8-67
X-polarized antennas, 1-28
XY-scanners, 33-19

Yagi-Uda antennas, 32-76
Yardarm, shipboard, simulation of, 20-93 to 20-94
Yield of crops, forecasting of, 22-15
Y-polarized antennas, 1-28

Zernike cylindrical polynomials, 16-13
Zero-bias diodes, 24-43
Zeroes of the Airy function, 4-59
Zeroes of Bessel functions, 1-45
Zeroes of W, 4-89
Zigzag antennas. *See also* Log-periodic zigzag antennas
  dipoles, as E-field probes, 24-39 to 24-40
  tv, 27-31 to 27-32
  wire, 9-33 to 9-38
Zone-plate lens antennas, 17-20, 17-22
Zones, FCC allocation and assignment 27-4
Zoning of lenses, 16-6, 16-28 to 16-30
  and bandwidth, 16-54 to 16-55, 21-16
  with constrained lenses, 16-43 to 16-45
  with dielectric lenses, 16-38 to 16-41
  with equal group delay lenses, 16-46
  with millimeter-wave antennas, 17-15, 17-19 to 17-21